ENGINEERING FUNDAMENTALS FOR PROFESSIONAL ENGINEERS' EXAMINATIONS

ENGINEERING FUNDAMENTALS FOR PROFESSIONAL ENGINEERS' EXAMINATIONS

LLOYD M. POLENTZ, P.E.

Consulting Engineer
Instructor, Engineering Fundamentals
University of California Extension Division

MCGRAW-HILL BOOK COMPANY

New York St. Louis San Francisco Auckland Bogotá Düsseldorf
Johannesburg London Madrid Mexico Montreal New Delhi
Panama Paris São Paulo Singapore Sydney Tokyo Toronto

Library of Congress Cataloging in Publication Data

Polentz, Lloyd M

Engineering fundamentals for professional engineers'
examinations.

Includes index.
1. Engineering—examinations, questions, etc.
I. Title.
TA159.P6   1979   620   78-21927
ISBN 0-07-050380-X

34567890 MUBP   865432

The editors for this book were Tyler G. Hicks and Joseph
Williams, the designer was Elliot Epstein, and the production supervisor was Sally Fliess. It was set in Century Expanded by Black Dot, Inc.

Printed by The Murray Printing Company and bound by
The Book Press.

To my parents
Gustav E. Polentz
Ethel M. Polentz

CONTENTS

4 THERMODYNAMICS

5 MECHANICS OF MATERIALS

6 ELECTRICITY AND ELECTRONICS

7 CHEMISTRY

PREFACE

This book is the outgrowth of some 20 years of teaching a course in engineering fundamentals for the Extension Division of the University of California. The purpose of the course, and of the first edition of this book, was to aid practicing engineers in reviewing their engineering. The purpose of this second edition remains the same—to help engineers review the fundamental principles of engineering.

New material has been added in each chapter, and some of the explanations in the first edition have been expanded and simplified. The new material includes the following:

Chapter 1, Mathematics—A discussion of the equations of conic sections, i.e., parabolas, ellipses, and hyperbolae; maxima and minima; rate problems, including salt or solution type; lengths of curves and areas of surfaces; numerical integration; worked-out examples of each subject are included.

Chapter 2, Mechanics—Cable supports, relative velocity, satellite problems with examples; simple harmonic motion with example.

Chapter 3, Fluid Mechanics—Expanded sections on hydrostatic force and fluid momentum forces, Toricelli's theorem, and jet propulsion, with examples.

Chapter 4, Thermodynamics—First, second, and third laws of thermodynamics; gas flow through nozzles; rocket thrust; and refrigeration; with worked-out sample problems.

Chapter 5, Mechanics of Materials—Shear stresses in beams, with examples.

Chapter 6, Electricity and Electronics—Expanded discussions on Kirchhoff's laws, resonant circuits, and phasors and phasor algebra, with worked-out problems.

Chapter 7, Chemistry—Solutions; concentration; normality, molarity, and molality; electrochemistry.

Chapter 8, Engineering Economics—Annuities, depreciation, and cash flow, with illustrative problems.

In addition to the above material, a group of multiple-choice problems (of the type now used in the fundamentals examinations) has been included at the end of each chapter.

Detailed solutions to many different types of problems, with explanations of the principles involved, have been included. And problems covering various different aspects of the subjects covered are given at the end of each chapter. Careful study of the text and completion of the problems will provide adequate preparation for successful completion of a fundamentals examination. The problems have been carefully selected to illustrate the various aspects of the different subjects covered in past examinations. Working these problems will automatically prepare a candidate for the multiple-choice part of the examination.

Most of the engineers who took the courses that led to the writing of this book were reviewing for the fundamentals portion of the examination for registration as a professional engineer. In almost every class, however, there were also a number of engineers who were reviewing either to gain a better grasp of the principles of engineering to aid them in their professional activities as practicing engineers, or to prepare for graduate study in engineering. Fortunately all these goals can be achieved in the same fashion, because there is no better guide to a review of engineering fundamentals than the material covered in the fundamentals portion of the professional engineering examinations.

There have been changes in both the examination format and extent of the material covered since this book was first published. The material in this revision has been modified to cover those changes. The majority of states and territories—more than 90%—now use the fundamentals examinations prepared by the National Council of Engineering Examiners. These examinations are open book with multiple-choice answers. It is expected that the few holdouts will also soon switch over to multiple-choice questions.

The fundamentals examination is 8 hours in length and is divided into two 4-hour sessions, morning and afternoon. The morning portion of the NCEE examination consists of 150 short multiple-choice questions and problems. Usually the examinee is required to answer 100 of these. Be careful in your selection. You will probably be graded on the first 100 you do. Those you do above the 100 count won't be included in your grade. But read the instructions on your examination paper. Your state may grade differently.

The afternoon portion of the examination contains three problems in each of seven subjects, for a total of 21 questions. The seven categories are:

Statics	Mechanics of Materials
Dynamics	Electricity
Fluid Mechanics	Economics and Investment
Thermodynamics	

You will be asked to do a total of five problems, probably in a minimum of four categories, with no more than one in economics and no more than two in any other category. Each of the afternoon problems contains 10 parts, with five multiple-choice selections for each part. Usually each subsequent part will depend upon an answer calculated previously; that is, the problems are ordinarily sequential. Part b depends upon part a, part c on part b, and so on. In some cases a single major problem may be built up of two or more relatively short problems treating the same general situation. By building a cluster of problems about one situation, it is possible to break the answer down into 10 different parts.

The material coverage has changed only slightly. The principal differences are in the emphasis placed on the different subjects and the inclusion of a few points on nuclear energy. The seven subjects covered in the afternoon session are discussed above. The coverage in the morning session is somewhat broader and includes the following subjects:

Subject	Approximate Percent of Points in Morning Session
Mathematics	15
Chemistry	9
Nuclear Energy	3
Economics and Investment	8
Thermodynamics	12
Mechanics of Materials	9
Statics	9
Dynamics	9
Fluid Mechanics	9
Electricity	12
Physics and Miscellaneous	5

The coverage of mathematics has been reduced in that there are no afternoon problems in mathematics. However, mathematics is still often needed to solve problems in the other disciplines. The chapter on mathematics has been expanded to provide a more complete coverage.

The emphasis on chemistry has also been reduced, with no problems in that subject in the afternoon. There is more emphasis on definitions, however, and the chapter on chemistry has been expanded to cover these.

Nuclear energy is given a minor emphasis in the examination. It is a very broad and involved subject, to the extent that it would be impossible to give adequate treatment to nucleonics in a book like this one. The same is true of the Physics and Miscellaneous category. These two categories are treated only in the morning session. The total amount of credit in the entire examination for them is small, and the chances of treating these categories meaningfully in a chapter or two are slight. Study time will be spent much more beneficially on the other, major subjects. All of the major subjects are treated in sufficient detail to enable a person who studies them to solve the majority of the problems in those categories that will be given in a fundamentals examination.

No effort has been made to cover all the subjects in sufficient detail to solve all the problems that might be given. That would require a library. But there is enough information included herein to adequately prepare a person to pass the fundamentals examination. This has been demonstrated by a number of people with no college background who have successfully passed fundamentals examinations based largely on what they learned from the first edition of this text.

The emphasis in all examinations has been on the application of the fundamental engineering principles, and this application has been stressed throughout the writing of this book. Memorization of formulas has served to get many a college student through a given course with a passing grade. Such a procedure is not practicable, however, for a comprehensive type of examination such as an engineer-in-training examination or the fundamentals portion of a professional engineering examination. This is also true in the solution of problems which confront an engineer in the practice of his or her profession. There are too many different cases and too many different types of problems to permit memorization of a formula for each case. It is actually easier, and certainly much more satisfactory, to learn how to apply the fundamental principles of engineering and to develop a formula to fit the case at hand from them.

In working with engineers, and helping them review their engineering fundamentals through the past decade, the author has found that the most effective method of teaching consists of essentially three steps.

1. Explain the engineering principle.

2. Illustrate the application of this principle by means of an appropriate problem or problems.

3. Supply problems for individual practice and review.

This three-step sequence—explain, illustrate, practice—is the pattern followed in this book. The principles are first discussed, then illustrated through the solution of sample problems given in past examinations. A group of sample problems is given at the end of each chapter for the reader to solve for practice. These sample problems have been selected from past examinations given by the various states and territories, or have been modeled after problems given in different fundamentals examinations.

The first groups of sample problems is not organized in quite the same way in which problems in those subjects are presented in the fundamentals examinations, in that they are not clustered in groups of ten with a central theme. Instead, they have been selected to illustrate the more important principles in the subject matter of that chapter. The ability to apply those principles is the important goal. Once you have mastered those principles, you will be able to solve problems singly or in the groups in which they are presented in fundamentals examinations. The attempt here has been to give a good cross-section of the type and complexity of the problems the examinee may expect to be confronted with.

Ten multiple-choice problems have been added at the end of each chapter. These are of the type you can expect in the morning session of the fundamentals examination as it is given at the present time.

The important thing in working these problems is not just obtaining the correct answer but understanding how you arrived at it. Method is of more importance in the review process than numerical answers alone. Learn how to work the problems, not just how to get the answers to a specific group of problems.

This brings us to a discussion of how one should complete such an examination most effectively. The first thing to do when you sit for an examination is to read the instructions *carefully*. When you have these in mind, read through the examination problems, checking those you are sure you can do as you go through. Then go back and do these easiest ones first. Then do the more difficult ones and those you think you can do. Then check your work and spend the rest of the time on the really difficult problems— the ones you are doubtful about being able to do. Doing the easy ones first is important. Too many people have fallen into the trap of becoming intrigued with a very difficult (for them) problem and have spent the major part of an examination period trying to solve it because of the challenge it presented. You can easily flunk an examination in this way. This brings up another important consideration: *You should budget your time during the examination.* An easy problem is worth just as many points as a difficult one, and extra time spent on a difficult problem will not pay for itself. In the morning session you are required to answer 100 questions. There are 240 minutes in four hours. If you knock off 20 minutes for reading instructions and taking care of personal needs, you still have 220 minutes, or 2.2 minutes per question. Based on this, if you spend more than two minutes, say, on any one 1-pointer, you are wasting time. Skip it and come back later.

Similarly, in the afternoon you are required to answer five cluster problems. This means that you can safely spend 40 minutes on each one. But don't run over that allotment. You will undoubtedly find that if you know how to work a cluster problem, you will be able to answer all of the parts in 30 minutes or less (maybe much less). If you don't know how to work it, you will just be wasting the extra time you spend on it. So, time yourself.

When you work a problem, the author recommends drawing a figure whenever possible and labeling the figure completely. This will be of considerable help in solving the problem. The figure should be drawn to scale and constructed neatly. It will then provide a rough graphical check of your solution. You should take a scale, straight edge, compass, and protractor to the examination for this purpose. When you write your solution, it should be put in a logical sequence of steps, with all steps written down. This will make it easy for you to check. And it is frequently necessary to check one's work, especially during an examination. A little extra time spent in putting your work down neatly and logically can pay big dividends, particularly when a problem is a sequential one where each part may depend upon the preceding part or parts.

This book has been written to help engineers review subjects with which they are already reasonably familiar. It has not been written to serve as an introduction. If you find that the text proceeds too rapidly, or if you are very rusty, it is recommended that you avail yourself of a book devoted exclusively to the subject causing you difficulty. In most cases your college texts will be adequate as references. If you don't have any reference texts yourself, it may be necessary to borrow or buy them. Be sure that you page a text before you buy it. The latest text is not necessarily the best. The fundamental principles of engineering have not changed in many years and an older book may be more clearly written. Engineering is not a great deal freer from faddism than are other branches of education.

The emphasis in this book is on technical material and on the coverage of the fundamentals portions of the various state examinations. No information has been included on registration procedures or requirements. This type of information has been well covered in the following books: LaLonde, William S.: "Professional Engineer's Examinations Questions and Answers," McGraw-Hill Book Company, Inc., New York; and Constance, John: "How to Become a Professional Engineer," McGraw-Hill Book Company, Inc., New York. The reader who is interested in becoming a registered professional engineer is advised to write to the board in charge of registration in the state or territory in which he or she wishes to become registered. The board should be contacted as far in advance of the time he or she wishes to become registered as possible, because it sometimes takes an appreciable length of time to satisfy all the necessary requirements. The state board of registration is the only official source and the only completely reliable source of information regarding registration. Contact it early.

There are, of course, many people who have been helpful in the preparation of this revision. Foremost among these is my wife Rose, whose gentle prodding kept me on the track and who cheerfully typed all of the new material. I should also like to express my appreciation to those who took the time to point out the occasional errors which crept into the first edition, especially those who did so in a helpful and constructive manner. And I should like to thank those instructors and students who offered suggestions as to how the coverage of the material might best be expanded and improved.

All of the material in this book has been checked and rechecked, especially the answers to the sample problems. There is still a chance, however, that some errors may have remained undetected. If you find any such errors, I would appreciate your bringing them to my attention. I am grateful to all those who have written to me in the past and hope that future users of this book will also send me their comments.

Lloyd M. Polentz

ENGINEERING FUNDAMENTALS FOR PROFESSIONAL ENGINEERS' EXAMINATIONS

1

MATHEMATICS

INTRODUCTION

One of the most difficult and certainly one of the most important factors in any review of mathematics is the "bounding" of the material to be covered. As a guide, a study was made of fundamentals examinations given during a period of more than 15 years. The study of these various examinations indicated that the examinee can expect problems requiring a knowledge of algebra with particular emphasis on exponents and logarithms, trigonometry, perhaps some analytic geometry, differential calculus, and a small amount of integral calculus.

1-1. ALGEBRA

The basic algebra requirement includes factoring and solution of quadratic equations. All engineers will recall the second-order-binomial expansion— $(a \pm b)^2 = a^2 \pm 2ab + b^2$ and the other common special product— $(a + b)(a - b) = a^2 - b^2$.

1-2. FACTORING

A past examination problem asked:

■ If 4 is a root of the equation $x^3 - 5x^2 - 2x + 24 = 0$, find the remaining roots.

If 4 is a root of the equation, then $x - 4$ will be a factor of the polynomial. We can find the other factor by long division.

$$\begin{array}{r}
x^2 - x - 6 \\
x - 4\overline{\smash{\big)}\, x^3 - 5x^2 - 2x + 24} \\
\underline{x^3 - 4x^2} \\
-x^2 - 2x \\
\underline{-x^2 + 4x} \\
-6x + 24 \\
\underline{-6x + 24}
\end{array}$$

The quotient $x^2 - x - 6$ can be further factored by inspection into $(x - 3)(x + 2)$. We then have that

$$(x - 4)(x - 3)(x + 2) = 0$$

and the roots are 4, 3, and -2.

1-3. QUADRATIC EQUATIONS

Quadratic equations can frequently be solved by factoring. However, only the general form of the quadratic equation will be discussed here and the solution derived. This general solution can then be used for all special cases as well. The general form of the quadratic equation may be written as $ax^2 + bx + c = 0$, where a, b, and c are constants. A solution to this equation may be obtained by completing the square. The procedure is as follows:

Transfer the constant term c to the right-hand side of the equation and divide through by the constant of the x^2 term, a. This gives $x^2 + (b/a)x = -c/a$. The left side of the equation can be made a perfect square by adding to it a constant equal to the square of one-half the constant part of the x term, which amount would equal $(\frac{1}{2}b/a)^2$. This gives the equation

$$x^2 + \frac{b}{a}x + \frac{b^2}{4a^2} = -\frac{c}{a} + \frac{b^2}{4a^2}$$

The left side is now a perfect square, and if we combine the terms on the right-hand side of the equation and take the square roots of both sides, the result is $x + (b/2a) = \pm\sqrt{b^2 - 4ac}/2a$; it should be remembered that a square root may be either positive or negative. Transferring the constant on the left side to the right side gives the familiar general solution to a quadratic equation

$$x = \frac{-b \pm \sqrt{b^2 - 4ac}}{2a}$$

As an example to illustrate the use of this equation, assume that a problem reduces to the relationship $17x^2 + 41x - 74 = 19$. This becomes

$$17x^2 + 41x - 93 = 0$$

Then
$$x = \frac{-41 \pm \sqrt{41^2 - 4 \times 17 \times (-93)}}{2 \times 17}$$
$$= -1.21 \pm 2.63$$
$$= -3.84 \quad \text{or} \quad 1.42$$

Note that both roots found in this case are real, but if the term $b^2 - 4ac$ is less than zero, both roots will be imaginary.

1-4. LOGARITHMS AND EXPONENTS

There are two systems of logarithms in use: the common, or Briggsian, logarithms, which have the base 10, and the natural, or Naperian, logarithms, which have the base e. The same principles of operation apply to both systems, since both are merely applications of the theory of exponents. A logarithm to the base 10 will be abbreviated "log," and a natural logarithm will be abbreviated "ln" throughout this work. We have then, by definition, that if $a = \log B$, $10^a = B$; and if $m = \ln P$, $e^m = P$.

An example taken from a past examination is:

■ What is the logarithm of 243 to the base 3? What is the number whose logarithm is 5 to the base 1?

For the first part, we have $\log_3 243 = A$; then $3^A = 243$, and $A \log 3 = \log 243$, giving $A = \log 243/\log 3 = 2.386/0.477 = 5$, or $3^5 = 243$, and the answer is 5. For the second part, $\log_1 B = 5$, $B = 1^5$, $B = 1$.

Another problem asked:

■ Find M if $\log_5 M = \log_{25} 4$.

Restating, we have $\log M$ to the base $5 = \log 4$ to the base 25. Let $\log_5 M = A$; then, by definition, $5^A = M$; $\log_{25} 4 = A$, and, by definition, $25^A = 4$. Taking the logs of this last relationship, we get $A \log 25 = \log 4$, or $A = \log 4/\log 25 = 0.602/1.398 = 0.431$. Then, from the other relationships, $M = 5^A = 5^{0.431} = 2$, which is the required answer. Or, since $25^A = 4$,

$$(5^2)^A = 5^{2A} = (5^A)^2 = 4 = 2^2$$

giving $5^A = 2$. Since $5^A = M$, $M = 2$.

1-5. TRIGONOMETRY

The examinee should review trigonometry rather thoroughly if he or she is at all rusty. Some of the more basic things with which the examinee should be particularly familiar are the signs of the functions in the various quadrants, the definitions of the various functions and their relationships to one another, the law of sines, the law of cosines, double-angle relationships, and the basic right triangles.

The various functions can be explained by means of the unit circle. Referring to Fig. 1-1, OP is the radius and is equal to 1;

$$\sin \theta = PM = \text{opposite side divided by the hypotenuse}$$
$$\cos \theta = OM = \text{adjacent side divided by the hypotenuse}$$
$$\tan \theta = \frac{\sin \theta}{\cos \theta} = \frac{PM}{OM}$$

By proportion $PM/OM = TA/OA = TA$; so $\tan \theta = TA$.

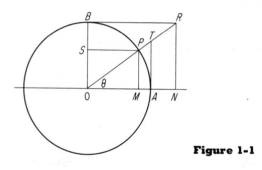

Figure 1-1

Also by proportion $OA/TA = BR/OB$, or $1/TA = BR$. The cotangent is equal to 1/tangent, so $BR = \cot \theta$.

Similarly, $OM/OP = OA/OT$, or $OM = 1/OT$, giving $\sec \theta = OT$ and $PM/OP = RN/OR = OB/OR$, or $PM = 1/OR$, giving

$$\csc \theta = OR$$

Summarizing, we have:

$$\sin \theta = PM \qquad\qquad \sec \theta = 1/\cos \theta = OT$$
$$\cos \theta = OM \qquad\qquad \csc \theta = 1/\sin \theta = OR$$
$$\tan \theta = \sin \theta/\cos \theta = TA \qquad \cot \theta = 1/\tan \theta = BR$$

It might also be noticed that the relationship $\sin^2 \theta + \cos^2 \theta = 1$ is apparent from the diagram.

The signs of the functions in the various quadrants can be determined easily by means of the unit circle, as can also the relationships of the functions of second-, third-, and fourth-quadrant angles to the functions of first-quadrant angles.

Second-quadrant angles, as shown in Fig. 1-2, will possess the following features:

$\sin \theta = PM$ positive $\qquad$ $\sin \theta = \sin (180° - \theta)$
$\cos \theta = SP$ negative $\qquad$ $\cos \theta = -\cos (180° - \theta)$
$\tan \theta = \sin \theta / \cos \theta$ negative $\qquad$ $\tan \theta = -\tan (180° - \theta)$

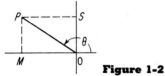

Figure 1-2

Third- and fourth-quadrant angles may be analyzed similarly (see Figs. 1-3 and 1-4).

Figure 1-3 $\qquad$ **Figure 1-4**

From these figures it can also be seen that $\sin (-\theta) = -\sin \theta$, $\cos (-\theta) = \cos \theta$, and $\tan (-\theta) = -\tan \theta$. In addition it can be noted that $\sin \theta = \cos (90 - \theta)$ and $\tan \theta = \cot (90 - \theta)$, or $\tan \theta = 1/[\tan (90 - \theta)]$.

Three types of right triangles which should be reviewed are the 30°-60° right triangle, the 45° right triangle, and the 3-4-5 right triangle.

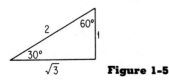

Figure 1-5 $\qquad$

Figure 1-6

The 3-4-5 right triangle (Fig. 1-7) is met quite commonly in mechanics and is a standard method for laying out right angles. The interior angles are almost 37° and 53°. Another common form of right triangle is the 5-12-13 right triangle.

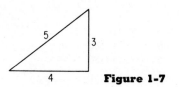

Figure 1-7

The law of sines, which holds for any triangle (Fig. 1-8), is easily remembered and need only be stated. It is:

$$\frac{\sin A}{a} = \frac{\sin B}{b} = \frac{\sin C}{c}$$

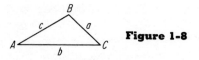

Figure 1-8

1-6. LAW OF COSINES

The law of cosines is more difficult to remember and will be derived (Fig. 1-9).

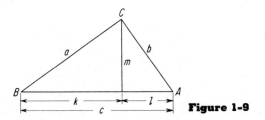

Figure 1-9

Given: Triangle *ABC* with the values of sides *b* and *c* and angle *A* known.

Find: The length of side *a*.

Solution: First construct line m from the vertex C perpendicular to base c.

$$a^2 = m^2 + k^2 \qquad k = c - l \qquad l = b \cos A$$
$$k^2 = c^2 - 2bc \cos A + b^2 \cos^2 A$$
$$m^2 = b^2 - l^2 = b^2 - b^2 \cos^2 A$$
$$a^2 = (b^2 - b^2 \cos^2 A) + (c^2 - 2bc \cos A + b^2 \cos^2 A)$$

which gives

$$a^2 = b^2 + c^2 - 2bc \cos A$$

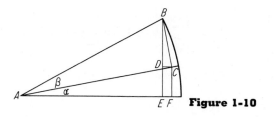

Figure 1-10

The trigonometry requirement also includes manipulation of the various functions algebraically with their various equivalents. No general set of rules can be laid down to cover all cases; the reader is urged to become familiar with the more important equalities and different methods of substitution.

Algebraic manipulation of functions can be demonstrated through solution of a past examination problem:

■ What is $\sec \theta \sin^3 \theta \cot \theta$ equal to?

Substitution gives $1/(\cos \theta) \times \sin^3 \theta \times \cos \theta/\sin \theta = \sin^2 \theta$, which also equals $1 - \cos^2 \theta$.

Another example is:

■ Simplify $1 - \tan^2 \theta \cos^2 \theta$.

Substitution gives $1 - (\sin^2 \theta/\cos^2 \theta) \times \cos^2 \theta = \cos^2 \theta$.

One other phase of trigonometry covered in past examinations is that of calculations involving angles concerned with surveying problems. An example of this is furnished by the following question taken from a past engineer-in-training examination:

■ Lay out a curve tangent at point A and tangent to line C (Fig. 1-11).

a. What is the radius of the curve?

b. What is the length of the curve? (Graphical solution not permitted.)

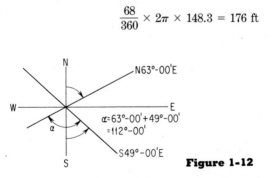

Figure 1-11

Whenever possible, the first step in such a problem is to construct the figure to scale. This gives a clearer picture of the problem (sometimes sketches are just enough out of scale to mislead the observer) and also provides a rough graphical check.

First determine the angle between lines B and C. This can be done easily from the figure by including the north-south line (Fig. 1-12). The center of the arc of the circle forming the curve will lie on the bisector of the angle (Fig. 1-13). This gives a right triangle, since a line drawn from the center of a circle to the point of tangency of a line tangent to that circle is perpendicular to the tangent line. With one leg 100 ft in length and the adjacent angle equal to $^{112}/_2 = 56°$, the other leg, which is the radius of the circle, must equal $100 \tan 56° = 100 \times 1.483 = 148.3$ ft. The length of the curve will equal $68°/360° \times$ circumference, or

$$\frac{68}{360} \times 2\pi \times 148.3 = 176 \text{ ft}$$

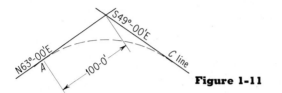

Figure 1-12

The reader should also remember that the sum of the internal angles of a polygon is equal to $(n - 2) \times 180$, where n is the number of sides of the polygon. This relationship is important in determining the error of closure of a polygonal survey. A problem selected from a past examination will serve to illustrate this.

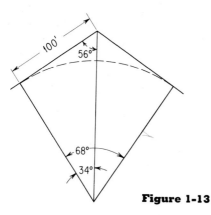

Figure 1-13

■ A surveyor measures the following angles on the property as shown in
Fig. 1-14. However, when he gets back to the office he finds that his
work does not check. Check the following data and show which two
angles do not check.

$$a = 21°47'10'' \qquad e = 36°17'41''$$
$$b = 40°43'40'' \qquad f = 26°13'09''$$
$$c = 71°10'30'' \qquad g = 87°36'40''$$
$$d = 46°18'50'' \qquad h = 29°52'30''$$

From Fig. 1-14 we see that the property can be broken into triangles.
Adding the angles of the individual triangles we have:

$$a + b + g + h = 180°$$
$$c + d + e + f = 180°0'10''$$
$$a + b + c + d = 180°0'10''$$
$$e + f + g + h = 180°$$

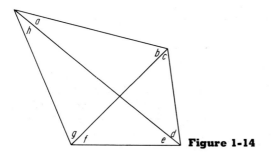

Figure 1-14

Since the sum of the interior angles of a triangle must equal 180°, two of these are in error. Since angles c and d are both in the two incorrect triangles and are not in any of the correct sums, while angles a, b, e, f, g, and h all appear in triangles adding exactly to 180°, the two angles which do not check are c and d.

1-7. FUNCTIONS OF ANGLES CLOSE TO ZERO

For angles which are close to zero an examination of Fig. 1-1 shows that the tangent of the angle will be very close in value to the sine of the angle. In fact, for angles less than 2°30′ the difference between the values of the tangent and the sine occurs in the fifth place (tan 2°30′ = 0.04366, sin 2°30′ = 0.04362); the difference is approximately 0.1 percent. The error of considering the tangent as equal to the sine does not exceed 1 percent until the angle exceeds 8°. In addition, the tangent and the sine of small angles are very nearly equal to the angle itself measured in radians. There are 2π radians in a circle. Then 2π radians equal 360°, and one radian equals $360/2\pi$ = 57.296°. The error obtained by considering sin $\alpha = \alpha$ is a little less than the error obtained by considering sin α = tan α. A 0.1 percent error occurs at about 5° (5° = 0.08727 radians, sin 5° = 0.08716) and 1 percent error occurs at an angle of approximately 14° (14° = 0.24435 radians, sin 14° = 0.24192).

An example of the usefulness of this relationship is given by a problem taken from a past examination:

■ The surveyor wishes to determine the distance AC in Fig. 1-15. How many minutes or seconds of arc would cause an error of 1.0 ft in the distance AC?

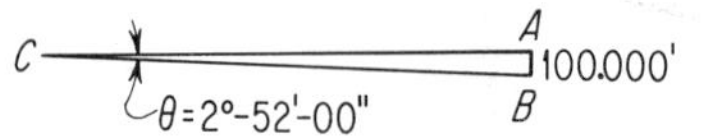

Figure 1-15

$$AC = \frac{100.000}{\tan \theta} = \frac{100.000}{0.05007} = 1997 \text{ ft}$$

If angle θ is smaller, then the distance AC will be longer for AB = 100.000 ft. A smaller angle would give the relationship

$$AC + \Delta AC = \frac{100.000}{\tan(\theta - \Delta\theta)}$$

For $\Delta AC = 1$ ft,

$$1/AC \times AC = (^1/_{1997}) \times AC = 0.0005 \times AC$$
$$\frac{\Delta AC}{AC} = {}^1/_{1997} = 0.00050\, AC \quad \text{and} \quad AC + \Delta AC = 1.00050\, AC$$

Then $1.00050 AC = 100.000/[\tan (\theta - \Delta\theta)]$. But $AC = 100.000/\tan \theta$, so $1.00050 \times 100.000/\tan \theta = 100.000/[\tan (\theta - \Delta\theta)] = 100.05/\tan \theta$.

We know that for very small angles $\tan \theta$ is very nearly equal to θ, so we can safely substitute the angles for their tangents in this case. This gives

$$100.050/\theta = 100.000/(\theta - \Delta\theta)$$

which reduces to

$$0.050\theta = 100.050\Delta\theta \quad \text{or} \quad \Delta\theta = 0.00050\theta$$

where θ and $\Delta\theta$ must be in the same units.

Converting θ to seconds of arc gives

$$\theta = 2 \times 60 \times 60 + 52 \times 60 = 10,320''$$

The number of seconds of arc which would cause an error of ± 1.0 ft in the distance AC would equal $0.0005 \times 10,300 = 5.15''$ or $5''$ of arc.

1-8. ANALYTIC GEOMETRY

A few of the fundamentals of analytic geometry should be reviewed, primarily those pertaining to straight lines and circles.

1-9. EQUATION OF A STRAIGHT LINE

The equation of a straight line through the origin (Fig. 1-16) can be written as

$$\frac{y}{x} = \text{constant} = \tan \theta \quad \text{or} \quad y = mx$$

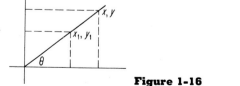

Figure 1-16

Alternatively, using points x and y, $y = (y_1/x_1)x$. The more general case is one in which the line does not pass through the origin. The line may then be specified by one point and the angle or by two points (Fig. 1-17).

If two points are given, the slope equals $(y_2 - y_1)/(x_2 - x_1) = m$, or $m = (y - y_1)/(x - x_1)$, so the equation of the line is

$$\frac{y_2 - y_1}{x_2 - x_1} = \frac{y - y_1}{x - x_1}$$

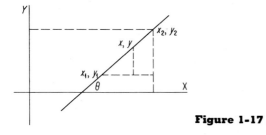

Figure 1-17

or $y = y_1 + [(y_2 - y_1)/(x_2 - x_1)](x - x_1)$, which is of the form $y = mx + c$.

If one point and the slope of the line are given, the equation of the line is $y - y_1 = m(x - x_1)$. The special cases where one intercept or more is given will be seen to fit easily one of the cases discussed.

The slope of a perpendicular to a line can be seen to equal the negative reciprocal of the slope of the line (Fig. 1-18).

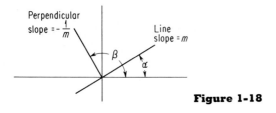

Figure 1-18

We know that $\tan \beta = -\tan (180 - \beta)$. From Fig. 1-18 we see that $180 - \beta = 90 - \alpha$ and the tangent of $(90 - \alpha) = 1/\tan \alpha$. Combining these gives us $\tan \beta = -(1/\tan \alpha)$, or the slope of a perpendicular to a line equals the negative reciprocal of the slope of the line.

1-10. EQUATION OF A CIRCLE

The equations of circles are as easily derived as those of a straight line. Given a circle with its center at the origin, the equation will be $x^2 + y^2 = r^2$, which can readily be seen from Fig. 1-19. If the center is not at the origin, the equation changes to $(x - a)^2 + (y - b)^2 = r^2$. This relationship follows readily from Fig. 1-20.

Figure 1-19

An example of the application of some of these relationships is afforded by a problem which asked:

■ 1. Find the points of intersection of:
$a.$ $x - 7y + 25 = 0$
$b.$ $x^2 + y^2 = 25$
2. Write the equation of a tangent to the circle (b), whose slope is $-3/4$.

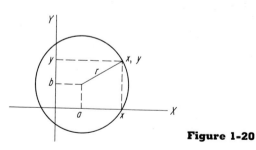

Figure 1-20

Equation (a) is that of a straight line and Eq. (b) is that of a circle. If the line crosses the circle, then the two points of intersection will satisfy both equations. If we solve them simultaneously, we can find both points.

$$x = 7y - 25$$
$$x^2 = 49y^2 - 350y + 625$$
$$\underline{x^2 = -y^2 \qquad\qquad + 25}$$
$$0 = 50y^2 - 350y + 600$$
$$0 = y^2 - 7y + 12$$
$$(y - 4)(y - 3) = 0 \qquad y = 4 \qquad y = 3$$
$$y = 4 \qquad x = 28 - 25 = 3$$
$$y = 3 \qquad x = 21 - 25 = -4$$

The two points are $x = 3$, $y = 4$; and $x = -4$, $y = 3$.

The given circle has its center at the origin, and the required tangent will be perpendicular to the diameter whose slope is the negative reciprocal of the slope of the tangent. The slope of the diameter would then be $4/3$, and the equation of the diameter would be $y = 4/3x$, since the diameter line would go through the origin.

The diameter line would intersect the circle at two points.

$$y^2 = 25 - x^2 \qquad y^2 = (16/9)x^2$$
$$(16/9)x^2 = 25 - x^2$$
$$x = \pm 3$$

The two points of intersection would be $x = 3$, $y = 4$; and $x = -3$, $y = -4$.

The tangent line selected would then go through point 3, 4, or point -3, -4 and have a slope of $-3/4$. Since the slope of a line also equals $(y - y_1)/(x - x_1)$, the equation of the required tangent is $(y - 4)/(x - 3) = -3/4$, or $(y + 4)/(y + 3) = -3/4$, which reduce to $4y + 3x = 25$ and $4y + 3x = -25$.

1-11. PARABOLAS

Another type of curve which has appeared in many of the fundamentals examinations is the parabola. A parabola is a conic section which is formed by the intersection of a plane and a cone when the plane is parallel to a straight line in the cone (see Fig. 1-21). The geometrical definition of a parabola is the locus of a point in a plane that is equidistant from a fixed point and a fixed line. The fixed point is called the "focus" and the fixed line is called the "directrix." The line perpendicular to the directrix through the focus is the axis of the parabola, and the intersection of the parabola and the axis is the vertex.

The general equation of a parabola is $y = Ax^2 + Bx + C$, which can also be written in the form $y = A(x + D)^2 + E$.

This is a quadratic curve, and there will be two values of x for each value of y. The curve will be concave downward if A is positive and concave

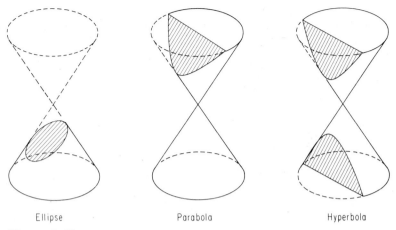

Ellipse Parabola Hyperbola

Figure 1-21

upward if A is negative. The curve will be symmetrical about its vertical axis.

If the equation of the parabola is of the form

$$x = Ay^2 + By + C$$

it will be symmetrical about its horizontal axis. See Fig. 1-22 for example.

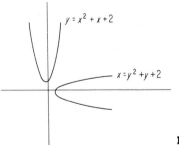

$y = x^2 + x + 2$

$x = y^2 + y + 2$

Figure 1-22

A problem in a past examination dealing with a parabola was of the following form:

■ A support cable in the shape of a parabola is attached to anchor posts

150 ft apart. The two attachment points are at the same elevation. The center point of the cable is midway between the two support points and is 30 ft below them. What is the equation of the parabola and how far is the cable below the elevation of the attachment points at distances of 25 and 50 ft from the attachment points measured on the horizontal?

First sketch the parabola, as shown in Fig. 1-23. Draw the X and Y axes through the center point of the cable. The coordinates of the three points will then be

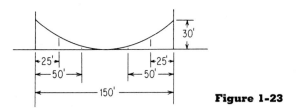

Figure 1-23

Point M: $x = -75$, $y = 30$
Point N: $x = 75$, $y = 30$
Point O: $x = 0$, $y = 0$

From the general form of the equation for a parabola $y = Ax^2 + Bx + C$, and the three given points, we can write the following three equations:

$$30 = 5625A - 75B + C$$
$$30 = 5625A + 75B + C$$
$$0 = C$$

The third equation says that $C = 0$ and the first two equations give us that $B = 0$ and that $A = 0.00533$. The required equation for the parabola is then

$$y = 0.00533x^2$$

At a distance of 25 ft from an attachment point $x = 50$ or -50 and $y = 13.33$ ft. The distance of the two points below the attachment points would then equal

$$30.00 - 13.33 = 16.67 \text{ ft}$$

At a distance of 50 ft from an attachment point $x = 25$ or -25 and $y = 3.33$ ft. The distance of the two points below the attachment points would then equal $30 - 3.33 = 26.67$ ft.

1·12. ELLIPSES AND HYPERBOLAS

There are three types of conic sections, as shown in Fig. 1-21, the ellipse, the parabola, and the hyperbola. An ellipse is formed by the intersection of a plane and one nappe of a cone when the plane is not perpendicular to the axis of the cone. Geometrically an ellipse is the locus of a point in a plane which moves so that the sum of its distances from two points (foci) is a constant which is greater than the distance between the two points. The general form of the equation of an ellipse is

$$\frac{(x - a)^2}{h^2} + \frac{(y - b)^2}{k^2} = 1$$

where the center is at (a, b) and the major axis is parallel with the X axis. The major axis is $2h$ long and the length of the minor axis is $2k$. If the center of the ellipse is at the origin, this equation reduces to the more common form of

$$\frac{x^2}{h^2} + \frac{y^2}{k^2} = 1$$

The general form of the equation can be reduced to

$$(x - a)^2 + c(y - b)^2 = d$$

This is the same form as the equation of a circle except for the numerical modifier (c) of the $(y - b)^2$ term.

A hyperbola is formed when a plane cuts both nappes of a cone (see Fig. 1-21). Geometrically it is the locus of a point in a plane that moves so that the difference of its distances from two fixed points (foci) is a constant. The equation of a hyperbola is similar to the equation of an ellipse except that the plus sign between the two terms becomes a minus sign. The general equation of a hyperbola is

$$\frac{(x - a)^2}{h^2} - \frac{(y - b)^2}{k^2} = 1$$

where the center is at a, b and the transverse axis is parallel to the X axis.

This equation can be modified to the form

$$(x - a)^2 - c(y - b)^2 = d$$

1·13. CALCULUS

Much of the engineer's use of calculus is devoted to determining the area under a curve and determining the maximum and/or the minimum point of a function.

1-14. DERIVATIVES

A few of the commoner derivatives should be recalled to memory. These include:

$$d\ ax = a\ dx \qquad a \text{ is a constant}$$
$$d\ (u + v) = du + dv \qquad u \text{ and } v \text{ are variable}$$
$$d\ (uv) = u\ dv + v\ du \qquad u \text{ and } v \text{ are variable}$$
$$d\ x^n = nx^{(n-1)}\ dx$$
$$d\ e^x = e^x\ dx$$
$$d\ \sin x = \cos x\ dx$$
$$d\ \cos x = -\sin x\ dx$$
$$d\ \ln x = dx/x$$

This list is the minimum that should be remembered. From these we can derive a few more important relationships; the derivations will obviate memorizing and will also provide drill in manipulating some of the listed derivatives.

The derivative of u/v is also important. $u/v = uv^{-1}$,

$$d\ (uv^{-1}) = ud(v^{-1}) + v^{-1}\ du$$
$$= u(-v^{-2}\ dv) + v^{-1}\ du$$
$$= -\frac{u\ dv}{v^2} + \frac{du}{v}$$

Combining terms gives $d\ (u/v) = (v\ du - u\ dv)/v^2$.

The derivative of $\tan x$ can be determined using this relationship.

$$\tan x = \frac{\sin x}{\cos x}$$
$$d\ \tan x = \frac{\cos x\ (\cos x\ dx) - \sin x\ (-\sin x\ dx)}{\cos^2 x}$$
$$= \frac{(\cos^2 x + \sin^2 x)\ dx}{\cos^2 x} = \frac{dx}{\cos^2 x} = \sec^2 x\ dx$$

1-15. INTEGRATION

Integration is the opposite of differentiation; thus we have:

$$\int a\ dx = ax + c \qquad \text{where } c = \text{the constant of integration}$$
$$\int x^n\ dx = (1/n + 1)x^{(n+1)} + c \qquad (\text{except for } n = -1)$$
$$\int dx/x = \ln x + c$$
$$\int e^x\ dx = e^x + c$$
$$\int \sin x\ dx = -\cos x + c$$
$$\int \cos x\ dx = \sin x + c$$

One more important integral which should be remembered is $\int u \ dv = uv - \int v \ du$, where u and v are both functions of the same variable.

An illustration of the usefulness of this relationship is provided by the examination problem which asked:

- *Find*: $\int x \cos x \ dx$.

Let $u = x$; then $du = dx$. $dv = \cos x \ dx$; then $v = \sin x$; $uv - \int v \ du = x \sin x - \int \sin x \ dx = x \sin x + \cos x + c$. To check, $d (x \sin x + \cos x) = \sin x \ dx + x \cos x \ dx - \sin x \ dx$, or

$$d (x \sin x + \cos x) = x \cos x \ dx$$

which checks.

Whenever we have an equation which gives y as a function of x [abbreviated $y = f(x)$], it can be represented by a curve [We shall assume that the function $y = f(x)$ is continuous and differentiable throughout the area in which we are interested.] when a series of points is plotted on a graph (Fig. 1-24). The derivative dy/dx is the limit of the fraction $\Delta y/\Delta x$ as Δx approaches zero.

1-16. MAXIMUM AND MINIMUM

As can be seen from Fig. 1-24, the quotient of $\Delta y/\Delta x$ is the slope of the hypotenuse of the indicated triangle or the tangent of the angle the hypotenuse makes with the abscissa. As Δx approaches zero, the triangle becomes smaller and smaller until, finally, it vanishes at the point x, y. At

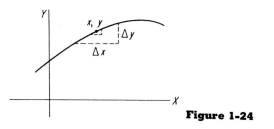

Figure 1-24

the point of extinction, the hypotenuse coincides with the tangent to the curve at the point x, y, and the slope of the tangent is given by the derivative of the function at that particular point. As we can see from Fig. 1-25, whenever the curve becomes a maximum or a minimum, the slope of the

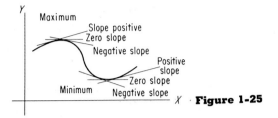

Figure 1-25

curve at the particular point of maximum or minimum is horizontal. In other words, the tangent to the curve at the particular point where the curve is either a maximum or a minimum will be parallel with the x axis and will, therefore, have a slope of zero, since the tangent of 0° (and of 180°) is zero. This means that at any point of maximum or minimum of the function $y = f(x)$ the derivative at that point, dy/dx, will equal zero. Conversely it can be shown that wherever $dy/dx = 0$, the tangent of the curve of the function $y = f(x)$ is parallel to the abscissa. In all but a few special cases this parallelism will indicate either a maximum or a minimum.

There are two general methods whereby we may determine whether the point at which $dy/dx = 0$ is a maximum or a minimum. The first is to take the second derivative of the function at the particular point; if d^2y/dx^2 is positive, the point is a minimum; if d^2y/dx^2 is negative, the point is a maximum.

Another method of determining whether $f(x)$ for $f'(x) = 0$ is a maximum or a minimum is to calculate the value of $f(x)$ a short distance on either side of the point of interest. If these values calculated for $f(x + \Delta x)$ and $f(x - \Delta x)$ are larger than the value calculated for $f(x)$, then $f(x)$ is a minimum. If they are smaller, then $f(x)$ is a maximum. Points on both sides are recommended to provide a check.

There have been a number of problems in past examinations in which it was necessary to determine a maximum or minimum value. Generally speaking it is necessary to form an equation in which there is one independent variable. Then differentiate with respect to the independent variable, set the resulting equation equal to zero, and solve.

Take as an example a type of problem which has been given in past examinations:

- It is desired to make a closed cylindrical container to hold a volume of 50 gal. The material to be used for the ends costs three times as much as the material to be used for the body of the container. What dimensions will give the most economical container?

The required volume is 50/7.48 = 6.68 ft³, a constant

Volume = 6.68 = $\pi r^2 h$
Surface area = $2\pi r^2 + 2\pi rh$
Cost = $2\pi r^2 \times 3M + 2\pi rh \times M = 6\pi r^2 M + 2\pi rhM$

We need an equation for cost in terms of either r or h. We have a relationship between r and h in the equation for volume, since the volume is a constant. We can solve for h in terms of r and then substitute in the relationship for cost.

$$h = \frac{V}{\pi r^2}$$

$$C = 6\pi r^2 M + 2\pi r\left(\frac{V}{\pi r^2}\right) M = 6\pi r^2 M + 2\left(\frac{V}{r}\right)M$$

Now differentiate with respect to the independent variable r and set equal to zero.

$$\frac{dC}{dr} = 12\pi rM - 2VMr^{-2} = 0$$

$$r^3 = \frac{V}{6\pi} = 0.35439 \qquad r = 0.7077 \text{ ft}$$

The diameter of the container would be 1.415 ft and the height would be 4.245 ft. The material cost would equal 28.316 × M.

Another problem which has been used was of the following type:

- A rectangle is to be constructed in the first quadrant with two sides coinciding with the X and Y axes. The fourth corner is to be on the curve $2x^2 + 3y^2 = 12$. Find the location of the fourth corner so that the area will be a maximum.

The area will equal the length of the base times the height. For this case $A = xy$. We must find an equation in which either x or y is the only independent variable. We have a relationship between x and y since the fourth corner must lie on the curve $2x^2 + 3y^2 = 12$. From the equation we find that $x = \sqrt{6 - 1.5y^2}$; then $A = y\sqrt{6 - 1.5y^2}$. Differentiate A with respect to y and set equal to zero.

$$\frac{dA}{dy} = y(\tfrac{1}{2})(6 - 1.5y^2)^{-1/2}(-3y) + (6 - 1.5y^2)^{1/2} = 0$$

$$\frac{1.5y^2}{(6 - 1.5y^2)^{1/2}} = (6 - 1.5y^2)^{1/2}$$

$$1.5y^2 = 6 - 1.5y^2 \qquad y = 1.4142 \qquad x = 1.7321$$

Area = 2.45 square units.

1-17. RATE PROBLEMS

There have been a number of rate problems in past examinations both of the radioactive-decay type and of the salt problem type. Many of the differential equations describing the rate of the action can be solved by separation of variables. However, some of the salt-solution-type problems require solution of the standard form of the first-order linear differential equation.

Radioactive-decay problems can be handled either by using the compound-interest law, as previously mentioned, or by setting up the differential equation and solving it. As an example, take the following problem.

- A radioactive substance has a half-life of 25 days. How long will it take to lose 90 percent of its radioactivity?

Solve first by using the compound-interest equation. The amount of radioactivity after n periods $= A_0 (1 - i)^n$. In this case the amount of radioactivity is decreasing, so the "interest rate" will be negative. Fifty percent remains after 25 days so

$$0.50 = (1 - i)^{25}$$
$$\ln 0.50 = 25 \ln(1 - i) \qquad 1 - i = 0.9726$$

so the time required for only 10% to remain would be given by the relationship

$$0.10 = 0.9726^n$$
$$n = \frac{\ln 0.10}{\ln 0.9726} = 83.0 \text{ days}$$

This problem can also be solved by means of a differential equation, as follows:

$$\frac{dA}{dt} = -kA \qquad \text{The rate of decay is proportional to the amount of radioactive material}$$

$$\frac{dA}{A} = -kdt \qquad \ln A = -kt + C$$

or $\qquad A = A_0 e^{-kt}$

Note that it is necessary to solve for k. When $A/A_0 = 0.50$, $t = 25$; so $0.50 = e^{-25k}$ and $k = 0.0277$, giving the equation for decay

$$A = A_0 e$$

for $A/A_0 = 0.10$

$$t = \ln 0.10/(-0.0277) = 83.0 \text{ days}$$

1-18. LIMIT OF AN INDETERMINATE FRACTION

In a past examination one of the problems was to determine the limit of the fraction $(5x^3 - \cos x + e^x)/(x^2 + 2x)$ as x approaches zero. If the value of zero is substituted for x, the indeterminate fraction of $0/0$ is obtained, which is not a solution to the problem. In the actual case the numerator may approach zero more rapidly than the denominator or vice versa. To solve this problem it is easiest to employ the rule of L'Hospital. This rule may be stated as follows: if $f(a) = \varphi(a) = 0$, and if the ratio $f'(x)/\varphi'(x)$ approaches a limit as $x \to a$, then $f(x)/\varphi(x)$ approaches the same limit. The same rule will hold if $f(a)$ and $\varphi(a)$ both approach infinity as $x \to a$. To apply this rule to the example given, since both numerator and denominator equal zero for $x = 0$, we differentiate both the numerator and denominator to obtain $(15x^2 + \sin x + e^x)/(2x + 2)$, which equals $1/2$ for $x = 0$. If the first differentiation results in another indeterminate fraction, then differentiate the numerator and denominator again, and again if necessary. As soon as a determinate fraction is obtained, however, the process must be discontinued; further differentiation will give an erroneous result.

1-19. DOUBLE INTEGRATION

A word about double integration should also be included. It was required to find A from the relationship

$$A = \int_0^\pi \int_0^R r \, dr \, d\theta$$

in a past fundamentals examination. The only point to remember here is that one variable must be considered constant while we integrate with respect to the other. The example can be rewritten as

$$A = \int_0^\pi \left(\int_0^R r \, dr \right) d\theta$$

which equals

$$\int_0^\pi \left(\frac{R^2}{2} \right) d\theta$$

which equals $(\pi R^2)/2$; this, incidentally, may be recognized as the area of a semicircle.

1-20. ADDITIONAL EXAMPLES AND SOLUTIONS

A few more problems selected from past fundamentals examinations are included with their solutions to help illustrate the principles discussed above.

■ 1. A box is to be constructed from a thin sheet of copper 10 in. square by cutting equal squares out of each corner and turning up the remaining parts of the sheet copper to form the four sides of the box. Since *this is a problem in mathematics, use calculus* and determine the *largest-volume* box that can be made from this material by the method described.

First, draw a figure (Fig. 1-26) illustrating the problem. Next, write an equation for the volume of the finished box in terms of the size of the cutout x.

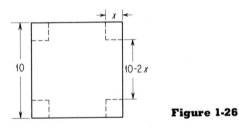

Figure 1-26

$$\text{Volume} = \text{height} \times \text{area of base} = x \times (10 - 2x)^2$$

which gives $V = 100x - 40x^2 + 4x^3$. Now that we have $V = f(x)$, we must differentiate V with respect to x and set $dV/dx = 0$ to find V_{max}. $dV/dx = 100 - 80x + 12x^2 = 0$. This may be factorable but can be solved easily by means of the general solution of a quadratic equation given previously under Quadratic Equations.

$$x = \frac{80 \pm \sqrt{6400 - 4800}}{24} = \frac{80 \pm 40}{24}$$
$$= 5 \text{ or } 1^2/_3$$

$x = 5$ is obviously a minimum, since this value for x would give a volume of zero. For illustration, however, let us take the second derivative of V with respect to x. $d^2V/dx^2 = -80 + 24x$; d^2V/dx^2 for $x = {}^5/_3$ equals -40, for $x = 5$ equals $+40$; so $x = 5$ gives V_{min} and $x = {}^5/_3$ gives V_{max}. The maximum volume would then equal 74.1 in.³.

■ 2. Find the entire area between the curve $y = x/(1 + x^2)^2$ and its asymptote.

First sketch the curve (Fig. 1-27). We see that the asymptote is $y = 0$. The area will then be equal to

$$A = \int_{x=-\infty}^{x=\infty} y \, dx = 2\int_0^\infty \frac{x}{(1+x^2)^2}dx$$

since the curve is symmetrical about both x and y axes. The derivative of $(1 + x^2)$ is $2x$; so if we let $(1 + x^2) = B$, we can rewrite the integral as

$$A = \int_{x=0}^{x=\infty} B^{-2} \, dB$$

which equals

$$-B^{-1}\bigg|_{x=0}^{x=\infty}$$

or

$$A = -\left(\frac{1}{1+\infty^2} - \frac{1}{1+0}\right) = 1.0$$

which is the entire area between the curve and its asymptote.

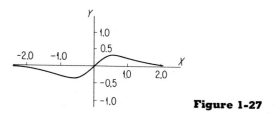

Figure 1-27

■ 3. Determine the equations of the tangent and the normal to the circle $(x - 4)^2 + (y + 3)^2 = 25$ at the point $(7,1)$.

The tangent to the curve equals dy/dx, and we can solve for this by differentiating the above equation and then solving for dy/dx. Differentiation gives $2(x - 4) \, dx + 2(y + 3) \, dy = 0$, and

$$\frac{dy}{dx} = -\frac{x-4}{y+3}$$

The slope of the tangent line at the point $(7,1)$ is $-3/4$, and the slope of the normal line at the point $(7,1)$ is $4/3$. Recalling the discussion of analytic geometry, we know that the equation will be of the form $y = mx + b$ and that the equation of the tangent line must have the slope $m = -3/4$ and pass through the point $(7,1)$; so we have $1 = -3/4(7) + b$; $b = 25/4$, and the equation of the tangent line is $y = -\frac{3}{4}x + \frac{25}{4}$, or $4y = -3x + 25$. Similarly, the equation of the normal line is found to be $3y = 4x - 25$.

■ 4. The equation of a parabola is $y^2 = 16x$. Find the equation of a chord through the points in the curve whose abscissas are 1 and 4 and whose ordinates are positive.

The two points will be $y = 4\sqrt{x}$ or $\begin{cases} y = 4 \\ x = 1 \end{cases}$ and $\begin{cases} y = 8 \\ x = 4 \end{cases}$. Required is the equation of the line which passes through two points. The slope of the line will be

$$\frac{y_2 - y_1}{x_2 - x_1} = \frac{8 - 4}{4 - 1} = \frac{4}{3}$$

The line will be of the form $y = mx + b$, and b can be determined by substituting the coordinates of either point. Using (1,4), $b = \frac{8}{3}$, and the required equation of the line is $3y = 4x + 8$.

■ 5. Two points P and Q on opposite sides of a stream are not intervisible because of an island. A straight line AB is run through Q, and the following measurements are made: $AQ = 824$ ft, $QB = 662$ ft, $QAP = 42.57°$, $QBP = 57.75°$. Calculate the distance PQ.

First draw and label the figure (Fig. 1-28). We can find the distances AP and BP by means of the law of sines.

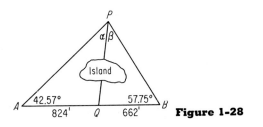

Figure 1-28

$$(\alpha + \beta) = 180 - (42.57 + 57.75) = 79.68°$$
$$PB = (AQ + QB)\frac{\sin 42.57°}{\sin 79.68°} = 1{,}022'$$

$AP = (AQ + QB)(\sin 57.75°/\sin 79.68°) = 1{,}278'$. Now, having two sides and the included angle of the triangles APQ and BPQ, we can find PQ in two ways, using the law of cosines.

$$(PQ)^2 = (PB)^2 + (QB)^2 - 2QB \cdot PB \cos 57.75° = (872 \text{ ft})^2$$

or $$(PQ)^2 = (AP)^2 + (AQ)^2 - 2 \cdot AP \cdot AQ \cos 42.57°$$

which also gives $PQ = 872$ ft.

■ 6. $\cos^2 x - 2 \tan^2 x = \sec^2 x$; find the principal values of x.

$\cos^2 x - 2(\sin^2 x/\cos^2 x) = 1/\cos^2 x$ gives $\cos^4 x - 2 \sin^2 x = 1$. $\sin^2 x = 1 - \cos^2 x$ gives $\cos^4 x - 2 + 2 \cos^2 x = 1$. Combining terms and rearranging gives $\cos^4 x + 2 \cos^2 x - 3 = 0$, which can be factored into $(\cos^2 x + 3)(\cos^2 x - 1) = 0$.

The factors are $\pm i\sqrt{3}$ and $\pm\sqrt{1}$. The principal values of x are, then, $x = \cos^{-1}(1) = 0°$ and $x = \cos^{-1}(-1) = \pi$.

■ 7. Find the coordinates of the point of intersection of the two tangents to the curve $y = x^2 + 4$ at the points (1,5) and (2,8).

The slope of the tangent to a curve equals dy/dx, and for this curve $dy/dx = 2x$. The first line will then have a slope equal to 2, and the second a slope equal to 4. Both lines will be of the form $y = mx + b$ and must pass through their respective points. For the first line $5 = 2 \times 1 + b$; $b = 3$, and the equation of the first line is $y = 2x + 3$. Similarly, the equation of the second line will be $y = 4x$. The intersection of these two lines will be at a point which will satisfy both these equations, and we can find that point by solving the two equations simultaneously.

$$y = 2x + 3$$
$$y = 4x$$

These two equations form one equation, $4x = 2x + 3$, which gives the answer, $x = 1\frac{1}{2}$ and $y = 6$.

■ 8. Determine the volume generated by the rotation about the x axis of the area bounded by the segment of the parabola $y^2 = 8x$ between the origin of the x axis and the ordinate $x = 6$.

First, draw the picture (Fig. 1-29). Since the volume to be determined is due to the rotation of a section of the parabola about the x axis, we can treat it as the volume occupied by a series of disks of thickness dx. The volume of each of these disks will equal $\pi r^2 \, dx$, where $r =$ the ordinate y.

We have, then, $V = \int_0^6 \pi y^2 \, dx$, or, substituting the equivalent value of y^2,

$$V = \int_0^6 \pi 8x \, dx = 8\pi (x^2/2) \Big|_0^6$$

$V = 8\pi(^{36}/_2) = 453$ cubic units.

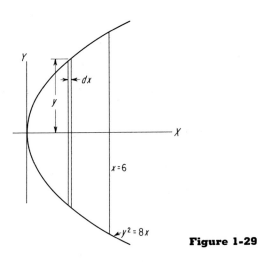

Figure 1-29

■ 9. Given: $y = wlx/2 - wx^2/2$. Determine the area between this curve and the x axis for values of x between $x = 0$ and $x = 1$.

The area will equal $\int_{x=0}^{x=1} y \, dx$, so we write

$$A = \int_0^1 \left(\frac{wlx}{2} - \frac{wx^2}{2} \right) dx = \frac{w}{2} \int_0^1 (lx - x^2) \, dx = \frac{w}{2} \left(\frac{lx^2}{2} - \frac{x^3}{3} \right) \Big|_0^1$$

$$A = \frac{w}{2} \left(\frac{l}{2} - \frac{1}{3} \right) = \frac{wl}{4} - \frac{w}{6}$$

■ 10. A roofer wishes to make an open gutter of maximum cross section whose bottom and sides are each 4 in. wide and whose sides have the same slope. What will be the width across the top?

To solve this problem we need an equation which will give the area as a function of the top dimension or of some other variable upon which the top dimension depends. If we draw a figure (Fig. 1-30), top dimension D is a function of the angle α. The area of the cross section can be equated as follows:

$$A = (4 \times 4 \sin \alpha) + (4 \cos \alpha \times 4 \sin \alpha) = 16 \sin \alpha + 16 \cos \alpha \sin \alpha$$

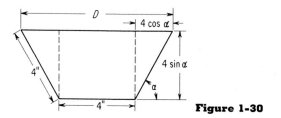

Figure 1-30

Next, find $dA/d\alpha$, set it equal to zero, and solve for α.

$$\frac{dA}{d\alpha} = 16 \cos \alpha + 16 \cos^2 \alpha - 16 \sin^2 \alpha = 0$$

Remembering that $\sin^2 \alpha = 1 - \cos^2 \alpha$, we obtain

$$2 \cos^2 \alpha + \cos \alpha - 1 = 0$$

Factoring gives $(2 \cos \alpha - 1)(\cos \alpha + 1) = 0$; $\cos \alpha = -1$ and $\cos \alpha = \frac{1}{2}$. These values give $\alpha = 180°$ and $\alpha = 60°$. For $\alpha = 180°$ the area would equal zero, so $\alpha = 60°$ is a maximum. The width of the top D equals $4 + 2 \times 4 \cos \alpha = 8$ in.

1-21. INTRODUCTION TO SAMPLE PROBLEMS

The emphasis on mathematics in the fundamentals examinations has diminished in the past few years. The problems devoted to mathematics per se are presently included in only the 150 short (one-point or less) questions in the morning session. The seven classifications of major problems in the afternoon portion of the fundamentals examination do not include mathematics. However, a good basis in mathematics is necessary to solve the problems given in the afternoon session. Moreover, it is always possible that mathematics may once again be elevated to a major classification in its own right in a future examination. In any event, a good grounding in mathematics is necessary for any sort of engineering competence.

The following problems have been selected to give a good general coverage of the types of problems that have been asked in past examinations. The problems have been divided into two groups. The first consists of complete problems which will require a few minutes (or, perhaps, a number of minutes) to solve. The second group is a series of multiple-choice problems of the type presently used in fundamentals examinations. Some of the problems in this latter group may require scratch-paper calculation; for the others the answers can be drawn from memory. These problems should not require more than $1\frac{1}{2}$ minutes each on the average—150 questions in the morning, totaling 4 hours, or 240 minutes.

SAMPLE PROBLEMS

1- 1 A bombardier is sighting on a target on the ground directly ahead. If the bomber is flying 2 miles above the ground at 240 mph, how fast must the sighting instrument be turning when the angle between the path of the bomber and the line of sight is 30°?

1- 2 A transformer core is to be built up of sheet-steel laminations, using two different strip widths, x and y, so that the resultant symmetrical cross section will fit within a circle of diameter D. Compute the values of x and y in terms of D so that the cross section of the core will have maximum value (see Fig. 1-P-2).

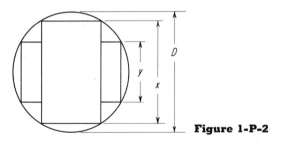

Figure 1-P-2

1- 3 It is desired to measure the width of a river. A base line 1000 ft long is laid out along one bank. An object on the other bank opposite the base line is sighted on from one end of the base line and found to make an angle with it of 60°. The same object is sighted on from the other end of the base line and found to make an angle with it of 30°. Calculate the width of the river.

1- 4 Find the equation of the parabola in the form $y = f(x)$ which passes through the points (1,2), (3,20), and (4,35).

1- 5 The base and altitude of a rectangle are 5 and 4 in., respectively. At a certain instant they are increasing continuously at the rate of 2 and 1 ips, respectively. At what rate is the area of the rectangle increasing at that instant? (Solve by calculus.)

1- 6 Write the equation of the straight line that passes through the point (0,6) and the point (−3,−8).

1- 7 A roll of tape is unrolled on a plane surface at the rate of 6 fps of length. Assuming a true circular roll, if the tape is $1/8$ in. thick, at what rate is the radius of the roll decreasing when the radius is 1 ft? (Solve by calculus.)

1- 8 A tract of land 400 by 300 ft is to be crossed diagonally (see Fig. 1-P-8) by a road 100 ft wide. What area is comprised in the right of way?

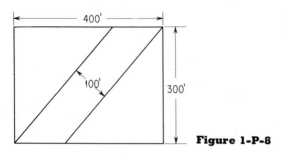

Figure 1-P-8

1- 9 $\sec^{-1} 1.40 = ?$

1-10 A gasoline tank consists of a horizontal closed cylinder 1 ft in diameter and 5 ft long. Determine the number of gallons in the tank when a gauge rod in the plane of the vertical diameter shows a depth of 4 in. in the tank. One gallon equals 231 cu in.

1-11 $\int_{\pi/2}^{\pi} \cos x\, dx = ?$

1-12 Ship A is sailing due south at 15 mph, and a second ship B 32 miles south of A is sailing due east at 12 mph.

    a. At what rate are they approaching or separating at the end of 1 hr?

    b. At the end of 2 hr?

    c. When do they cease to approach each other, and how far apart are they at that time?

1-13 Given: $y = wlx/2 - wx^2/2$. Determine the area between this curve and the x axis for values of x between $x = 0$ and $x = 1$.

1-14 Integrate: $\displaystyle\int_0^L \frac{dx}{\sqrt{7 + 6x - x^2}}$

1-15 *a.* Determine the points of intersection and the angle between the curve $(x + 2)^2 + (y - 3)^2 = 95$ and the line $2x + 3y = 12$.

    b. Simplify: $\cos x/(1 - \sin x) - (1 - \sin x)/\cos x$.

1-16 From one corner of a pasture, one fence bears due east and the other bears S62°E. The owner wishes to build a new fence diagonally across the corner starting at a point 800 ft due east of it. If he has only 1200 ft of fencing, where will the new fence intersect the one that runs southeast?

1-17 The equation of a parabola is $y^2 = 16x$. Find the equation of a chord through the points on the curve whose abscissas are 1 and 4 and whose ordinates are positive.

1-18 The equation of a curve is $4a^2y = (x^3/3) - 2ax^2 + 5a^2$.
  a. Determine the slope of $x = 0$ and $x = 2a$.
  b. Determine the points where the curve is parallel to the x axis.
  c. Find the points at which the slope is 1.

1-19 Determine the volume generated by the rotation about the x axis of the area bounded by the segment of the parabola $y^2 = 8x$ between the origin of the x axis and the ordinate $x = 6$.

1-20 A rectangular plot of ground has two adjacent sides along Highways 20 and 32. In the plot is a small lake, one end of which is 256 ft west from Highway 20 and 108 ft north from Highway 32.
  Find the length of the *shortest* straight path which cuts across the plot from one highway to the other and passes by, i.e., is just tangent to or touches, the end of the lake.

1-21 A woman is buying two square lots of unequal size which lie adjacent to each other. Together the lots have a total area of 12,500 ft². To embrace the two lots in a single enclosure would require 500 ft of fence. Determine:
  a. Dimensions of each lot
  b. Area of each lot in square feet

1-22 Prove that $\sec^2 \theta = 1 + \tan^2 \theta$.

1-23 Evaluate the following definite integral:

$$\int_0^{\pi/2} \frac{\cos \theta \, d\theta}{1 + \sin^2 \theta}$$

1-24 Determine the length of the catenary from $x = 0$ to $x = x_1$, using the exponential equation $y = \frac{1}{2}a(e^{x/a} + e^{-x/a})$.

1-25 Along the bank of a stream of water two points A and B are laid out 100 yd apart. A tree is visible from points A and B across the stream and very near to the opposite bank of the stream. The angle at A subtended by the distance from B to the tree is found to be 30°, and the angle at B subtended by the distance from A to the tree is 105°. With these data calculate:
  a. Angle at the tree subtended by the distance AB
  b. Width of the stream

1-26 Water is flowing into a conical reservoir 20 ft deep and 10 ft across the top, at the rate of 15 cfm. Find how fast the surface is rising when the water is 8 ft deep.

1-27 In the expression $R = re^{\tan x}$, where r is a constant and e is the natural logarithm base,
  a. What value of x will give $R = r$?
  b. What value of x will give $R = 0$?

1-28 Two ships leave the same port at the same time, one sailing due northeast at a rate of 6 mph, and the other sailing due north at a rate of 10 mph. Find the distance between the two ships after 3 hr of sailing.

1-29   The equation of a parabola is $x^2 = (b^2/a)y$, where b is the limit of x and a is the limit of y. Show by calculus that the area B equals $\frac{1}{3}ab$, and that the area C equals $\frac{2}{3}ab$ (Fig. 1-P-29).

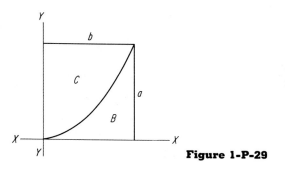

Figure 1-P-29

1-30   How would you determine logarithmic functions of an angle greater than 90°?

1-31   Simplify the following mathematical expressions:

$a.\ \dfrac{3}{5x^{-2}} + \dfrac{5}{3x^{-2}}$

$b.\ (8y^6Z^9)^{-4/3}$

$c.\ \dfrac{a^{-1}\,x^{-1}}{a^{-1} + x^{-1}}$

$d.\ \dfrac{\sqrt{10}}{\sqrt[3]{20}}$

$e.$ Solve for x: $\log_4 3 + \log_4 [x - (5/3)] = 2$

1-32   Two sides of a triangle are 84 and 48 ft. The angle opposite the 84-ft side is 35°20′. Find the other two interior angles and the third side.

1-33   A trapezoidal trough has a bottom twice as wide as the slant length of a side. What angle does the side make with the bottom for maximum capacity of the trough?

1-34   Solve for x by logarithms and *show each step*: $x = 5.29(0.896)^{0.435}$.

1-35   The half-life of a radioactive element is the time required for half a given quantity of that element to disintegrate into a new element. For example, it takes 1600 years for half a given quantity of radium to change into radon. In another 1600 years, half the remainder will have disintegrated, leaving one-quarter of the original amount. The half-life of radium is therefore said to be 1600 years. If the rate of disintegration of a radioactive material is 35 parts per 100 every hour, what is the half-life of this material?

1-36   Find the coordinates of the maximum ordinate on the curve

$$y + 10 = 3x^2 - 2x$$

1-37 *a.* Plot the curve $y = x^3/100 - 5x/4$ for the interval $x = 0$ to $x = 9$.
   b. Write the equation od the tangent to the curve at $x = 5$.
   c. Calculate the area between the tangent defined in (*b*), the *x* axis, and the line $x = 5$.

1-38 A cylindrical tank having a flat top and bottom is to have a capacity of 10,000 cu ft. What should be its diameter and length if the surface area is to be a minimum?

1-39 A surveyor runs a line $AC = 380.00$ ft. With 0 as the midpoint on the line AC, he lays out a 20° curve CB. What is the distance between A and B? (See Fig. 1-P-39.)

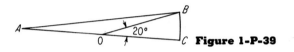

Figure 1-P-39

1-40 If x approaches zero in the expression $y = (5x^3 - \cos x + e^x)/(x^2 + 2x)$, then what does y approach as a limit?

1-41 Find the value of $\displaystyle\int_0^{\pi/2} \sin \theta \cos \theta \, d\theta$.

MULTIPLE-CHOICE PROBLEMS

For each question, select the correct answer from the five given possibilities.

1M- 1 The $\log_{10}$ of 2 is 0.301030. The $\log_{10}$ of ¹/₂ is:
   (*a*) $9.301030 - 10$ (*d*) $9.69897 - 10$
   (*b*) $0.301030/2$ (*e*) $1/0.301030$
   (*c*) $1 - 0.301030$

1M- 2 Which of the following is incorrect?
   (*a*) $\cos^2 A = \tan A \cot A$
   (*b*) $\sin A = \cos A \tan A$
   (*c*) $\cos 2A = \cos^2 A - \sin^2 A$
   (*d*) $2\sin^2 A = 1 - \cos 2A$
   (*e*) $\sin (A + B) = \sin A \cos B + \cos A \sin B$

1M- 3 The logarithm of the number -0.342 is:
   (*a*) positive   (*d*) A complex number
   (*b*) negative   (*e*) not a real number
   (*c*) zero

1M- 4 The value of $\tan (A + B)$, where $\tan A = $ ¹/₃ and $\tan B = $ ¹/₄ is (A, and B are acute angles):
   (*a*) ⁷/₁₂   (*d*) ⁷/₁₃
   (*b*) ¹/₁₁   (*e*) none of the above
   (*c*) ⁷/₁₁

1M- 5 If $x^{3/4} = 8$, x equals:
- (*a*) 6   (*d*) 16
- (*b*) 9   (*e*) 20
- (*c*) -9

1M- 6 If $\log_A 10 = 0.250$, *then* $\log_{10} A$ equals:
- (*a*) 4   (*d*) 0.25
- (*b*) 0.50   (*e*) 10
- (*c*) 2

1M- 7 The csc of 960° is equal to:
- (*a*) $-2\sqrt{3}/3$ (*d*) -2
- (*b*) 1   (*e*) $-^3/_2$
- (*c*) $^1/_2$

1M- 8 If $i = -1$, then i^{27} equals:
- (*a*) 0   (*d*) 1
- (*b*) i (*e*) -1
- (*c*) $-i$

1M- 9 If $x = \log_B N$, then:
- (*a*) $x = N^B$ (*d*) $N = B^x$
- (*b*) $B = x^N$ (*e*) $N = x^B$
- (*c*) $x = B^N$

1M-10 If x equals $+6$ and -4, the equation satisfying both these values would be:
- (*a*) $2x^2 + 3x - 24 = 0$ (*d*) $x^2 - 4x - 32 = 0$
- (*b*) $x^2 + 10x - 24 = 0$ (*e*) $x^2 - 3x - 18 = 0$
- (*c*) $x^2 - 2x - 24 = 0$

2

MECHANICS

Mechanics is the most basic of all the different subjects which combine to make up the composite of engineering. For this reason many of the fundamentals examinations place more emphasis on mechanics than on any of the other subjects.

This has been especially true in the past few years since the NCEE examinations have split the subject of mechanics into two parts—Statics and Dynamics. Each part has been treated as a major subject in recent past examinations. Thus, mechanics has been considered twice as important as any of the other subjects.

In recent fundamentals examinations, as previously noted, the single subject of mechanics has been divided into two subdivisions—statics and dynamics. Each of these has been treated as a major subject. However, many of the same principles apply to both branches of the general subject of mechanics, and there is no clear line of demarcation. Therefore, in this review, mechanics will be treated as a whole, and statics and dynamics will be treated as subdivisions of the major subject rather than as major subjects in their own right.

2-1. STATICS

The subject of statics can be summarized by the statement of Newton's first law of motion. "When a body is at rest or moving with constant speed in a straight line, the resultant of all of the forces exerted on the body is zero." Mathematically this can be written as:

$$\Sigma F_x = 0 \qquad \Sigma F_z = 0$$
$$\Sigma F_y = 0 \qquad \Sigma M = 0$$

or the summation of all the forces in the X, Y, and Z planes must equal zero and the summation of all moments must also equal zero. These conditions are both necessary and sufficient.

The study of statics consists merely of the development of different methods of satisfying the above four equations, which reduce to three equations for the two-dimensional case.

2-2. VECTORS

Forces are vectors, and both the magnitude of the force and its direction of application must be given before the force is completely defined. A vector quantity is one which has both magnitude and direction. If only the magnitude is given, the quantity is not a vector but a scalar quantity. A vector quantity can be represented graphically by an arrow of the correct length, pointed in the proper direction. Vector quantities can be added vectorially by moving the vector to be added parallel to itself until the tail or starting point of the vector to be added (the second vector) just touches the head or ending point of the vector to which it is to be added (the first vector). The vector sum will then be a vector starting from the tail of the first vector and ending at the head of the second vector. This process can be repeated for any number of vectors; the tail of the third vector should just touch the head of the second, etc. The final sum of all the vectors will be the vector starting at the tail (or beginning) of the first vector and ending at the head (or ending) of the last vector. This is shown graphically in Fig. 2-1. Note that we have added only the vector forces and have ignored any moments that may be present. These must be handled separately and will be discussed later.

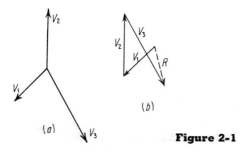

Figure 2-1

The described method, when two forces are considered, is sometimes termed the "triangle law," which may be stated as follows: "The resultant of two concurrent forces is represented in magnitude and in direction by the third side of a triangle, the other two sides of which represent the two forces in magnitude and direction."

A slightly different method of describing the same thing is by the "parallelogram law," which can be stated as: "The resultant of two forces which act on a rigid body is represented in magnitude and in direction by the diagonal of a parallelogram, the sides of which represent the magnitudes and directions of the two forces." These are illustrated in Fig. 2-2.

All the above methods are essentially the same, and any number of vectors can be added by either method by first adding vector 1 to vector 2, getting resultant 1, and then adding R_1 and V_3, getting R_2, which is equal to $V_1 + V_2 + V_3$, etc. (Fig. 2-1).

If two vectors can be added to equal a third vector, then it should also

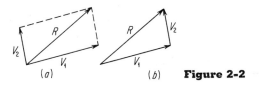

(a)   (b)   **Figure 2-2**

be possible to divide any single vector into two smaller vectors; this can be done. Any vector can be divided into two or more components. For mathematical calculations it is usually convenient to divide the vector into its components in the X and Y directions for two-dimensional problems and into its X, Y, and Z components for three-dimensional problems. The two-dimensional case is illustrated in Fig. 2-3. Here we see that each of the individual vectors is broken down into its X and Y components, and then all the X components and all the Y components are added. The resultant component forces obtained are the X and Y components of the resultant of these three forces and are easily calculated as shown, with $R = \sqrt{R_x^2 + R_y^2}$ and the angle equal to $\tan^{-1}(R_y/R_x)$.

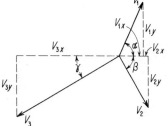

Figure 2-3

$$\Sigma F_x = V_{1x} + V_{2x} + V_{3x} =$$
$$V_1 \cos \alpha + V_2 \cos \beta - V_3 \cos \gamma$$

$$\Sigma F_y = V_{1y} + V_{2y} + V_{3y} =$$
$$V_1 \sin \alpha - V_2 \sin \beta - V_3 \sin \gamma$$

Total angles, measured from 0 to 360°, may be used, but it is usually easier to sketch the problem and use angles of 90° or less.

Often the summation of the component forces alone will not be sufficient to give the required answer. In such an event it will be necessary to use the summation of the moments as well. An example is shown in Fig. 2-4. Here we have a beam loaded as shown. There is no component in the X direction, and there are two unknown quantities R_1 and R_2. Two equations are required to determine the two unknowns; these equations result from the two relationships that the summation of the forces in the Y direction must equal zero and the summation of moments about any point in the system must also equal zero for the system to be in equilibrium.

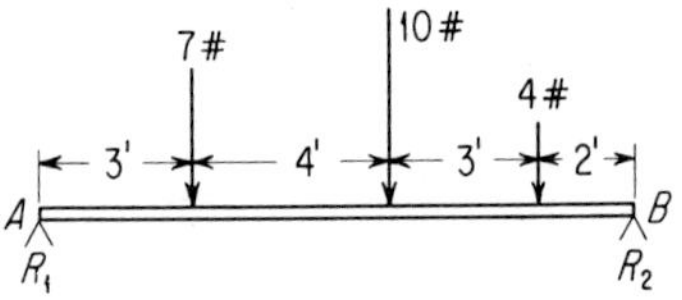

Figure 2-4

$$\Sigma F_y = 0 = R_1 + R_2 - 7 - 10 - 4$$
$$\Sigma M_A = 0 = -3 \times 7 - 7 \times 10 - 10 \times 4 + 12R_2$$
$$R_2 = \frac{21 + 70 + 40}{12} = 10.9\,\text{lb}$$
$$R_1 = 21 - 10.9 = 10.1 \text{ lb}$$

Check by taking
$$\Sigma M_B = 0 = 2 \times 4 + 5 \times 10 + 9 \times 7 - 12R_1.$$
$$R_1 = \frac{8 + 50 + 63}{12} = 10.1 \text{ lb}$$

2-3. COPLANAR FORCES

All determinate statics problems can be solved by the application of the two requirements that the summation of all forces and the summation of all moments must equal zero. A considerable saving in time and effort can often be made, however, by the application of the principle that if three forces are in equilibrium the forces must be coplanar and must also be either parallel or concurrent. This means that if we have three forces acting in a static condition, either these three forces must be parallel or else the lines of action of these three forces, when extended, must intersect at a single point. This principle can be extended to include the case of four coplanar forces by obtaining the resultant of two of the four forces, which then gives us three forces. The cases of five or more coplanar forces can also be treated similarly by the method of combining two or more forces into a single resultant. Examples of the application of this principle are shown in Fig. 2-5, where it is required to find the reactions of the planes on a 100-lb sphere held as shown.

This problem can be solved graphically by constructing the force triangle as shown, since we know both the magnitude and the direction of the weight force W, and the directions of the forces R_1 and R_2. The mathematical solution follows directly from the figure by the application of the law of sines.

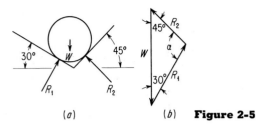

(a)   (b)   **Figure 2-5**

$$\frac{W}{\sin \alpha} = \frac{R_1}{\sin 45°} = \frac{R_2}{\sin 30°} \qquad \alpha = 180 - (30 + 45) = 105°$$

$$R_1 = \frac{\sin 45°}{\sin 105°} \times W = \frac{0.707}{0.966} \times 100 = 73.2 \text{ lb}$$

$$R_2 = \frac{\sin 30°}{\sin 105°} \times W = \frac{0.500}{0.966} \times 100 = 51.8 \text{ lb}$$

2-4. FREE-BODY DIAGRAMS

When working a problem concerning the equilibrium of a body or a juncture of forces, it is necessary to consider *all* the forces acting on the particular body or point under investigation. It is just as important to omit any force which does not act on the body being investigated, since one force too many or one force too few would destroy the conditions of equilibrium. The first step in attacking a problem in statics should then be the determination of just which forces are acting upon the body or juncture under investigation. The diagram of all these forces is called a free-body diagram.

A "free-body" is any system in static equilibrium. When a system is in static equilibrium, all internal forces are in balance, so no effect of any internal force can penetrate the boundary of a free body. The internal forces can thus all be ignored. Only the external forces must be considered. It should be noted that weight is always an external force since weight is a force caused by the gravitational attraction of the earth.

First a decision must be made as to what constitutes the body to be used. The writer then finds it convenient to draw a line around that part of the system to be considered as the free body. All known external forces acting on the free body should then be clearly shown on the diagram with their correct magnitudes and directions. The known external forces generally include the forces applied to the body and the weight of the body. The weight of a body will act as if it were concentrated at the center of mass of the body. The reactions to be solved for are exerted at the points where the free body is supported or connected to other bodies. The points of application of these unknown reactions will be known, but the directions of the forces may or may not be known, and each force may or may not be accompanied by a couple. Some samples of free-body diagrams are shown in Fig. 2-6. The unknown forces should be labeled and the senses of these unknown forces assumed. If the assumed sense of an unknown force is incorrect, subsequent calculations will show the correct direction. All the internal forces in a free-body diagram will be balanced within the free body and will have no effect on any of the external forces acting. See also Sec. 2-6.

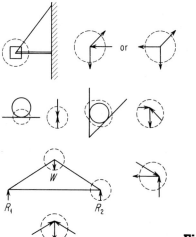

Figure 2-6

2-5. METHOD OF JOINTS

Figure 2-7 is an example of a statics problem taken from a past engineer-in-training examination and serves to illustrate some of the above points.

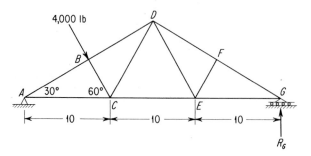

Figure 2-7

Here it is required to determine the reactions at A and G and the forces in the various members of the truss.

The three external forces acting on the framework are the reactions R_A and R_G and the applied force of 4000 lb. The vertical force at G, R_G, may be found by taking moments about point A.

$$\Sigma M_A = 0 = -4{,}000 \times 10 \cos 30° + 30 R_G$$
$$R_G = \frac{34{,}600}{30} = 1156 \text{ lb}$$

From $\Sigma F_y = 0$,

$$A_y + 1153 - 4000 \cos 30° = 0$$
$$A_y = 2308 \text{ lb}$$
$$A_x = 4000 \sin 30° = 2000 \text{ lb}$$

Construct a free-body diagram about point A (Fig. 2-8). Here we have two unknowns, the forces in AB and AC. These can be determined by satisfying the two equilibrium relationships: $\Sigma F_x = 0$ and $\Sigma F_y = 0$.

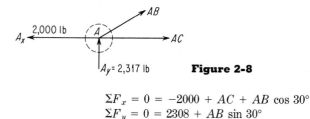

Figure 2-8

$$\Sigma F_x = 0 = -2000 + AC + AB \cos 30°$$
$$\Sigma F_y = 0 = 2308 + AB \sin 30°$$

$AB = -2308/\sin 30° = -4616$ lb, so the assumed direction of action of the force in member AB was incorrect, and AB is in compression, not in tension as tentatively expected.

$AC = 2000 - AB \cos 30° = 6000$ lb tension. We now have two unknown forces meeting at point B and three unknown forces meeting at point C. Since we have only two equations at our disposal ($\Sigma F_x = 0$ and $\Sigma F_y = 0$), we next solve for the forces acting at point B (Fig. 2-9).

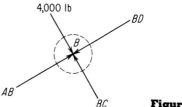

Figure 2-9

B is taken as a pin joint, so it can withstand no torque. Therefore, the applied force of 4000 lb must be resisted entirely by member BC. The force in BC will then equal 4000 lb compression. Similarly, the force in BD will equal 4616 lb compression.

There are now three unknown forces at point D and two at point C, so the forces acting at point C can now be determined (Fig. 2-10).

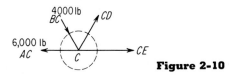

Figure 2-10

$$\Sigma F_x = 0 = -6000 + 4000 \cos 60° + CD \cos 60° + CE$$
$$\Sigma F_y = 0 = -4000 \sin 60° + CD \sin 60°$$

which gives CD = 4000 lb tension. Then

$$CE + 6000 - 2000 - 2000 = 2000 \text{ lb tension}$$

Now two of the four forces acting at point D are known and the remaining two unknown forces may be solved for (Fig. 2-11).

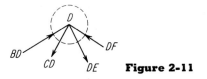

Figure 2-11

$$\Sigma F_x = 0 = BD \cos 30° - CD \cos 60° + DE \cos 60° - DF \cos 30°$$
$$\Sigma F_y = 0 = BD \sin 30° - CD \sin 60° - DE \sin 60° + DF \sin 30°$$

Solution of these two equations gives DE = 0 and DF = 2310 lb. Examination of the forces at point F indicates that since F is a pin joint, the force in EF must equal zero and $FG = DF$ = 2310 lb compression.

Examination of the forces acting at point E (Fig. 2-12) shows that $EG = CE$ = 2000 lb tension.

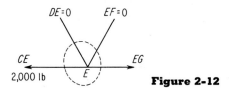

Figure 2-12

The forces acting at point G must total to zero if the analysis is correct (Fig. 2-13).

$$\Sigma F_x = 2310 \cos 30° - 2000 = 0$$
$$2000 - 2000 = 0$$
$$\Sigma F_y = 1156 - 2310 \sin 30° = 0$$
$$1156 - 1155 = 1 \text{ (slide-rule discrepancy)}$$

The check indicates that the calculations are correct.

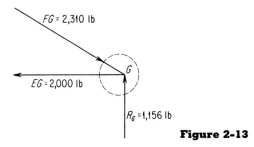

FG = 2,310 lb

EG = 2,000 lb

G

R_G =1,156 lb

Figure 2-13

The method used in determining the forces in the members of the truss is termed the "method of joints."

Another selected example is shown in Fig. 2-14. The problem is to determine:

■ *a*. The vertical components of the reactions at M and G
   b. The horizontal components of the reactions at M and G
   c. The stress in member DH

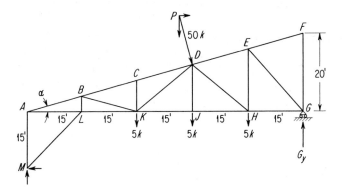

Figure 2-14

Since point G is on rollers, it can withstand no horizontal force. All the horizontal force must, then, be resisted at point M. The 50-kip load must first be resolved into its x and y components. Call the 50-kip load P. Then

$$\tan^{-1} (^{20}/_{75}) = \tan^{-1} 0.267 = 15°$$
$$P_x = 50 \text{ kips } (\sin 15°) = 12.9 \text{ kips}$$
$$P_y = 50 \text{ kips } (\cos 15°) = 48.3 \text{ kips}$$

Taking moments about point M gives $DJ = {}^3/_5 \times 20 = 12$ ft.

$$\Sigma M_M = 0 = -30 \times 5 \text{ kips} - 45 \times 5 \text{ kips} - 60 \times 5 \text{ kips} - 45$$
$$\times 48.3 \text{ kips} - (12 + 15) \times 12.9 \text{ kips} + 75G_y$$
$$G_y = \frac{3198}{75} = 42.6 \text{ kips}$$

From $\Sigma F_y = 0$,

$$M_y = 5 \text{ kips} + 5 \text{ kips} + 5 \text{ kips} + 48.3 \text{ kips} - G_y = 20.7 \text{ kips}$$

This can be checked by taking moments about point G.

$$\Sigma M_G = 0 = 15 \times 5 \text{ kips} + 30 \times 5 \text{ kips} + 45 \times 5 \text{ kips} + 30$$
$$\times 48.3 \text{ kips} - 12 \times 12.9 \text{ kips} - 75 M_y - 15M_x$$
$$M_y = \frac{1551}{75} = 20.7 \text{ kips}$$

which checks. To complete this check, we must know M_x, which equals P_x since $G_x = 0$.

$$M_x = 12.9 \text{ kips}$$

acting to the left from $\Sigma F_x = 0$. The answers to parts (a) and (b) are, then,

a. $M_y = 20.7$ kips $\quad G_y = 42.6$ kips
b. $M_x = -12.9$ kips (or 12.9 kips to the left) $\quad G_x = 0$

2-6. METHOD OF SECTIONS

We can determine the force in member DH by the method of joints as illustrated in the previous example, starting at either point M or point G. A quicker way is the method of sections. First, cut a section through members DE, DH, and JH, and draw the resulting free-body diagram, showing the actions of all the external forces (Fig. 2-15).

The section of the truss shown in Fig. 2-15 is a free-body diagram. All of the forces are in equilibrium since this section is statically stable. That means that all of the internal forces are in balance. This being the case, we can ignore the mutually balanced internal forces. Since the system is in

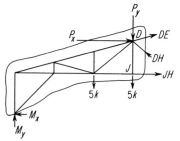

Figure 2-15

equilibrium, the external forces are also balanced. We then need only equate the external forces and solve for the unknowns.

Figure 2-15 shows that nine different forces are applied to the section under consideration. P_x and P_y are the components of the applied force P. M_x and M_y are the components of the reaction force at M. The two $5k$ forces are weight forces. It might be noted here that a weight, or mass, will always act to apply an external force to any system or free body on the earth which does not include the center of mass of the earth. This follows from the fact that weight is really the force of gravity produced between a mass on the surface of the earth and the mass of the earth which acts to draw them together. Thus any system or "free body" which contains a mass will have an external force exerted on that mass by the mass of the earth. In a static system this weight force will act at the center of mass of the object and will be directed toward the center of mass of the earth.

The remaining three forces applied to the section, or "free body," shown in Fig. 2-15 are the forces exerted on this subsystem by the three structural members DE, DH, and JH. These members extend outside of the "free body" and thus the forces in them act upon it, the free body.

The directions of the forces in members DE, DH, and JH are assumed; if these directions are later found to be in error, this will be shown by the signs of the answers. There are three unknown quantities; thus, three equations are necessary. Since the conditions of $\Sigma F_x = 0$, $\Sigma F_y = 0$, and $\Sigma M = 0$ must all be met for equilibrium to exist, three equations can be written.

First, the angle between DH and the horizontal must be determined (Fig. 2-16).

$$\beta = \tan^{-1}(^{12}/_{15}) = 38.7° \qquad \sin \beta = 0.625$$
$$\cos \beta = 0.781$$
$$\alpha = \tan^{-1}(^{20}/_{75}) = 15° \qquad \sin \alpha = 0.258$$
$$\cos \alpha = 0.966$$

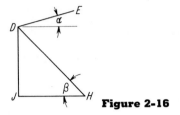

Figure 2-16

$$\Sigma F_x = 0 = -M_x + P_x + DE \cos \alpha + JH - DH \cos \beta$$
$$0 = 0.966DE - 0.781DH + JH$$
$$\Sigma F_y = 0 = M_y - 5 \text{ kips} - 5 \text{ kips} - P_y + DH \sin \beta - DE \sin \alpha$$
$$37.6 = 0.625DH - 0.258DE$$
$$\Sigma M_M = 0 = -30 \times 5 \text{ kips} - 5 \text{ kips} \times 45 - 45P_y - 27P_x - DE$$
$$\times 15 \cos \alpha - 15JH + 45DH \sin \beta + 27DH \cos \beta$$
$$2.896 = 49.2DH - 15JH - 14.5DE$$

Simultaneous solution of these three equations gives $DH = 77.2$ kips. Note that the direction of the force in DE was incorrectly assumed and that this member is actually in compression instead of in tension as initially assumed.

The same result can be obtained more easily by taking the section on the right end of the truss (Fig. 2-17).

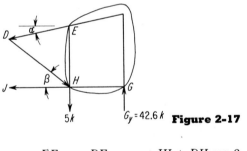

$G_y = 42.6 k$ **Figure 2-17**

$$\Sigma F_x = -DE \cos \alpha - JH + DH \cos \beta = 0$$
$$-0.966DE - JH + 0.781DH = 0$$
$$\Sigma F_y = 42.6 \text{ kips} - 5 - DH \sin \beta - DE \sin \alpha$$
$$37.6 \text{ kips} = 0.625DH + 0.258 DE$$
$$\Sigma M_G = 0 = 5 \text{ kips} \times 15 + 20 \cos \alpha \times DE + 15 \sin \beta \times DH$$
$$75 \text{ kips} = -19.3DE - 9.38DH$$

Simultaneous solution of these three equations also gives $DH = 77.1$ kips, which checks.

2-7. CABLE SUPPORTS

The case of a weight suspended from a (frictionless) pulley requires a slightly different approach. As an example take the problem illustrated in Fig. 2-18. It is required to find the tension in the cable and the distances H and S.

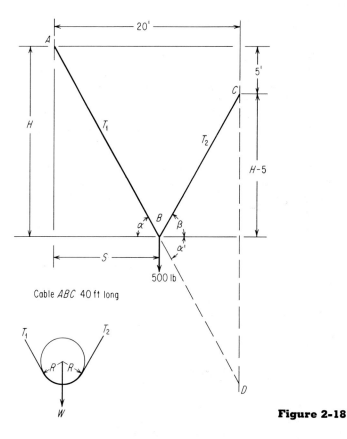

Figure 2-18

Take the pulley as a free body in static equilibrium. The summation of moments about the axis of the pulley must then equal zero, or $T_2 R - T_1 R = 0$. Thus $T_1 = T_2$, and the tension is the same over the length of the cable. Similarly, since the summation of forces in the X direction must equal zero,

$$T \cos \beta - T \cos \alpha = 0 \quad \text{and} \quad \alpha = \beta$$

Vertical angles are equal so $\alpha = \alpha'$. Then triangles BCE and BDE are congruent and the distance AD equals 40 ft. Then $\alpha = \cos^{-1}(20/40) = 60°$.

From the geometry of the figure $40 \sin \alpha = 2H - 5$, which gives $H = 19.82$ ft. Since $S = H/\tan \alpha$, $S = 11.44$ ft. The length of cable section T_1 equals 22.89 ft and the length of section T_2 equals 17.11 ft.

2-8. CENTER OF GRAVITY

The location of the center of gravity of a body or of a system is important because the weight of the body or system acts at its center of gravity. Later, when we review the subject of dynamics, the location of the center of gravity will again prove important because the inertia of a body acts at the center of gravity (or the center of mass). By definition, the sum of all the moments due to the weight (or mass) of a body about its center of gravity must equal zero, which gives the mathematical expression $\Sigma r \, \Delta W = 0$, where r is the distance to the infinitesimal weight ΔW, and the summation of all the ΔW's equals the total weight.

2-9. CENTROID

For a plane surface, which has no weight, the term center of gravity is replaced by the "centroid of the area," which gives the mathematical expression $\Sigma r \, \Delta A = 0$. Since we know that the weight of a body or system will act as if it were concentrated at the center of gravity, the moment of a body or system about any point 0 will equal WS, where S is the horizontal distance from point 0 to the center of gravity. In addition, the moment of the body or system about the point 0 will also equal $\Sigma a \, \Delta W$, where ΔW is an increment of the weight of the system and a is the horizontal distance from 0 to the center of gravity of that increment of weight. The distance from the point 0 to the center of gravity can then be determined by setting these two equations equal to each other and solving for S, or $(\Sigma a \, \Delta W)/W = S$. Referring to Fig. 2-19, this can be illustrated as follows:

Figure 2-19

$$S = \frac{\Sigma a \, \Delta W}{\Sigma \, \Delta W} = \frac{a_2 W_2 + a_3 W_3}{W_1 + W_2 + W_3}$$

Taking moments about the center of gravity of W_1, if we take $a_2 = 5$ ft, $a_3 = 12$ ft, $W_1 = 3$ lb, $W_2 = 5$ lb, $W_3 = 7$ lb, then the distance from the center of gravity of W_1 to the center of gravity of the system is:

$$S = (5 \times 5 + 12 \times 7)/(3 + 5 + 7) = 7.27 \text{ ft}$$

To prove this, calculate the sum of all the moments about the center of gravity of the system.

$$\Sigma M_{CG} = W_1 \times 7.27 + W_2 \times 2.27 - W_3 \times 4.73$$
$$= 21.81 + 11.35 - 33.11 = 0.05$$

Another illustration is afforded by the determination of the centroid of a triangle.

First orient the triangle so that the apex lies on the Y axis and the base is parallel to the Y axis (Fig. 2-20).

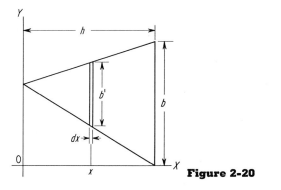

Figure 2-20

$$S = \frac{\displaystyle\int_0^h x \, dA}{A} \qquad \begin{aligned} dA &= b' \, dx \\ b' &= (x/h) \times b \\ A &= \tfrac{1}{2}bh \end{aligned} \qquad \text{giving } dA = (b/h) \cdot x \, dx$$

$$S = \frac{\displaystyle\int_0^h x \, b/h \, x \, dx}{\tfrac{1}{2}bh} = \frac{b/h \displaystyle\int_0^h x^2 \, dx}{\tfrac{1}{2}bh} = \frac{\tfrac{1}{3}h^3}{\tfrac{1}{2}h^2} = \tfrac{2}{3}h$$

or the centroidal axis of a triangle is two-thirds the distance from the apex to the base. By repeating the above calculations for different orientations it may be shown that the centroid is at the intersection of the three lines drawn parallel to each of the three sides at distances one-third the way from the side to the opposite vertex. If the triangle had thickness and weight, the distance to the center of gravity would equal

$$\frac{\int x\, dW}{W} = \frac{\int x\, dA \cdot t \cdot w}{A \cdot t \cdot w} = \frac{\int x\, dA}{A}$$

which is the same as the calculation for the centroid of the area. Another past examination problem was as follows:

- Find the location of the horizontal centroidal axis of the built-up beam shown in Fig. 2-21.

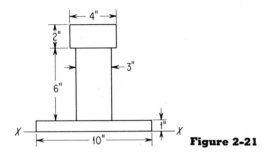

Figure 2-21

$$\bar{y} = \frac{\Sigma s\, \Delta A}{A} = \frac{\frac{1}{2}(10 \times 1) + (1 + 3)(6 \times 3) + (1 + 6 + 1)(2 \times 4)}{10 \times 1 + 6 \times 3 + 4 \times 2}$$
$$= 3.91 \text{ in.}$$

Another problem was stated as follows:

- A horizontal plate rests on three posts: A, B, C. Material of uniform density is piled on the plate at a depth increasing uniformly from zero at the right edge to a maximum at the left edge. If the total weight of the material is 1000 lb, what are the reactions at A, B, and C? Disregard the weight of the plate. (See Fig. 2-22.)

This problem can be solved by determining the location of the center of gravity and then proceeding as if all the weight were concentrated at that point. The center of gravity will be located two-thirds of the distance from end C to end AB and midway between the 9-ft edges A and B.

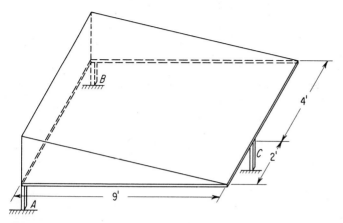

Figure 2-22

From

$$\Sigma F_y = 0 = -1000 + A + B + C$$
$$\Sigma M_{A-B} = 0 = -1000 \times 3 + 9C$$

which gives

$$C = 333 \text{ lb}$$
$$\Sigma M_{\text{edge }B} = 0 = -3 \times 1000 + 4C + 6A$$
$$A = \frac{1667}{6} = 278 \text{ lb}$$
$$B = 1000 - 278 - 333 = 389 \text{ lb}$$

or, taking

$$\Sigma M_{\text{edge }A} = 0 = -1000 \times 3 + 2C + 6B$$
$$B = \frac{2334}{6} = 389 \text{ lb}$$

which checks.

One more example, illustrating that the center of gravity or centroid need not always lie within the piece, is afforded by a past examination problem:

- Find the location of the centroid of the figure shown in Fig. 2-23.

Find the distance of the centroidal axis from the left edge.

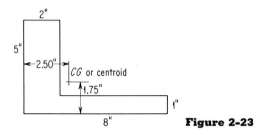

Figure 2-23

$$\frac{(5 \times 2) \times 1 + (6 \times 1)(2 + 3)}{(5 \times 2) + (6 \times 1)} = 2.50 \text{ in.}$$

Next determine the distance of the centroidal axis above the base

$$\frac{(5 \times 2) \times 2\frac{1}{2} + (6 \times 1) \times \frac{1}{2}}{(5 \times 2) + (6 \times 1)} = 1.75 \text{ in.}$$

To summarize, the resultant of the weight of a body or system will pass through one point in the body, or the body extended, for all orientations of the body; this point is defined as the center of gravity of the body or system. The centroid of an area corresponds to the center of gravity of a thin plate and may be determined as the center of gravity is determined, by assuming the area to be a thin plate.

2-10. FRICTION AND FRICTION ANGLE

The next important topic to be discussed is the effect of friction. A frictional force will always be in such a direction as to resist motion. It is a passive force and can never exceed whatever force may be applied to an object. It can never cause motion; it may vary from zero to a maximum. The maximum frictional force which can be exerted on an object is equal to the coefficient of friction times the force *normal* to the surface along which the frictional force is exerted. This is easily demonstrated in its simplest form by the forces acting on a block resting on an inclined plane (Fig. 2-24).

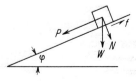

Figure 2-24 μ = coefficient of friction between block and plane.

In this figure we see that the frictional force is equal to $W \cos \varphi \times \mu$, or the normal force times the coefficient of friction between the block and the plane. The force parallel to the face of the plane tending to cause the block to slide down the plane is equal to $W \sin \varphi$. If the angle of the plane is adjusted so that slipping impends, or if the block is pushed to start it moving and the velocity of the block down the plane remains constant, the value of the angle φ is termed the "friction angle." In this case the frictional force resisting motion will be just equal to the force component $W \sin \varphi$. This gives the relationship $W \sin \varphi = \mu W \cos \varphi$, which reduces to $\mu = \tan \varphi$, or the coefficient of friction between the block and the plane is equal to the tangent of the friction angle. This concept of a friction angle is extremely useful and considerably simplifies the working of the more complicated friction problems. It is possible to add the friction angle to whatever mechanical angle is being considered (or subtract the friction angle from the mechanical angle) to obtain an equivalent angle of a frictionless system, and the answer can be readily calculated. The method is illustrated as follows:

- Referring to Fig. 2-25, how large a horizontal force F must be applied to a block of weight W resting on an incline of angle α to cause motion to impend up the slope? The coefficient of friction between the block and the slope is μ.

Figure 2-25

The force acting up the plane parallel to it equals $F \cos \alpha$. The forces resisting motion are the force down the slope due to the weight of the block, equal to $W \sin \alpha$, and the frictional resistance equal to the force exerted by and on the block normal to the plane—N times the coefficient of friction μ. N is made up of two components, $W \cos \alpha$ and $F \sin \alpha$.

Adding the forces acting, we have

$$F \cos \alpha = W \sin \alpha + (W \cos \alpha + F \sin \alpha)\theta$$

and if we substitute $\tan \varphi$ for its equivalent, μ, we get

$$F \cos \alpha = W \sin \alpha + (W \cos \alpha + F \sin \alpha) \frac{\sin \varphi}{\cos \varphi}$$

or

$$F \cos \alpha \cos \varphi = W \sin \alpha \cos \varphi + W \cos \alpha \sin \varphi + F \sin \alpha \sin \varphi$$

which gives

$$F(\cos \alpha \cos \varphi - \sin \alpha \sin \varphi) = W(\sin \alpha \cos \varphi + \cos \alpha \sin \varphi)$$

which reduces to

$$F \cos (\alpha + \varphi) = W \sin (\alpha + \varphi)$$

or
$$F = W \tan (\alpha + \varphi)$$

which is the same result we should have obtained had we considered the slope angle to be α + the friction angle, or $(\alpha + \varphi)$, and considered the surfaces to be frictionless.

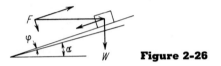

Figure 2-26

Utilizing the concept of friction angle, resketch the problem as shown in Fig. 2-26. In this case we get:

$$F \cos (\alpha + \varphi) = W \sin (\alpha + \varphi)$$
$$F = W \tan (\alpha + \varphi)$$

as before.

Similarly it can be shown that to find the force to just hold the block in place we can consider the friction angle as being subtracted from the slope angle and again treat the system as if it were frictionless.

From Fig. 2-27,
$$F \cos (\alpha - \varphi) = W \sin (\alpha - \varphi)$$
or
$$F = W \tan (\alpha - \varphi)$$

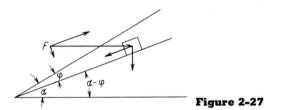

Figure 2-27

By adding or subtracting the friction angle, it is possible to incorporate the effects of friction without treating the frictional components separately. One implication of the friction-angle concept is immediately discernible;

i.e., if the friction angle is greater than the angle of the slope, the system will be statically stable.

Figure 2-28 shows the condition of a statically stable wedge where $\varphi > \alpha$ and the weight will not slip.

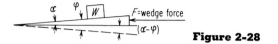

Figure 2-28

On the other hand, Fig. 2-29 shows a wedge which is not statically stable. The weight would slide as if on a frictionless plane with a slope of $\alpha - \varphi$; or the wedge would shoot out if the force F were removed. Also, if the sum of the slope angle plus the friction angle equals or exceeds 90°, it would be impossible to push the block up the slope with a force which is applied horizontally.

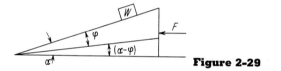

Figure 2-29

A word of caution should be given about the use of the friction angle: Do not change the direction of the application of the force. As an example, if it is desired to determine the force required to push a block up a plane, the direction of the force must not be changed when the friction angle is added to the slope angle to determine the resisting force (Fig. 2-30).

$$F \cos \varphi = W \sin (\alpha + \varphi)$$
$$F = \frac{W}{\cos \phi} \sin (\alpha + \varphi)$$

where F is the force parallel to the face of the plane required to push the block up the plane.

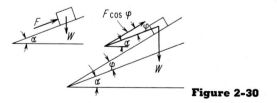

Figure 2-30

A more involved type of wedge and friction problem is illustrated in Fig. 2-31.

■ For the wedge system shown, find the force F required to raise the weight W. The coefficient of friction for all surfaces is equal to $\mu =$ $\tan \varphi$.

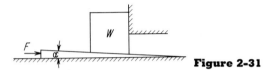

Figure 2-31

Applying the principle of friction angle to the problem, we can construct a new figure which will incorporate the effects of the friction angles and give us an equivalent frictionless system (Fig. 2-32). Here we can determine the relationship between F and W by means of the conservation of energy (or equivalence of work); i.e.,

$$FS = Wh_{\text{effective}}$$

$h_{\text{effective}}$ will be made up of two parts:

$$h_{\text{eff}} = S[\tan (\alpha + \varphi) + \tan \varphi] + h'$$

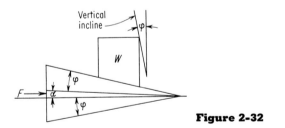

Figure 2-32

where h' is the additional elevation due to the movement to the left which results from the movement of the upper right corner of the weight in the figure along the back angle of the vertical incline. These quantities are shown in Fig. 2-33.

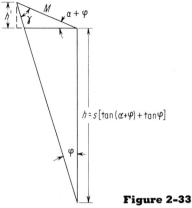

$$h = s[\tan(\alpha+\varphi) + \tan\varphi]$$

Figure 2-33

$$\frac{M}{\sin \varphi} = \frac{h}{\sin \gamma} \qquad h' = M \sin (\alpha + \varphi)$$
$$\gamma = 90 - (\alpha + 2\varphi)$$
$$\sin \gamma = \cos (\alpha + 2\varphi)$$

$$h' = \frac{\sin \varphi}{\sin \gamma} \times h \times \sin (\alpha + \varphi)$$

or

$$h' = \frac{\sin \varphi \times \sin (\alpha + \varphi)}{\cos (\alpha + 2\varphi)} \times h$$

$$h_{\text{eff}} = S[\tan (\alpha + \varphi) + \tan \varphi]\left[1 + \frac{\sin \varphi}{\cos (\alpha + 2\varphi)} \sin (\alpha + \varphi)\right]$$

which gives

$$F = W[\tan (\alpha + \varphi) + \tan \varphi]\left[1 + \frac{\sin \varphi}{\cos (\alpha + 2\varphi)} \sin (\alpha + \varphi)\right]$$

If we take $W = 1000$ lb, $\mu = 0.3$, and $\alpha = 5°$, this will give

$$F = W(\tan 21.7° + \tan 16.7°)\left(1 + \frac{\sin 16.7°}{\cos 38.4°}\sin 21.7°\right)$$
$$= W(0.398 + 0.300)\left(1 + \frac{0.288}{0.784} \times 0.370\right)$$
$$= 793 \text{ lb}$$

The same results may be obtained by a combination of the principles of friction angle and concurrent forces. To use this procedure it is first necessary to draw the free-body diagrams of the block and wedge and to show all the forces acting at their effective angles (Fig. 2-34).

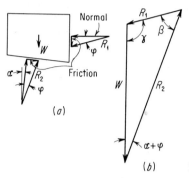

Figure 2-34

Referring to the figure,

$$\frac{W}{\sin \beta} = \frac{R_2}{\sin \gamma}$$

$$\gamma = 90 + \varphi$$
$$\sin \gamma = \cos \varphi$$
$$\beta = 90 - (\alpha + 2\varphi)$$

$$R_2 = \frac{\cos \varphi}{\sin \beta} W = \frac{0.958}{0.784} W = 1.223W$$

The concurrent-force diagram for the wedge is shown in Fig. 2-35.

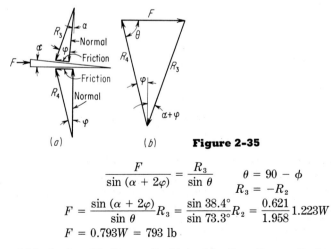

Figure 2-35

$$\frac{F}{\sin (\alpha + 2\varphi)} = \frac{R_3}{\sin \theta}$$

$$\theta = 90 - \phi$$
$$R_3 = -R_2$$

$$F = \frac{\sin (\alpha + 2\varphi)}{\sin \theta} R_3 = \frac{\sin 38.4°}{\sin 73.3°} R_2 = \frac{0.621}{1.958} 1.223W$$

$$F = 0.793W = 793 \text{ lb}$$

which checks with the result obtained by the other method.

2-11. JACKSCREW

One other example is the derivation of the mechanical efficiency of a jackscrew (Fig. 2-36). First assume a square thread, with a pitch P.

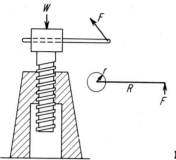

Figure 2-36

$$\text{Pitch} = \frac{1}{\text{No. of threads/in.}}$$

Also consider the screw to be single-threaded (lead equal to pitch). The coefficient of friction is equal to μ, and the mean thread diameter is D_m. From Fig. 2-37 we can see that if we unwind the thread, the lead angle λ is easily determined; it is equal to $\tan^{-1}(\text{lead}/\pi D_m)$. The friction angle φ is equal to $\tan^{-1}\mu$. The equivalent angles of frictionless screws are $\lambda + \varphi$ if the load is being raised and $\lambda - \varphi$ if the load is being lowered. A torque T_q, lb-in., applied to the jackscrew will supply a force F equal to $2T_q/D_m$ at the thread.

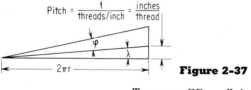

Figure 2-37

$$\text{Torque} = FR \qquad \text{lb-in.}$$

The equivalent force at a distance of 1 in. from the axis is then equal to the torque, and the work done per revolution of the screw equals $2\pi \times$ torque. This also equals the work done in raising the weight a distance equal to the pitch P plus the work done against friction, or

$$2\pi Tq = \pi 2r \tan(\lambda + \varphi)W$$
$$Tq = \frac{D_m \tan(\lambda + \varphi)}{2}W = r\tan(\lambda + \varphi)W = FR$$

If $\varphi > \lambda$, the jackscrew will be statically stable and will require no force on the handle to maintain the load in its position. On the other hand, if $\lambda > \varphi$, the weight will lower by itself if all force is removed from the handle. This case is similar to that of the wedge, and the effective angle is $\lambda - \varphi$ (Fig. 2-38).

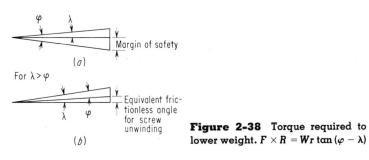

For $\lambda > \varphi$

Figure 2-38 Torque required to lower weight. $F \times R = Wr \tan (\varphi - \lambda)$

The torque in the preceding example was calculated from the equivalence of the work done on the system and the energy of the system. This is a very important relationship based on the theory of the conservation of energy. The change in energy of a system—potential energy plus kinetic energy (and heat energy resulting from friction)—will equal the work done on the system, or by the system. This will be discussed further in the next section.

2-12. DYNAMICS

Just as the subject of statics is based on Newton's first law of motion, the subject of dynamics is based on Newton's second law plus effects of Newton's other two laws. The first law has already been stated in the section devoted to statics, but we shall repeat it here with the other two of Newton's laws so that we shall have them all together. The three laws may be stated as follows:

1. Every body persists in its state of rest or of uniform motion in a straight line unless it is compelled by some force to change that state.
2. The rate of change of the momentum of a body is proportional to the force acting upon it and it is in the direction of the force ($F = ma$).
3. Action and reaction are equal and opposite.

2-13. EQUATIONS OF LINEAR MOTION

Linear motion, or motion along a straight line, is called "velocity," where velocity is a vector quantity (it has both magnitude and direction) and is equal to the rate of change of distance with time. Average velocity equals displacement/elapsed time, and if we take the distance covered as $s - s_0$ in the time interval $t - t_0$, the average velocity

$$v_{avg} = \frac{s - s_0}{t - t_0}$$

This can also be written as $v = \Delta s/\Delta t$, which, as Δt approaches zero, has the limiting value $v = ds/dt$. Similarly, the rate of change of velocity with time is termed "acceleration," and the limiting value of the acceleration, $a = dv/dt = d^2s/dt^2$. If we start with a constant acceleration, $a = dv/dt$, and integrate, we get $v = at + c_1$, where c_1 is the constant of integration. Replacing v by its equivalent, we have $ds/dt = at + c_1$, which, upon integration, gives us $s = \frac{1}{2}at^2 + c_1t + c_2$, where c_2 is the constant resulting from the second integration. If $s = s_0$ and $v = v_0$ at time $t = 0$, we can solve for the constants of integration. Substitution of these gives us the general relationship for distance: $s = s_0 + v_0t + \frac{1}{2}at^2$.

If $s = 0$ at $t = 0$, this gives us the more familiar relationship $s = v_0t + \frac{1}{2}at^2$. If the equation $v = v_0 + at$ is solved for t and the value substituted in the equation for s, we get $v^2 = v_0^2 + 2as$. Thus we can easily derive the four basic equations of linear motion:

$$v = v_0 + at$$
$$v^2 = v_0^2 + 2as$$
$$\text{avg } v = \frac{v + v_0}{2}$$
$$s = v_0t + \frac{1}{2}at^2 \quad \text{or} \quad s = v_{avg}t$$

An example of the use of these equations is afforded by a past examination problem, which asked:

■ An automobile starting from rest increased its speed in 18 sec to 66 fps. What was its final speed in miles per hour? What was its average speed in miles per hour during the 18 sec its speed was increasing? Assume uniform acceleration. Find also the acceleration and the distance traveled during the 18 sec.

From $v = v_0 + at$, $v = 66$ fps, $v_0 = 0$, $t = 18$ sec, and

$$a = \frac{66}{18} = 3.67 \text{ ft/sec}^2$$

the final speed in miles per hour equals

$$66 \text{ fps} \times 3600 \text{ sec/hr} \times \frac{1}{5280} \text{ miles/ft}$$

which equals 45 mph.

The average speed = $(45 + 0)/2 = 22.5$ mph, and the distance traveled equals average $v_{avg} \times t = {}^{66}/_2 \times 18 = 594$ ft; or the distance may be calculated from

$$s = {}^1/_2 at^2 = {}^1/_2 \times 3.67 \times 18^2 = 594 \text{ ft}$$

2-14. RELATIVE VELOCITY

Velocity is defined as the rate of change of distance (from some arbitrarily selected point) with time, or $v = ds/dt$. Thus, a velocity possessed by an object must be in reference to some other object or reference system. That is, *all velocities are relative*. Usually we think of a velocity (i.e., any noncelestial velocity) in reference to the surface of the earth and tend to think of this as *the* velocity of an object. That can become confusing. Take, for example, the following problem:

■ A runaway railroad car is moving along a (frictionless) horizontal stretch of track at a constant velocity of 30 mph. A locomotive starts out in pursuit and reaches a speed of 60 mph when it is exactly 1 mile behind the runaway car. The engineer (locomotive engineer, that is) starts to decelerate the engine at a constant rate so that when it touches the car to couple with it, the locomotive will be going at the same rate of speed as the car. What should be the rate of deceleration and what additional distance would the runaway car travel before being caught?

This problem can be worked by relating all velocities to the surface of the earth. If this approach is used, then, referring to Fig. 2-39, the following relationships can be written:

$$D_L = D_c + 1$$
$$60t + {}^1/_2 at^2 = 30t + 1 \qquad \text{giving } 30t + {}^1/_2 at^2 = 1$$

With constant deceleration the average velocity of the locomotive would equal $(60 + 30)/2 = 45$ mph.

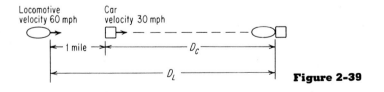

Locomotive velocity 60 mph   Car velocity 30 mph

1 mile D_c D_L

Figure 2-39

$$D_L = D_c + 1 \qquad \text{gives } 45t = 30t + 1 \qquad \text{or } t = \text{}^1/_{15} \text{ hr.}$$
$$30 \times (^1/_{15}) + ^1/_2 a(^1/_{15})^2 = 1 \text{ which gives}$$
$$a = -450 \text{ mph/hr}$$
$$D_L = 30 \times (^1/_{15}) = 2 \text{ miles}$$

It is more straightforward and simple to use relative velocities. The initial relative velocity equals

$$60 - 30 = 30 \text{ mph} \qquad \text{or} \qquad u_0 = 30 \text{ mph}$$

The final relative velocity equals zero.

$$u^2 = u_0{}^2 + 2aS$$
$$0 = 30^2 + 2a \times 1 \qquad \text{giving } a = -450 \text{ mph/hr}$$
$$S = u_0 t + ^1/_2 at^2 \qquad \text{or} \qquad S = u_{\text{avg}} \times t$$
$$u_{\text{avg}} = {}^{30}/_2 = 15 \text{ mph} \qquad S = 1 \text{ mile} \qquad t = ^1/_{15} \text{ hr}$$

so the runaway car would travel an additional 2.0 miles before being caught. The deceleration rate of 450 mph/hr sounds high, but is actually only 0.183 ft/sec², or 0.0057g.

2-15. FLIGHT OF A PROJECTILE

One other example of the use of the equations of linear motion is afforded by a past examination problem which asked for details regarding the flight of a projectile. The problem and its analysis are as follows:

- A bullet leaves a gun muzzle at a velocity of 2700 ft/sec at an angle of 30° with the horizontal.
  a. What is the maximum height to which the bullet will travel?
  b. What is the maximum distance the bullet will travel horizontally measured along the same elevation as the gun muzzle? Disregard air resistance.

First sketch the figure and determine the velocity components (Fig. 2-40).

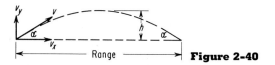

Figure 2-40

The vertical component of the velocity $v_y = v \sin \alpha$, and the component $v_x = v \cos \alpha$. In the vertical direction the bullet will be subjected to the acceleration of gravity, and if air resistance is disregarded, the acceleration will equal g. The time required for the bullet to reach its apogee can be determined from the fact that at the top the vertical velocity will equal zero. Then, from $v = v_0 + at$, we have $0 = v_y - gt$, or

$$\text{Time to apogee} = \frac{v_y}{g} = \frac{v \sin \alpha}{g}$$

The time for the bullet to fall back to the earth will equal the time to rise to the apogee, or the total time of flight will equal $2 \times [(v \sin \alpha)/g]$. The maximum height can be calculated from the relationship $v_{y\text{avg}} \times t = h$, where $v_{y\text{avg}} = (v_y + 0)/2$. The time t to rise to the height h has already been determined, and

$$h = \frac{v_y}{2} \frac{v_y}{g} = \frac{(v_y)^2}{2g}$$

The height h can also be determined from the relationship

$$v^2 = v_0{}^2 + 2as$$

which gives $(v_y)^2 = 2gh$, or $h = v_y{}^2/2g = (v^2 \sin^2 \alpha)/2g$. The range R will equal the distance traveled horizontally during the total time of flight at the constant horizontal velocity $v_x = v \cos \alpha$. $R = v \cos \alpha \times (2v \sin \alpha)/g = 2[(v^2 \cos \alpha \sin \alpha)/g]$. This last relationship could be reduced to $R = (v^2 \sin 2\alpha)/g$, by substituting equivalent angle-function relationships, but this additional step is no advantage to the occasional user.

The numerical answers will be:

$$h = \frac{v^2 \sin^2 30°}{2g} = \frac{2700^2 \times 0.500^2}{2 \times 32.2} = 28{,}300 \text{ ft}$$

$$R = 2\frac{v^2 \cos \alpha \sin \alpha}{g} = 2\frac{2700^2 \times 0.866 \times 0.500}{32.2} = 196{,}000 \text{ ft}$$

Note that the only equations necessary for the solution of this problem are the basic velocity relationships: $v = v_0 + at$, $s = v_{\text{avg}}t$, and $v^2 = v_0{}^2 + 2as$.

2-16. MOMENTUM AND IMPULSE

The momentum of a body is equal to the product of its mass times its velocity, or $M = mv$. Newton's second law of motion states that $F = k(dM/dt)$, where k is the constant of proportionality. If we consider the mass as being constant, dM/dt reduces to $m(dv/dt)$, and if we select the proper units k is equal to 1. This gives us the familiar relationship $F = ma$.

A falling body has a force acting on it equal to its own weight, and the acceleration equals the acceleration of gravity; thus $W = mg$, or mass $= W/g$. Newton's second law of motion can then be stated as $F = ma$, where $m = W/g$. This is a very important relationship and is the foundation of the study of dynamics.

In addition to this, we see from Newton's second law of motion that if no external force acts, the momentum of a system will remain constant. This leads to the concept of the conservation of momentum. Since momentum is equal to $m \times v$, or is the product of a vector and a scalar, it is a vector quantity and thus must possess both magnitude and direction.

From $F = ma$ and $a = dv/dt$, we obtain $F \, dt = m \, dv$, from which we see that $F \, dt = mv_2 - mv_1$. The integral of $F \, dt$ is termed the "impulse," and the quantity $m(v_2 - v_1)$ is the change in momentum of mass m. This gives the relationship that impulse equals change in momentum, where impulse is also a vector quantity.

To illustrate, let us take as an example a golf ball which is struck by a club. High-speed photography shows that the club is in contact with the ball for approximately one-half-thousandth of a second and the velocity of the ball is 250 fps immediately after being struck. What force is exerted by the club if the ball weighs 1²/₃ oz? Assume that the force is constant during the period of contact, so the impulse will equal $0.005F$. The change in momentum of the golf ball will equal mv, or $[1.67/(16 \times 32.2)]250 = 0.810$ lb-sec, or 0.810 slug-ft/sec. Since the change in momentum is equal to the impulse, $F = 0.810/0.0005 = 1620$ lb.

This problem could also have been worked using the relationships $F = ma$ and $a = \Delta v/\Delta t$, giving

$$F = \frac{1.67}{16 \times 32.2} \frac{250}{0.0005} = 1620 \text{ lb}$$

Another example of the conservation of momentum is illustrated by the following problem:

■ A plastic body weighing 100 lb is moving with a velocity of 20 ft/sec and overtakes another plastic body, weighing 150 lb, moving in the same direction with a velocity of 15 ft/sec. What is the final velocity, and what is the loss in kinetic energy?

From the principle of the conservation of momentum we have

$$m_1v_1 + m_2v_2 = (m_1 + m_2)v_r$$

The resulting velocity

$$v_r = \frac{m_1 v_1 + m_2 v_2}{m_1 + m_2} = \frac{100 \times 20 + 150 \times 15}{100 + 150}$$
$$= 17 \text{ fps}$$

Note that a $1/g$ factor was canceled out of each term.

$$\text{KE initially} = \frac{1}{2} m_1 v_1{}^2 + \frac{1}{2} m_2 v_2{}^2$$
$$= \frac{100}{2g} \times 20^2 + \frac{150}{2g} \times 15^2$$
$$= 1{,}144 \text{ ft-lb}$$
$$\text{KE final} = \frac{1}{2}(m_1 + m_2) v_r{}^2$$
$$\frac{100 + 150}{2g} \times 17^2$$
$$= 1120 \text{ ft-lb}$$

The loss in kinetic energy = 24 ft-lb.

Note that energy is not a vector quantity and that the direction has no bearing on the amount of kinetic energy possessed by a moving body. Momentum, on the other hand, is a vector quantity, and direction of motion is very important. It might also be mentioned that even though 24 ft-lb of mechanical energy has been lost in the impact, the total energy of the system is constant, and this lost mechanical energy will have been converted into heat because of the force of the impact and the subsequent deformation of both bodies.

The impact in this sample problem is a plastic impact or a completely inelastic impact with a coefficient of restitution equal to zero. If a small body rebounds from a very large one, such that the change of velocity of the large body is negligible, the coefficient of restitution $e = -(v_2/v_1)$. (It must be remembered that velocity is a vector quantity; thus v_2 will be opposite in sign to v_1.) If a ball rebounds or bounces up after being dropped onto a fixed plate, then

$$e = -\frac{v_2}{v_1} = \frac{\sqrt{2gh_2}}{\sqrt{2gh_1}} = \sqrt{\frac{h_2}{h_1}}$$

or the coefficient of restitution is equal to the square root of the height of bounce divided by the height of drop. If the impact is between two objects which do not differ greatly in mass, then

$$e = -\frac{v_2 - u_2}{v_1 - u_1}$$

where $v_1 - u_1$ = relative velocity before impact
$v_2 - u_2$ = relative velocity after impact

As an example of the use of this relationship, determine the velocities after impact of an 8-lb ball traveling with a velocity of 20 ft/sec which overtakes and strikes a 12-lb ball traveling in the same direction with a velocity of 8 ft/sec if the coefficient of restitution is 0.75 (Fig. 2-41).

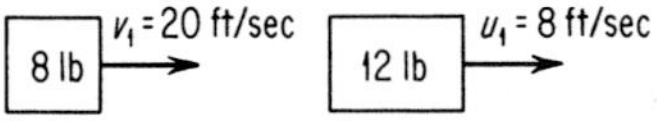

$v_1 = 20$ ft/sec   8 lb $u_1 = 8$ ft/sec   12 lb

Figure 2-41

$$e = 0.75 = -\frac{v_2 - u_2}{20 - 8}$$

or

$$9 = u_2 - v_2$$

but also

$$m_1 v_1 + m_2 u_1 = m_1 v_2 + m_2 u_2$$

$$\frac{8}{g} \times 20 + \frac{12}{g} \times 8 = \frac{8}{g} v_2 + \frac{12}{g} u_2$$

$$32 = v_2 + 1.5 u_2$$

$$9 = -v_2 + u_2$$

$$u_2 = 16.4 \text{ ft/sec}$$

$$v_2 = 7.4 \text{ ft/sec}$$

The velocity of the 8-lb ball reduces from 20 to 7.4 ft/sec, and the velocity of the 12-lb ball is increased from 8 to 16.4 ft/sec.

If the coefficient of restitution is 1, we have a perfectly elastic collision and no energy is lost. Referring to the first example in this section, if the plastic bodies had been given as elastic bodies, we should have had

$$m_1 v_1 + m_2 u_1 = m_1 v_2 + m_2 u_2$$
$$\tfrac{1}{2} m_1 v_1^2 + \tfrac{1}{2} m_2 u_1^2 = \tfrac{1}{2} m_1 v_2^2 + \tfrac{1}{2} m_2 u_2^2$$

which reduces to

$$42.5 = v_2 + 1.5 u_2$$
$$737.5 = v_2^2 + 1.5 u_2^2$$

which can be solved simultaneously to give $v_2 = 14$ ft/sec and $u_2 = 19$ ft/sec.

The same result can be obtained by utilizing the relationship that

$$e = -(v_2 - u_2)/(v_1 - u_2)$$

The two cases considered have concerned bodies moving along the same line. Should the bodies be moving at an angle to each other, the same relationships would apply, but it would be necessary to consider the components of velocity in the X and Y directions and the components of momentum in the X and Y directions. This is a more complex problem, involving four equations, and will not be discussed further here.

One other important concept of linear momentum is that the linear momentum of any moving-mass system is the product of the mass of the whole system and the velocity of the mass center of the system, and its direction agrees with that of the velocity of the mass center.

The equations of equilibrium, $\Sigma F_x = 0$ and $\Sigma F_y = 0$, state that if the sum of the external forces acting on a system equals zero, the acceleration of the center of mass of that system equals zero. This means that if there is no external force acting on a system, the center of mass of the system remains at rest or continues its motion in a straight line at a constant speed. This is an example of the conservation of momentum and may be stated as follows: "If the resultant of the external forces which act on a body equals zero, then the linear momentum of the body remains constant."

Note that the forces considered are external forces; it can, therefore, be stated that the action of an internal force cannot result in a displacement of the center of mass of a system, or if no work is done upon (*not* inside) a system, the location of the center of mass of the system cannot change in a horizontal direction. As an example, a 200-lb man is standing at the rear of a 450-lb barge 18 ft in length. The front of the barge is 2 ft away from a pier, but the barge is floating freely in the water and is not tied to anything. The man walks forward in the barge. Will he be able to reach the pier? Assume no frictional drag of any kind on the boat, i.e., no external force acting.

Since no external force is acting, the location of the center of mass of the system must remain constant. Then the moment of the system about point A must also remain constant (Fig. 2-42).

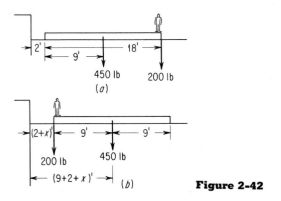

Figure 2-42

$$M_A = -11 \times 450 - 20 \times 200 = -8950 \text{ ft-lb}$$

The moment of the system about point A will be the same after the man has walked the 18 ft to the front end of the barge, so the barge must move outward x ft.

$$-8950 = -450(11 + x) - 200(2 + x)$$

$x = 5.54$ ft, or the barge will move 5.54 ft away from the pier. The man would then be 7.54 ft from the pier—a long reach; *or*, assuming the man moves and then the system (barge and man) moves, the 200-lb man moves 18 ft, so the 650-lb system must move

$$\frac{200}{650} \times 18 = 5.54 \text{ ft}$$

or man impulse equals barge impulse; that is, $200(18 - x) = 450x$, giving $x = 5.54$ ft.

One further example is the classical problem of the monkey and the bananas (Fig. 2-43). Monkey and bananas weigh the same, the rope and pulley are massless, and the system is frictionless. What happens as the monkey climbs the rope?

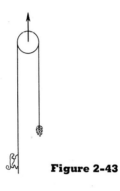

Figure 2-43

The tension in the rope must be the same over its entire length. Thus the impulse produced by the monkey is the same as the impulse acting on the bananas, and since their masses are the same the velocities must be equal. The distance between the monkey and the bananas will, then, stay the same—at least until the bananas jam the pulley.

2-17. CONSERVATION OF ENERGY

Another important relationship, which has been mentioned before, is that of the conservation of energy. The work done on a system must equal the change in energy of the system plus the energy lost in the process (e.g., energy lost in overcoming friction), if we disregard any effects of nuclear fission or change in mass. Furthermore, the work done by a force F on a body while the body undergoes a displacement is defined as the product of the displacement and the component of the force in the direction of the displacement, or $W = Fs \cos \theta$. The magnitude of the force need have little bearing on the amount of work done. As an example, take the case of a man pushing a crate across a loading platform. The sum of the forces acting on the crate includes the force of the platform acting upward, the retarding force of friction, and the force exerted by the man causing motion. Disregarding, for the moment, the frictional force, the total force acting on the crate = $\sqrt{50^2 + 500^2} = 502$ lb (Fig. 2-44). But the work is equal to $R \times S \times \cos \theta$, or $50 \times S$. Another way of looking at it is that any work done must appear either as work against something (friction) or as an increase in kinetic or potential energy.

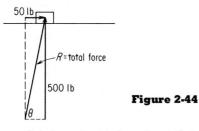

Figure 2-44

It is important to keep in mind that work is equal to the product of the distance times the component of the acting force in the direction of the distance. The mechanical energy of a body is the energy it possesses because of its position and its velocity. The energy due to position, termed the "potential energy," is equal to the weight of the body times its height above some arbitrarily selected datum plane. The change in potential energy is equal to the weight of the object times the change in elevation, or PE = $W \Delta h$, ft-lb. The kinetic energy is the energy possessed by a body by virtue of its velocity and equals $\frac{1}{2}mv^2$. The equality between these two can easily be seen by taking as an example a weight W at a height h with an initial velocity of zero. After the weight falls the distance h, it will experience a loss in potential energy equal to $W \times h$. The force acting through the distance h will have been W, and the object will have been subjected to a constant acceleration g. The velocity at the end of the fall will then be equal to $\sqrt{2gh}$ (from $v^2 = v_0^2 + 2as$, where $a = g$ and $s = h$). The total energy is constant, so ΔKE equals $Wh = \frac{1}{2}mv^2$, which gives $Wh = \frac{1}{2}m(2gh) = mgh$, so the loss in potential energy equals the increase in kinetic energy. Another way of looking at it is that the increase in potential energy is exactly equal to the work done on the body.

As an illustrative example, let us take the case of an object which slides down a chute onto a horizontal table, as shown in Fig. 2-45.

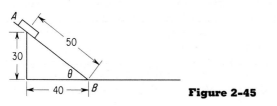

Figure 2-45

■ Assume that the initial horizontal velocity is the same as that gained in sliding from A to B. What distance S along the horizontal plane will the object slide before coming to rest if the coefficient of friction is 0.3?

We can work this problem a number of different ways; let us look at two.

First find the answer using the velocity and acceleration relationships. The force parallel to the surface of the plane tending to cause the object to slide is $0.60W$. The resisting frictional force f is equal to the normal force $(0.80W)$ times the coefficient of friction, which gives $f = 0.24W$. The resultant force causing the object to accelerate down the plane is then $0.60W - 0.24W = 0.36 \ W$. Since $F = ma$, the acceleration is $a = 0.36W/(W/g) = 0.36g$. The velocity at the point B is then found to equal $\sqrt{2 \times 0.36g \times 50} = 6\sqrt{g}$, from the relationship $v^2 = v_0^2 + 2as$. The force acting on the object as it slides horizontally is equal to $-0.3W$, giving an acceleration of $-0.3g$. Since the final velocity will be zero,

$$S = \frac{v_0^2}{-2a} = \frac{36g}{0.6g} = 60 \text{ ft}$$

The second method of determining the desired distance is by means of the theory of the conservation of energy. In sliding down the plane the block will lose $30W$ ft-lb of potential energy. This must all be dissipated in work done in overcoming friction. The frictional work will equal $0.24W \times 50$ while sliding down the plane and $0.30W \times S$ while sliding along the horizontal. Then

$$30W = 0.24W \times 50 + 0.30W \times S$$

and
$$S = \frac{30 - 12}{0.30} = 60 \text{ ft}$$

2-18. D'ALEMBERT'S PRINCIPLE

The subject of dynamics can be considerably simplified by the application of D'Alembert's principle. By this means problems in dynamics can be considered as problems in statics and can be treated by the methods of statics. D'Alembert's principle is based upon the use of a fictitious "inertial force," which is equal and opposite to the force causing the acceleration.

The existence of this imaginary inertial force follows from Newton's third law—action and reaction are equal and opposite. This states that the object pushes just as hard against the force as the force pushes against the object. These two forces must be equal in magnitude and opposite in sense. The equation $F = ma$ can then be written $F - ma = 0$, where ma equals the resisting "inertial force," and F is the resultant of all real external forces acting on the body. The sum of the forces acting on the body is zero, and the problem reduces to an equilibrium which can be treated by the methods of statics.

An example of the use of this principle is illustrated in Fig. 2-46. In this figure is shown a 50-lb block resting on a horizontal plane. The block is pulled by a cord which goes over a frictionless, inertialess pulley and is then attached to a 25-lb weight, which is allowed to fall. The forces on both weights, including the inertial forces, are added, and the problem is seen to reduce to a problem in statics which can be solved by means of the relationship that the summation of all forces acting must equal zero.

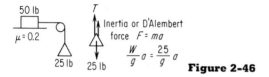

Figure 2-46

The force acting downward is the weight of the block, or 25 lb. This is balanced by two forces upward, the tension in the cord, T, and the D'Alembert or inertia force, which equals mass times acceleration, giving $T = 25 - (25/g) \times a$. The forces acting on the 50-lb block are next added, and since the tension in the cord is the same over its entire length, the force acting to cause motion of the 50-lb weight is equal to T. This is balanced by two resisting forces, the frictional force—$50 \times 0.2 = 10$ lb—and the D'Alembert force—$(50/g) \times a$, which gives $T = (50/g) \times a + 10$ (Fig. 2-47). Since $T = T$, $25 - (25/g) \times a = (50/g) \times a + 10$; $a = 0.2g$, and $T = 20$ lb.

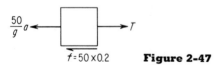

Figure 2-47

2-19. ROTATIONAL MOTION

The concepts of linear velocity, displacement, and acceleration all have their exact counterparts in rotational motion. So, too, have the concepts of force and mass. These concepts and their counterparts are listed in the following table:

Linear concepts	Rotational concepts
Displacement, s, ft	θ, radians
Velocity, v, ft/sec	ω, radians/sec
Acceleration, a, ft/sec^2	α, radians/sq sec
Force, pounds	Torque, ft-lb
Mass $= W/g$, slugs	I = moment of inertia, slug-ft^2

The corresponding relationships are:

$$s = v_0 t + \tfrac{1}{2}at^2 \qquad \theta = \omega_0 t + \tfrac{1}{2}\alpha t^2$$
$$v = v_0 + at \qquad \omega = \omega_0 + \alpha t$$
$$F = ma \qquad T = I\alpha$$
$$\mathrm{KE} = \tfrac{1}{2}mv^2 \qquad \mathrm{KE} = \tfrac{1}{2}I\omega^2$$

The relationships $T = I\alpha$ and $\mathrm{KE} = \tfrac{1}{2}I\omega^2$ are somewhat different, and the moment of inertia I is also different. These can be related to the linear relationships by some brief reasoning.

2-20. MOMENT OF INERTIA

If a force F is applied to a mass m which is connected to a weightless, rigid rod of length r (Fig. 2-48), the other end of which is pivoted, the work done by the force will equal $F \times r\, d\theta$, where $r\, d\theta$ is the distance through which the force travels. $F \times r$ is the torque about the pivot point, so we have $dW = T_q\, d\theta$. The velocity attained by the mass will equal v, where $v = \omega r$. The work done on the mass will result in an increase in kinetic energy, or $dW = d(\tfrac{1}{2}mv^2)$, which becomes

$$T\, d\theta = mv\, dv = m(\omega r)(rd\omega) = mr^2 \omega\, d\omega$$

or $T_q = mr^2 \omega (d\omega/d\theta)$. $\omega = d\theta/dt$, and $\alpha = d\omega/dt$, so $\omega(d\omega/d\theta) = \alpha$ (obtained by substituting $\omega/d\theta = 1/dt$ from $\omega = d\theta/dt$, the equation for angular velocity).

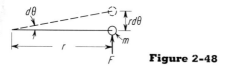

Figure 2-48

Substituting this relationship in the former equation gives $T = (mr^2)\alpha$, which corresponds to $F = ma$. The quantity mr^2 is termed the moment of inertia, and for a body where the mass is not concentrated at a distance r

from the center point $I = \Sigma \, \Delta m r^2$, or the moment of inertia of a body equals the sum of the products of all the point masses which make up the total mass of the body times the square of their distances from the axis about which the moment of inertia is desired. To return for a moment to the equation for kinetic energy, we see, for the concentrated mass used in the illustration, that $\frac{1}{2}mv^2$ becomes $\frac{1}{2}mr^2\omega^2$, which equals $\frac{1}{2}I\omega^2$.

2-21. RADIUS OF GYRATION

From the equation $I = mr^2$ comes also the relationship $r = \sqrt{I/m}$, where r is termed the radius of gyration or the distance from the axis at which the mass m could be concentrated to give the same moment of inertia. The radius of gyration is usually indicated by k rather than r.

$$I = k^2 m$$

Note that, in general, the mass of a body cannot be considered as acting at its center of mass for the purpose of calculating the moment of inertia. This can be illustrated easily by calculating the moment of inertia of a uniform, slender rod about an axis through one end and perpendicular to the rod (Fig. 2-49). The mass of the rod will equal $\rho A L$, where ρ is the density in slugs/cubic foot, A the cross-sectional area in square feet, and L the length in feet. The dm of the relationship $I = \int r^2 \, dm$ will then be $\rho A \, dl$, and

$$I = \int_0^L l^2 \rho A \, dl$$

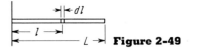

Figure 2-49

$\frac{1}{3}\rho A L^3 = \frac{1}{3}mL^2$, since $m = \rho A L$. Similarly, the moment of inertia of a slender, uniform rod about an axis through its center and perpendicular to the rod will equal

$$\int_{-L/2}^{L/2} l^2 \, dm = \frac{1}{12}ML^2 \qquad \text{(Fig. 2-50)}$$

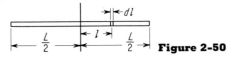

Figure 2-50

In the case of the rod pivoted at one end, the center of mass is at a distance of $L/2$ but the radius of gyration equals $L/\sqrt{3}$. The moment of inertia of the plane of a rectangular surface about an axis through its center and lying in the plane can be determined similarly (Fig. 2-51).

$$I = \sum x^2 \, \Delta A = \int_{-h/2}^{h/2} x^2 b \, dx = \frac{bh^3}{12}$$

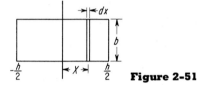

Figure 2-51

The moment of inertia about an axis in the plane of an area is called the "plane moment of inertia"; the moment of inertia of an area about an axis perpendicular to the area is termed the "polar moment of inertia."

2-22. POLAR MOMENT OF INERTIA

The polar moment of inertia of an area with respect to any axis is equal to the sum of the moments of inertia of the area with respect to any two rectangular axes in the plane of the area which intersect on the given polar axis.

$$J_z = \sum r^2 \, \Delta A = \sum (x^2 + y^2) \, \Delta A$$

where J_z = polar moment of inertia about the perpendicular axis 0 (Fig. 2-52)

$$J_z = \int x^2 \, dA + \int y^2 \, dA = I_x + I_y$$

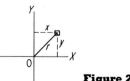

Figure 2-52

Application of this principle aids in simplifying a calculation of some of the moments of inertia of the different areas. As an example, determine the moment of inertia of a circular section about the Y axis (Fig. 2-53).

$$I_y = \int_{-r}^{r} x^2 \, dA = \int_{-r}^{r} x^2(2y \, dx) = \int_{-r}^{r} (r^2 - y^2)(2y \, dx)$$

which becomes involved.

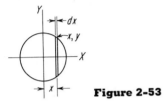

Figure 2-53

If we utilize the relationship that $J_z = I_y + I_x = 2I_y$, we can determine I_y much more easily (Fig. 2-54).

$$J_z = \int_{0}^{r} \rho^2 \, dA = \int_{0}^{2\pi} \int_{0}^{r} \rho^3 \, dp \, d\theta = 2\pi \int_{0}^{r} \rho^3 \, dp$$

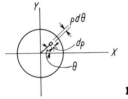

Figure 2-54

Since $dA = (\rho \, d\theta) \times d\rho$,

$$J_z = 2\pi \frac{r^4}{4} = \frac{\pi r^4}{2}$$

$$J_y = \tfrac{1}{2}J_z = \frac{\pi r^4}{4}$$

or by a different method (Fig. 2-55),

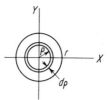

Figure 2-55

$$J_z = \int_0^r \rho^2 \, dA \qquad dA = 2\pi\rho \, d\rho$$

$$J_z = \int_0^r 2\pi\rho^3 \, d\rho = \frac{\pi r^4}{2}$$

as before.

2-23. PARALLEL-AXIS THEOREM

If it is desired to calculate the moment of inertia about an axis different from the moment of inertia about the centroid of an area, this can be done by means of the "parallel-axis" theorem. As an example, determine the moment of inertia of an area about an axis $X - X$ given that the moment of inertia about an axis $0 - 0$ through the centroid of the area equals I_0. Axis $X - X$ is parallel with axis $0 - 0$ (Fig. 2-56).

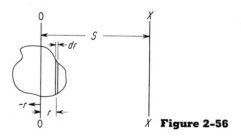

Figure 2-56

$$I_x = \int_{-r}^r (s - r)^2 \, dA = \int_{-r}^r r^2 \, dA - \int_{-r}^r 2rs \, dA + \int_{-r}^r s^2 \, dA$$

$$= \int_{-r}^r r^2 \, dA - 2s \int_{-r}^r r \, dA + s^2 \int_{-r}^r dA = I_0 + As^2$$

since $\int_{-r}^{r} r^2 \, dA$ equals the moment of inertia about the axis $0 - 0$. $\int_{-r}^{r} r \, dA$ about the centroid of the area must equal zero, since that defines the location of the centroid, as discussed previously.

2-24. CENTRIPETAL ACCELERATION

It can be shown that an object which is moving in a circular path has an acceleration toward the center of the circle of motion equal to the product of the square of the angular velocity times the radius. From Fig. 2-57 we can see that $v = r(d\theta/dt) = r\omega$. The change of velocity Δv equals the distance CB, since velocity is a vector quantity.

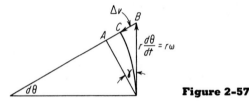

Figure 2-57

As the angle $d\theta$ gets smaller and smaller, point A approaches point C and angle γ approaches $d\theta$. At the instant $d\theta$ vanishes we have $dv = r(d\theta/dt)$ $\sin \, d\theta = r(d\theta/dt)d\theta$. Acceleration toward the center $= dv/dt = r(d\theta/dt)(d\theta/dt) = r\omega^2 = a_c = $ centripetal acceleration ($d\theta/dt = \omega = $ constant, so the derivative is zero).

From the figure we also see that

$$v = r\frac{d\theta}{dt} = r\omega$$

$$a_c = \frac{v^2}{r}$$

since

$$\omega = \frac{v}{r} \qquad \frac{d\omega}{dt} = \frac{dv}{dt}\frac{1}{r}$$

or

$$\alpha = \frac{a}{r} \qquad \text{and} \qquad a = \alpha r$$

where $\alpha = $ angular acceleration
$a = $ circumferential acceleration

2-25. CENTRIPETAL FORCE

Since we have an acceleration, if we also have a mass then we must have a force from $F = ma$, where the force equals $m \times \omega^2 r$ or $m(v^2/r)$. This force is exerted toward the center and is termed the "centripetal" force. It is balanced by an equal and opposite outward force termed a "centrifugal" force, centrifugal meaning "fleeing a center."

Two points should be noted in connection with centrifugal force. The first is that the forces and accelerations discussed are instantaneous; even if the path of the body is not a true circle, it will have an instantaneous center of curvature, and there will thus be an instantaneous centripetal acceleration (and an instantaneous centripetal force) toward this center of curvature which can be calculated from the relationships given. If the object is moving in essentially a straight line, then the distance to the center of curvature will be so large that the term v^2/r will be so small as to be negligible. The second point of importance is that if the body moving in a curved path is too large to be considered as a point, then the mass should be considered as concentrated at the center of mass and the location of the center of mass will determine the values of the radius of curvature and the tangential velocity.

An example of the application of these principles is afforded by the past problem:

■ What superelevation (slope of roadway) is required on a highway for 60-mph traffic on a 2500-ft-radius curve so that a 200-lb driver will exert no sideward force on the car seat? What force will the driver exert perpendicular to the seat?

First, draw a figure showing the forces acting (Fig. 2-58). The forces acting on the car (or any object in the car) are shown in the sketch. Any variation due to a variation in the radius will be negligible, since 2500 ft is so

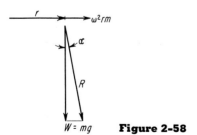

Figure 2-58

much greater than any distance which would be measured inside the car. The centrifugal force equals $\omega^2 rm$, or $(v^2/r) \times m$, and the force of gravity acting downward equals W, $= mg$.

$$\tan^{-1}\left[\frac{(v^2/r)m}{mg}\right] = \tan^{-1}\left(\frac{v^2}{rg}\right)$$

$$60 \text{ mph} = \frac{60 \times 5{,}280}{60 \times 60} = 88 \text{ ft/sec}$$

$$\alpha = \tan^{-1}\left(\frac{88^2}{2500\ g}\right) = 5.5°$$

The superelevation of the roadway should be 5.5°.

$$R = \sqrt{[(v^2/r)m]^2 + (mg)^2} = m\sqrt{9.6 + 1036.8} = 32.35m$$

$$R = \frac{200}{g}32.35 = 201 \text{ lb}$$

which is the force exerted by the driver perpendicular to the seat.

2-26. SATELLITE

A few years ago there was quite a bit of interest in satellites and the forces that hold them up. Some of this interest was reflected in fundamentals examinations by the inclusion of satellite problems. One such problem asked:

■ If the moon circles the earth once in every 28 days, approximately how far is it from the earth?

The earth-moon system is in equilibrium; therefore, the forces acting on the moon are balanced. That is, the force acting to hold the moon away from the earth is exactly equal and opposite to the force attracting it toward the earth.

To calculate these forces, simplify the problem slightly by making two assumptions: (1) the orbit of the moon about the earth is circular; (2) no influence is exerted on the moon by any members of our solar system other than the earth. These assumptions will affect the accuracy of our answer slightly, but only slightly, and since we have been asked to approximate only the actual distance, they are permissible.

The two forces acting on the moon are (1) the centrifugal force outward and (2) the gravitational force toward the earth. These two forces are equal.

The centrifugal force equals $\omega^2 rm$ where individual quantities are as previously discussed. The speed of rotation is once in 28 days so

$$\omega = \frac{2\pi}{28 \times 24 \times 3600} = 2.60 \times 10^{-6} \text{ rad/sec}$$

r is the radius of rotation and is thus the distance between the center of the moon and the center of the earth. The centrifugal force, or force holding the moon away from the earth is, then,

$$F_c = 6.75 \times 10^{-12} \, rm_m$$

The gravitational attraction between two masses is equal to the gravitational constant times the product of the two masses divided by the square of the distance between the two centers of mass. This equates to

$$F_g = K\frac{m_m m_e}{r^2}$$

where K is the gravitational constant, m_m is the mass of the moon, m_e is the mass of the earth, and r is the distance between the two centers of mass.

We can calculate a usable value for K by utilizing the fact that the gravitational force acting on an object at the earth's surface equals mg, where g is the acceleration of gravity. The radius of the earth is approximately 4000 miles, so the distance between the centers of mass of the object and the earth would equal 4000×5280 ft. If we substitute these values into the equation for gravitational force, we have

$$F_g = m_0 \times g = m_0 \times 32.2 = K\frac{m_e \times m_0}{(4000 \times 5280)^2}$$
$$K = 1.44 \times 10^{16}/m_e$$

This enables us to calculate the gravitational force between the earth and the moon.

$$F_g = \frac{1.44 \times 10^{16}}{m_e} \times \frac{m_e m_m}{r^2} = \frac{1.44 \times 10^{16} m_m}{r^2}$$

We know the two forces F_g and F_c are equal. Equating these gives

$$\frac{1.44 \times 10^{16} m_m}{r^2} = 6.57 \times 10^{-12} rm_m$$
$$r = 1.286 \times 10^9 \text{ ft, or } 244,000 \text{ miles}$$

If we subtract from this the radius of the earth (4000 miles) and the radius of the moon (1080 miles), we obtain 238,920 miles as the distance between the surfaces of the earth and the moon. It is interesting to note that this is very close to the reported value of 238,857 miles for the mean distance between the earth and the moon.

To be perfectly correct, the centrifugal force acting upward on an object at the surface of the earth at the Equator should also be included in the calculation to obtain the gravitational constant. The summation of the forces acting on a body at the surface of the earth at the equator would be

$$mg = K\frac{m_{earth} \times m}{r^2_{earth}} - \omega^2 r_{earth} \times m$$

or

$$g = K\frac{m_{earth}}{r^2_{earth}} - \omega^2 r_{earth}$$

$$\omega = \frac{2\pi}{24 \times 3600} = 7.272 \times 10^{-5}$$

$$\omega^2 = 5.288 \times 10^{-9}$$

$$r_{earth} = 4000 \times 5280 = 2.11 \times 10^7 \text{ ft} \qquad r^2_{earth} = 4.46 \times 10^{14}$$

$$\omega^2 r_{earth} = 0.111$$

$$K = \frac{r^2_{earth}}{m_{earth}}(32.17 + 0.11) = 1.439 \times 10^{16}/m_{earth}$$

compared with the value of $K = 1.44 \times 10^{16}/m_e$ which was obtained without including the effect of the centrifugal force.

Another problem was as follows:

■ An armature is keyed to a shaft 4 in. in diameter, which rests transversely on two parallel steel rails inclined to the horizontal with a slope of 1:12. The shaft rolls, without slipping and without rolling friction, a distance of 6 ft along the rails from rest in 1 min. What is the radius of gyration of the armature and its shaft about the longitudinal axis?

Make a sketch (Fig. 2-59). The energy possessed by the armature at the end of 1 min equals $\frac{1}{2}mv^2 + \frac{1}{2}I\omega^2$.

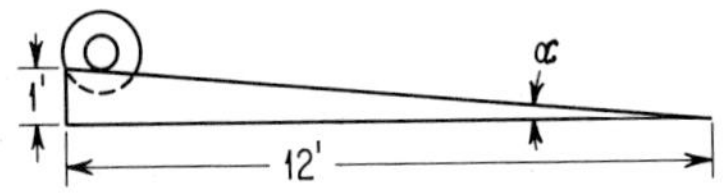

Figure 2-59

$$v_{avg} = \frac{6}{60} = 0.1 \text{ ft/sec} = \frac{(0 + v)}{2}; v = 0.2 \text{ ft/sec}, \omega = \frac{v}{r}$$

$$KE = \frac{1}{2}m(0.2)^2 + \frac{1}{2}I\frac{v^2}{r^2} \qquad r = 2 \text{ in.} = 0.167 \text{ ft}$$
$$I = k^2 m$$
$$k = \text{radius of gyration}$$

$$KE = \frac{1}{2}m(0.04) + \frac{1}{2}(k^2m)\frac{0.04}{(1/6)^2} = m(0.02 + 0.72k^2)$$

This increase in kinetic energy must equal the decrease in potential energy, since no work is done against friction. $\alpha = \tan^{-1}(1/12)$, giving $\sin \alpha = 0.083$.

$$\Delta PE = W \sin \alpha \times 6 = mg \times 0.083 \times 6 = 16.05m$$
$$16.05m = m(0.02 + 0.72k^2)$$
$$k = \sqrt{\frac{16.05 - 0.02}{0.72}} = 4.72 \text{ ft}$$

The same result can be obtained by determining the acceleration of the armature (Fig. 2-60).

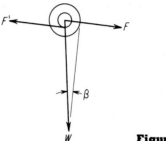

Figure 2-60

$$s = v_0 t + \frac{1}{2}at^2$$
$$6 = 0 + \frac{1}{2}a(60)^2 \qquad a = \frac{1}{300} \text{ ft/sec}^2$$
$$F = W \sin \beta = 0.083W$$
$$T_q = I\alpha = I\frac{a}{r}$$
$$T_q = F' \times r = F \times r$$

since there is no slipping and $F' = F$.

$$F \times r = I\frac{a}{r} = k^2 m \frac{a}{r}$$
$$mg \sin \beta \times r = k^2 m \frac{a}{r}$$
$$k^2 = \frac{0.083gr^2}{a}$$
$$= \frac{0.083g(1/6)^2}{1/300}$$
$$= 22.3$$
$$k = 4.72 \text{ ft radius of gyration}$$

A word might be in order here about the summation of the forces acting on the armature. The force F (Fig. 2-61) is just equal to the frictional force f

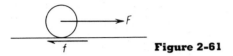

Figure 2-61

and is balanced by it. There is, then, no excess force to cause purely linear acceleration of the armature as a whole, and the total acceleration is angular.

If F exceeded f and the armature slipped as well as rolled, then there would be a linear acceleration in addition to that resulting from the angular acceleration.

Another problem asked:

- A 64.4-lb cylinder of radius $R = 2$ ft is pushed by a moving bulldozer. The coeficient of friction at points A and B is 0.20. Determine (a) the minimum acceleration of the bulldozer if the cylinder is to move to the right without rotating (b) the maximum acceleration of the bulldozer if the cylinder is to roll without slipping.

Make a sketch (Fig. 2-62). The maximum frictional force at point A will equal $0.20W$. The net torque acting on the cylinder will equal $f_A \times R - f_B \times R$, so if $f_B > f_A$ (Fig. 2-63), the cylinder will not roll, and if $f_B < f_A$, the cylinder will roll. If $f_B = f_A$, then the inertial force, or D'Alembert's force, must equal $W = mg = ma$, so if the cylinder is to move to the right without rotating, the acceleration must be greater than g, and if the cylinder is to roll without sliding, the acceleration must be less than g.

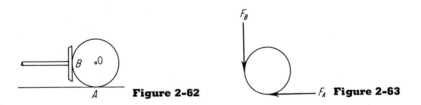

Figure 2-62   **Figure 2-63**

There are a few other phases of mechanics which have been emphasized in past examinations; they will be discussed here briefly.

2-27. SPRINGS

It has, upon occasion, been necessary to determine the energy absorbed by a spring because of the impact of a weight. The energy lost by the weight will, of course, be matched by an increase in the energy contained in the spring. The energy required to compress a spring a distance dx equals $F\ dx$, where the force will be equal to kx; k is the spring constant, and x is the distance the spring has been compressed. The total work to compress the spring a distance S will then equal $\int_0^S kx\ dx$, which equals $\frac{1}{2}kS^2$.

An example illustrating this is:

- A mass weighing 25 lb falls a distance of 5 ft upon the top of a helical spring with a spring constant of 20 lb/in.
  - *a.* What will the velocity of the mass be after it has compressed the spring 8 in.?
  - *b.* Calculate the maximum compression of the spring.

The fall of the weight will be 5 ft plus 8 in. when it has compressed the spring 8 in. At that time the spring will have absorbed energy equal to $\frac{1}{2}kS^2 = \frac{1}{2}(20\ \text{lb/in.}) \times (8\ \text{in.})^2 = 640$ lb-in., or 53.3 lb-ft. The change in potential energy equals

$$W \times h = 25(5 + 0.67) = \Delta PE = 142 \text{ ft-lb}$$

Part of this energy has been absorbed by the spring; the remainder, $142 - 53.3 = 88.7$ ft-lb, must remain as kinetic energy of the mass $= \frac{1}{2}mv^2$.

$$v = \sqrt{\frac{88.7 \times 2}{m}} = \sqrt{\frac{88.7 \times 2g}{25}} = 15.1 \text{ fps}$$

When the spring has been fully compressed, the velocity of the mass will be zero and all the change in potential energy of the mass will have been converted into potential energy of the spring.

$$W(5 \text{ ft} + S) = \frac{1}{2}kS^2$$

If S is in inches,

$$W(5 \times 12 + S) = \frac{1}{2}20S^2$$
$$2S^2 - 5S - 300 = 0$$
$$S = \frac{5 \pm \sqrt{25 + 2{,}400}}{4} = \frac{5 \pm 49.3}{4}$$
$$= 13.6 \text{ in.}$$

which is the maximum compression of the spring.

2-28. SIMPLE HARMONIC MOTION

Simple harmonic motion (SHM) is defined as the motion of a point in a straight line such that the acceleration of the point is proportional to the distance of the point from its equilibrium position, or $a = -kx = d^2x/dt^2$.

Simple harmonic motion can be shown to be the same motion to which a suspended mass is subjected by the force of a spring (Fig. 2-64). The static

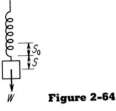

Figure 2-64

position of the weight is that position in which the spring is stretched the distance S_0. $S_0 \times k = W$, where k is the spring constant. If the weight is pulled down farther, a distance S, then the sum of the forces acting on the weight will equal

$$k(S_0 + S) - W = kS$$

Since the distance S is down and the force is up, $F = -kS$, giving $ma = -kS$, or $a = (-k/m) \times S$, where k/m is a constant, and we have the defining relationship for simple harmonic motion.

The velocity and time relationships for simple harmonic motion are usually determined by means of a "reference circle."

It can be shown that if a point moves with constant speed in a circular path, the motion of the projection of the point on a diameter of this reference circle is a simple harmonic motion (Fig. 2-65).

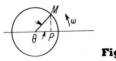

Figure 2-65

$$x = r \cos \theta = r \cos \omega t$$

where $\omega = $ a constant

$$v_p = \frac{dx}{dt} = -\omega r \sin \omega t = -\omega y$$

$$a_p = \frac{dv_p}{dt} = -\omega^2 r \cos \omega t = -\omega^2 x$$

This last equation is the equation defining simple harmonic motion. The time for P to make one complete oscillation is the same as the time for the point M to make one complete revolution, or $T = 2\pi/\omega$, and the frequency equals $1/T = \omega/2\pi$. A numerical example illustrating simple harmonic motion is as follows:

- A truck body lowers 5 in. when a load of 25,000 lb is placed on it. What is the frequency of vibration of the truck if the total spring-borne weight is 40,000 lb?

The frequency $= 1/T = \omega/2\pi$, and the acceleration of the weight $a = -\omega^2 x$, but $F = ma$, and the force acting on the mass $= -kx$, so $-kx = -\omega^2 m x$, and $\omega = \sqrt{k/m} = \sqrt{kg/W}$; then

$$f = \frac{\omega}{2\pi} = \frac{1}{2\pi}\sqrt{\frac{kg}{W}}$$

The spring constant $k = 25{,}000$ lb/ 5 in. $= 5000$ lb/in.

$$f = \frac{1}{2\pi}\sqrt{\frac{5000 \text{ lb/in.} \times 12 \text{ in./ft} \times 32.2 \text{ ft/sec}^2}{40{,}000 \text{ lb}}} = 1.105 \text{ oscillations/sec}$$

The restoring force acting on the truck body equals $-kx$, where x is the distance the spring is compressed past the neutral point (Fig. 2-66).

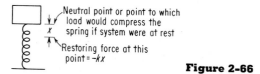
Neutral point or point to which load would compress the spring if system were at rest

Restoring force at this point $= -kx$

Figure 2-66

Another example of simple harmonic motion is afforded by the following problem:

- If the lever is assumed to be weightless in the system shown in Fig. 2-67, at what frequency would the weight move up and down, after it had been displaced vertically? The spring constant equals 5 lb/ft.

This is an example of undamped free vibration with a single degree of freedom. For small displacements the system will vibrate in conformance with the relationship for simple harmonic motion in the same way as the suspended mass shown in Fig. 2-66. Thus the motion will satisfy the differential equation

$$\frac{d^2 y}{dt^2} + \left(\frac{k}{m}\right) y = 0$$

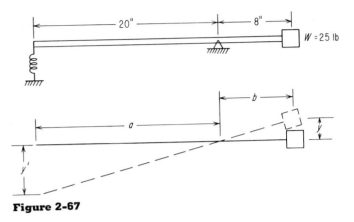

Figure 2-67

The general solution of this equation gives $\omega^2 = k/m$ where ω is measured in radians per second. A circular function repeats itself in 2π radians, so one cycle of vibratory motion is completed when $\omega\tau = 2\pi$ where τ is the period of motion, or the time required for a single cycle. The frequency, in cycles per second, would equal $1/\tau$, or $f = \omega 2\pi$.

In addition, $d^2y/dt^2 =$ acceleration. From Newton's second law, $F = ma$, we obtain $a = F/m$. From this $d^2y/dt^2 = F/m$, which, in turn, gives

$$\frac{F}{m} + \frac{k}{m}y = 0, \text{ or } F = -ky$$

where the minus sign is explained by the fact that the force is opposite in sense to the displacement of y.

The force acting on the weight (see Fig. 2-67) would equal $-(a/b)(k_{sp}y')$ (for small displacements) where $y' =$ displacement (extension or compression) of spring and $k_{sp} =$ spring constant. $b/y = a/y'$ so $y' = (a/b)y$ and the force acting on the weight would equal

$$F = -(a/b)^2 k_{sp}y$$

But force also equals $F = -ky$, so $k = (a/b)^2 k_{sp}$, $\omega^2 = k/m$, and $f = \omega/2\pi$.

Then
$$f = \frac{\sqrt{k/m}}{2\pi} = \frac{\sqrt{(a/b)^2 k_{sp}/m}}{2\pi}$$

$$\left(\frac{a}{b}\right)^2 = \left(\frac{20}{8}\right)^2 = 6.25 \qquad m = 25/32.2 = 0.776 \text{ slugs}$$

$$f = \frac{\sqrt{6.25 \times 5/0.776}}{2\pi} = 1.01 \text{ cycles per second}$$

STATICS—SAMPLE PROBLEMS

2S- 1   What are the maximum and minimum weights of the box that can be held in place by the force as shown in Fig. 2S-1?

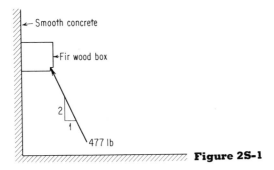

Figure 2S-1

2S- 2   A cable 25 ft long is connected between two points 20 ft apart horizontally and 5 ft apart vertically. If a 5000-lb load is hung on the cable by use of a frictionless pulley, where will the pulley come to rest, and what will be the total stress in the cable?

2S- 3   Locate the centroid along the center line from side A of the area shown in Fig. 2S-3.

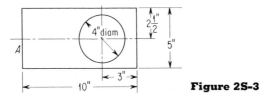

Figure 2S-3

2S- 4   A jackscrew having eight threads to the inch, i.e., $1/8$-in. threads, is operated manually by a handle. The point at which the handle is gripped moves through a 12-ft circumference.

   a. If friction is disregarded, what force must be applied at the grip on the handle to just raise a load of 2300 lb?

   b. If a shorter handle three-fourths as long is used as in (a), what force would be required at the grip on the short handle to just raise the same load as in (a)?

2S- 5 For the truss loaded as shown in Fig. 2S-5, find the force in member U_4L_4 and in U_3U_4.

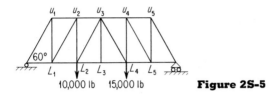

Figure 2S-5

2S- 6 The bracket shown in Fig. 2S-6 is fastened to a vertical wall at points B, C, and D, point D being 3 ft higher than C, with member AB horizontal. Find the stress in members AB, AC, and AD.

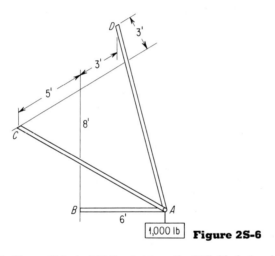

Figure 2S-6

2S- 7 The coefficient of friction between the 10-lb block A and the 20-lb block B, shown in Fig. 2S-7, is 0.40. The coefficient of friction between B and the horizontal plane is 0.15. Determine the force P required to cause motion of B to impend to the right.

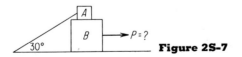

Figure 2S-7

2S- 8 A solid sphere weighing 1000 lb is held at rest on a 45° inclined plane by means of cables a, b, and c, of which a is attached to the center of the sphere. Solve for the tension in each cable and for the reaction of the plane on the sphere. (See Fig. 2S-8.)

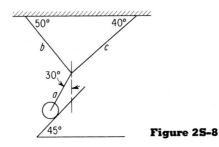

Figure 2S-8

2S- 9 Locate the centroid of the shaded area shown in Fig. 2S-9.

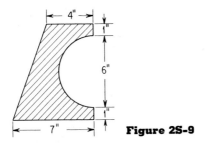

Figure 2S-9

2S-10 A steel block is 8 in. square and 12 in. high. A cylindrical hole 6 in. deep is to
be drilled from the upper end. Compute the necessary diameter of the hole
required so that the center of gravity of the block will be 5 in. from the
bottom of the block.

2S-11 Find stress in members CH and CD of the truss shown in Fig. 2S-11.

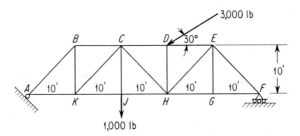

Figure 2S-11

2S-12 Solve for the magnitude and direction of the stresses in the members
AC, BC, and DC of the frame shown in Fig. 2S-12. Distances between
lettered points are as indicated. Plane ABC is horizontal, and plane CDE is
vertical.

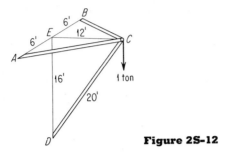

Figure 2S-12

2S-13 Figure 2S-13 shows how a weight can be raised by an ideal frictionless wedge.
    a. Find the relation between the load W and the applied force F and the wedge angle.
    b. Find the forces transmitted by the three rows of balls.
    c. For what angle can a weight W be raised that is ten times as large as the applied force F?

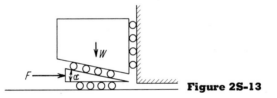

Figure 2S-13

2S-14 Calculate the stress in member L_1-U_2 in the truss shown in Fig. 2S-14.
    b. Calculate the stress in member U_0-U_1.

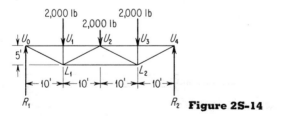

Figure 2S-14

2S-15 A weight of 4000 lb is supported at the midpoint of cable HNK, which is 18 ft long (Fig. 2S-15). If the safe tension in the cable is 2600 lb, determine:

a. Maximum distance between H and K so that the 2600-lb tension will not be exceeded

b. Angle θ for the condition in (*a*)

c. Direction and magnitude of the horizontal forces at H and K

d. Distance BN (B is midway between H and K)

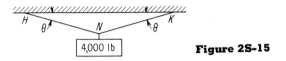

Figure 2S-15

2S-16   Total length of rope $A + B = 60$ ft (Fig. 2S-16). Weight on sheave is free to move on rope. Disregard the weight of rope.

a. Find tension in rope A and in rope B.

b. Find moment at base of pole, point C.

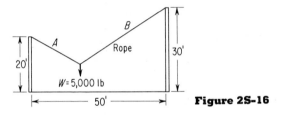

Figure 2S-16

2S-17   *a.* Find center of gravity of area of figure shown in Fig. 2S-17.

b. What is moment of inertia about Y-Y axis through the center of gravity?

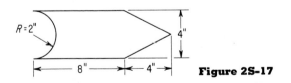

Figure 2S-17

2S-18   A wooden beam is made up of full-dimension timbers, as shown in Fig. 25.18.

a. What is the moment of inertia about axis x-x?

b. What load concentrated at center of such a beam 12 ft long can be supported with a stress of 1200 lb?

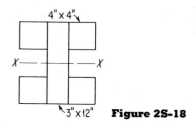

4" x 4"

X———X

3" x 12"   **Figure 2S-18**

2S-19 Refer to Fig. 2S-19 which shows a Fink truss, and determine the stresses in members labeled a and b. Express your final answer in pounds and state whether the stress is *tension* or *compression*.

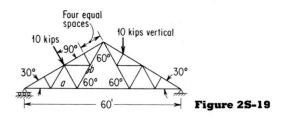

Figure 2S-19

2S-20 A cantilever beam is hinged to the wall as shown in Fig. 2S-20. What are the amount and direction (compression or tension) of stress in members A and B?

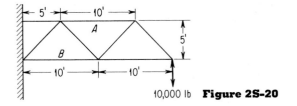

10,000 lb   **Figure 2S-20**

2S-21 *a.* A tractor, as shown in Fig. 2S-21, is on level ground. What pull X will remove all weight from the front wheels?

b. Assume the same tractor is going up a slope 20° from horizontal: what will X be in pounds?

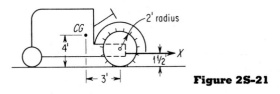

Figure 2S-21

2S-22 Given a truss as shown in Fig. 2S-22. Assume the horizontal components of the reactions at A and B are equal.

    a. Determine the horizontal and vertical components of the reactions at A and B.

    b. Determine the stress in AE, AD, BF, and BD. Be sure to state the kind of stress in each case.

    c. Check your results by analyzing joints D and C.

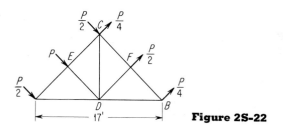

Figure 2S-22

2S-23 Find the least force P and the angle α just sufficient to move the wheel shown in Fig. 2S-23, radius r, which weighs W, over the curb of height h.

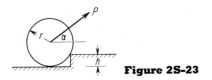

Figure 2S-23

2S-24 Find the location of centroid and moment of inertia about axis X-X through centroid of welded plate and angle girder shown in Fig. 2S-24.

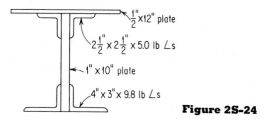

Figure 2S-24

2S-25 The ends of a cable 21 ft long are fastened to two rigid overhead points 15 ft apart and in the same horizontal plane. A weight of 200 lb is attached to the cable 9 ft from one end.

    a. Determine the stress in each segment of the cable.

       b. If the weight were attached to a roller pulley and permitted to move along the cable, what would be the position of equilibrium, and what would be the stress in each segment of the cable?

2S-26 Find the magnitude, direction, and distance from point A of the force required to act on the rigid body shown in Fig. 2S-26 to hold it in equilibrium.

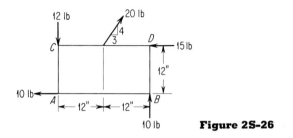

Figure 2S-26

2S-27 Given the truss as shown in Fig. 2S-27, determine the stresses in L_0U_2, L_0L_2, U_2U_4, and U_2L_4.

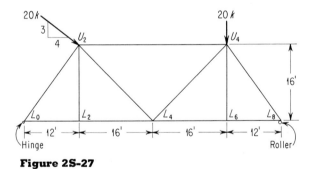

Figure 2S-27

2S-28 The weights A and B (Fig. 2S-28) are supported by a continuous rope which is attached at points C and D and passes around frictionless pulleys as shown. Disregard the weight of the rope and pulleys. Find the position of the pulley P relative to the point 0 for equilibrium.

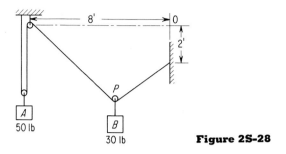

Figure 2S-28

2S-29 A homogeneous rod weighing 50 lb is hinged at A and rests on a block at B (Fig. 2S-29). The block, weighing 3 lb, rests on a horizontal floor. If the coefficient of friction for all surfaces is one-third, find the horizontal force P necessary to cause the block to slide to the left. Consider the hinge at A to be frictionless.

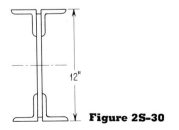

Figure 2S-29

2S-30 What is the moment of inertia about its horizontal neutral axis of a section assembled from four 3- by 3- by 0.5-in. angles and an 11- by 0.5-in. plate as shown in Fig. 2S-30.

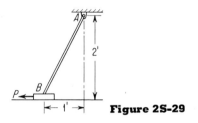

Figure 2S-30

2S-31 Two smooth cylinders rest as shown in Fig. 2S-31 on an inclined plane and a vertical wall. Compute the value of the force F exerted by the vertical wall.

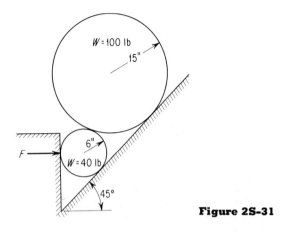

Figure 2S-31

2S-32 *a.* Draw a free-body diagram of the slider arm shown in Fig. 2S-32, indicating all the forces acting on the slider arm when the slider arm has a constant velocity due to the force F.
b. In a case where $d = 4$ in., $a = 2$ in., and the coefficient of friction is 0.20, what is the minimum dimension of x so that the force F will result in no movement of the slider arm?

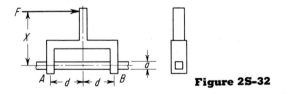

Figure 2S-32

2S-33 *a.* Three springs, having spring constants of 5 lb/in., 10 lb/in., and 15 lb/in., respectively, are connected in series as shown in Fig. 2S-33. What is the magnitude of the combined spring constant?
b. If the springs were arranged in parallel in such a way that a load deflects them equally, what is the magnitude of the deflection for a 100-lb load?

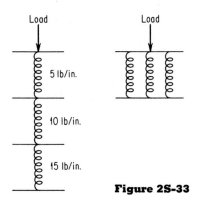

Figure 2S-33

2S-34   Three men are carrying a heavy 12-ft log of circular cross section 8 in. in diameter at the small end and 12 in. in diameter at the large end. One man takes the small end of the log and the other two place a bar under the log at a point so that each of the three carries an equal load. How far from the small end was the bar placed?

2S-35   The weightless beam shown in Fig. 2S-35 is suspended at each end by cables at (1) and (2). Find:

a. Value of tension in the cable at (2), which is at an angle θ_2 of (60°) with the horizontal
b. Angle θ_1
c. Tension F_1 in the cable supporting the beam at (1)

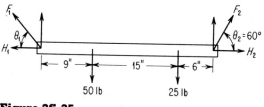

Figure 2S-35

2S-36   The hoisting rig in Fig. 2S-36 is composed of two symmetrical column members bd and cd with pin joints at b and c and a steel cable ad. A vertical load L of 8 tons is supported at d. The distance ae is 21 ft and ef is 10.5 ft. The angle *daf* is 30°, *def* is 60°, *dbc* is 60°. The distance *af* = 31.5 ft. Determine:

a. Tension in the cable F_{ad}

b. Tension or compression in the legs *cd* and *bd*, F_{bd} and F_{cd}

c. Horizontal forces along the line *bc*, F_{bc}

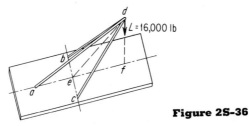

Figure 2S-36

STATICS—MULTIPLE-CHOICE PROBLEMS

For each question select the correct answer from the five given possibilities.

2SM- 1   Which of the following is true with respect to the vector diagram?

(a) $\vec{D} = \vec{A} + \vec{B} - \vec{C}$

(b) $\vec{2B} = \vec{A} + \vec{2C}$

(c) $\vec{2A} = \vec{B} + \vec{D}$

(d) $\vec{C} = \vec{D} - \vec{A} - \vec{B}$

(e) $\vec{A} - \vec{B} = \vec{D} - \vec{C}$

Figure 2SM-1

2SM- 2   The reaction at point *A* is:

(a) zero

(b) 40 lb ↑

(c) 40 lb ↓

(d) 40 lb ↑ plus 400 ft-lb

(e) 40 lb ↓ plus 400 ft-lb

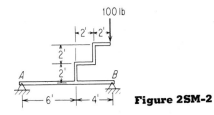

100 lb

A

B

Figure 2SM-2

2SM- 3 One newton equals:
(a) 10 ergs   (d) 10^2 ergs
(b) 10^3 ergs   (e) 10^4 ergs
(c) 10^5 ergs

2SM- 4 For a system to be in equilibrium, the sum of the external forces acting on the system must be:
(a) equal to unity   (d) a maximum
(b) indeterminant   (e) zero
(c) infinite

2SM- 5 The vector which represents the sum of a group of force vectors is called the:
(a) magnitude   (d) resultant
(b) sum   (e) phase angle
(c) force polygon

2SM- 6 Two weights are suspended on an inextensible weightless cord from frictionless pulleys as shown in the figure. Which of the following most nearly equals the angle θ at equilibrium?
(a) 60°   (d) 53°
(b) 37°   (e) 82°
(c) 75°

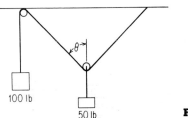

100 lb

50 lb

Figure 2SM-6

2SM- 7 A 50-lb pulley, supported as shown, carries an inextensible cable supporting an additional 50-lb load. The force the beam exerts on the pulley shaft is:

(*a*) 50 lb up   (*d*) 150 lb up

(*b*) 100 lb up   (*e*) 150 lb down

(*c*) 100 lb down

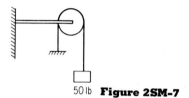

50 lb   **Figure 2SM-7**

2SM- 8   A 100-lb weight hangs by a string from the ceiling. It is pulled by a horizontal force until the string makes an angle of 30° with the vertical. The tension in the string most nearly equals:

(*a*) 100 lb   (*d*) 110 lb

(*b*) 125 lb   (*e*) 120 lb

(*c*) 115 lb

2SM- 9   An 80-lb body lies on a 30° slope. The frictional force most nearly equals:

(*a*) 35 lb   (*d*) 40 lb

(*b*) 50 lb   (*e*) 60 lb

(*c*) 45 lb

2SM-10   An object weighing 100 lb is held by two strings 4 ft long, the fixed ends of which are fastened to a ceiling 4 ft apart. What is the tension in each string?

(*a*) 58 lb   (*d*) 50 lb

(*b*) 62 lb   (*e*) 48 lb

(*c*) 60 lb

DYNAMICS—SAMPLE PROBLEMS

2D- 1   A chain hangs over a smooth peg, 8 ft of its length being on one side and 10 ft on the other. If the force of friction is equal to the weight of 1 ft of the chain, find the time required for the chain to slide off.

2D- 2   A falling weight of 1000 lb is used to drive a pile into the ground. If the weight of the pile is 800 lb and the weight is dropped 20 ft, what will be the depth of penetration of the pile, assuming an average resistance to penetration of 30,000 lb? Assume the impact between the weight and pile to be perfectly inelastic.

2D- 3   A horizontal force is applied to a block moving in a horizontal guide. When the block is at a distance *x* from the origin, the force is given by the equation

$F = x^3 - x$. What work is done by moving the block from 1 ft to the left of the origin to 1 ft to the right of the origin?

2D- 4 In the system shown in Fig. 2D-4, $W = 20$ lb and is observed to vibrate ten times in 15 sec. Disregard the weight of the lever arm and find the stiffness coefficient of the spring S.

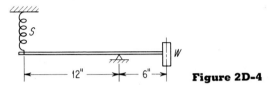

Figure 2D-4

2D- 5 An unbalanced force in pounds that varies with time in seconds as follows— $F = 800 - 120t$—acts on a body with a weight of 3220 lb for 5 sec. If the body had an initial velocity of 10 fps in the direction of the unbalanced force, what final velocity would the body have? What would be the initial and final power, in horsepower, required?

2D- 6 A wheel 2 ft in diameter weighting 64.4 lb has a radius of gyration of 9 in. If it starts from rest and rolls without slipping down a 30° plane, what will be its speed when it has moved 20 ft?

2D- 7 A rectangular door swings horizontally, is 3 ft wide, and weighs 60 lb uniformly distributed. The door is supported by frictionless hinges installed along one of its vertical sides. It is controlled by a spring which exerts on it a torque proportional to the angle through which it is turned from its closed position. If the door is opened 90° and released, the torque due to the spring is then 3 lb-ft. Determine the angular velocity at the time it has swung back and is passing through the 45° position.

2D- 8 A body is moving in a straight line according to the law

$$S = \frac{1}{4}t^4 - 2t^3 + 4t^2$$

When $S =$ distance traveled in time t,
a. Find its velocity formula and its acceleration formula.
b. When is its velocity at a maximum?
c. During what interval is it moving backward?

2D- 9 A locomotive is traveling round a curve of mean radius 1000 ft at a speed of 40 mph. Find the correct amount of superelevation so that there shall be no side thrust on the flanges of the wheels at that speed. If the gauge of the railway track is 5 ft and the height of the center of gravity of the locomotive is 4 ft above rail level, find the greatest speed at which the locomotive could travel round the curve without actually overturning.

2D-10 A belt 8 in. wide and $\frac{1}{4}$ in. thick drives a pulley 12 in. in diameter at 1650 rpm. The maximum tension the belt may have is 200 psi. If the slack side has three-fourths the tension of the tight side, what horsepower may be transmitted?

2D-11   A hoisting cable has a safe working strength of 40,000 psi of cross-sectional area. What is the greatest weight that can be raised with an acceleration of 8.05 ft/sec^2 by a cable $1^1/_2$ in. in diameter? Disregard friction.

2D-12   A kite is 120 ft high, with 150 ft of cord out. If the kite moves horizontally 4 mph directly away from the boy flying it, how fast is the cord being paid out in feet per second?

2D-13   A bullet leaves a gun muzzle at a velocity of 2700 fps at an angle of 45° to the horizontal.

   a. What is the maximum height to which the bullet will travel?

   b. What is the maximum distance the bullet will travel horizontally, measured along the same elevation at the gun muzzle?

   Disregard air resistance.

2D-14   An airplane has a true air speed of 240 mph and is heading true north. A 40-mph head wind is blowing 45° from true north from the northeast.

   a. Draw a velocity vector diagram for the conditions outlined above.

   b. Solve for the ground speed.

   c. Solve for the true course the airplane is making with reference to the ground.

2D-15   The rifling in a 30-caliber rifle causes the bullet to turn one revolution per 10 in. of barrel length.

   a. If the muzzle velocity of the bullet is 2900 fps, how many rpm is the bullet rotating at the instant it leaves the muzzle?

   b. Assuming constant acceleration for the bullet for the 22 in. of barrel travel, what is the magnitude of the acceleration?

2D-16   A faster-moving body is directly approaching a slower-moving body going in the same direction. The speed of the faster body is V_1; the speed of the slower body is V_2. At the instant when the separation of the two bodies is d, the faster body is given a constant deceleration of a. Determine the formula for the minimum value of d, so that there will be no collision.

2D-17   A parachutist is falling with a speed of 176 fps when his parachute opens. If the air resistance is $Wv^2/256$ lb, where W is the total weight of the man and parachute, find his speed as a function of time after the parachute is opened.

2D-18   At the beginning of the drive, a golf ball has a velocity of 170 mph. If the club stays in contact with the ball for $^1/_{25}$ sec, what is the average force on the ball? The weight of the ball is 1.62 oz.

2D-19   An artificial satellite is circling the earth at a radius of 4500 miles. Assume that the pull of gravitation is the same at that elevation as at the earth's surface.

   a. What speed in radians per second is required to maintain the satellite at that radius?

   b. What is total kinetic energy in the satellite if it weighs 50 lb?

2D-20   The fire door shown in Fig. 2D-20 is hung from a horizontal track by means of wheels at A and B. The door weighs 200 lb. The wheel at A has rusted on its bearing and slides, rather than rolls, on the track, the coefficient of friction being one-third. Wheel B rolls freely, and friction is negligible. Determine the force P necessary to give the door an acceleration of 8 fps. Compute the horizontal and vertical components of the reactions at A and B.

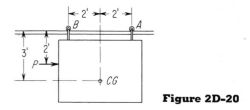

Figure 2D-20

2D-21 A truck of dimensions and weights shown in Fig. 2D-21 is rolling across a 30-ft bridge.
   a. What is the distance X for maximum moment?
   b. Calculate the maximum moment and where it is located.

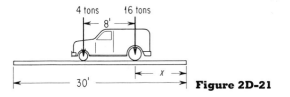

Figure 2D-21

2D-22 A balloon is ascending vertically at a uniform rate for 1 min. A stone let fall from it reaches the ground in 5 sec. Find:
   a. Velocity of the balloon
   b. Height from which the stone fell
2D-23 A body takes just twice as long to slide down a plane at 30° to the horizontal as it would if the plane were smooth. What is the coefficient of friction?
2D-24 How far from the center of a phonograph record turning at 78 rpm will a pickle lie without being thrown off, if the coefficient of friction is 0.30?
2D-25 Block A (Fig. 2D-25) weighs 200 lb and is placed 5 ft from the edge of the table. A string passing over a smooth pulley connects blocks A and B as shown. Block B weighs 50 lb and is 2 ft from the floor. The coefficient of

sliding friction for block A is 0.15. Block B is released from rest in this position.

a. With what velocity does B strike the floor?

b. How far from the edge of the table will block A come to rest?

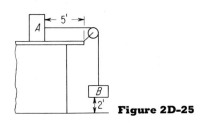

Figure 2D-25

2D-26 A cast-aluminum drum, 2 ft in diameter and 1 ft in length, is supported by an axial shaft on smooth bearings. A light cord wrapped on the drum is pulled by a constant force of 20 lb until 15 ft of cord is unwrapped and then let go. What is the rpm of the drum when the pull on the cord has ceased?

2D-27 A 50-ton car moving at a speed of 3 mph strikes a bumping post equipped with a 40,000-lb spring. Assuming that all the shock is absorbed by the spring, how far will the spring be compressed?

2D-28 An automobile which weighs 3700 lb approaches a curve on a concrete paved highway. The radius of the curve is 1000 ft and the pavement is flat (no superelevation). If the coefficient of friction between the tires and the pavement is 0.15, what is the maximum speed at which the driver may round the curve without skidding sidewise? What should be the transverse slope of the pavement to drive at 60 mph on the curve and keep the resultant forces perpendicular to the pavement? Assume the pavement is 24 ft wide and the distance between treads is 6 ft.

2D-29 An elastic body weighing 100 lb is moving with a velocity of 20 fps and overtakes another elastic body, weighing 150 lb, moving in the same direction with a velocity of 15 fps. What is the final velocity, and what is the loss in kinetic energy?

2D-30 A body weighing 200 lb starts from rest at the top of a plane which makes an angle of 60° with the horizontal. After sliding 8 ft 4 in. down the plane, the body strikes a 100-lb/in. spring. The coefficient of friction is 0.25. The body remains in contact with the plane throughout. Determine the compression of the spring in inches.

2D-31 A 100-lb wheel 18 in. in diameter, rotating at 150 rpm in stationary bearings, is brought to rest by pressing a brake shoe radially against its rim with a force of 20 lb. If the radius of gyration of the wheel is 7 in., and if the coefficient of friction between the brake shoe and rim is steady at 0.25, how many revolutions will the wheel make in coming to rest?

2D-32 A solid cylinder 4 ft in diameter and weighing 12,880 lb is rolling at a linear
speed of 40 fps. How far will it roll up a 10 percent grade before coming to a
stop?

2D-33 The friction surface on a friction disk type of clutch has an outside diameter
of 10 in. and an inside diameter of 3 in. The coefficient of friction for the
surfaces in contact may be assumed as 0.3. What pressure must be applied to
the plate if the clutch is to transmit 100 hp at 3300 rpm?

2D-34 For the conditions shown in Fig. 2D-34, compute the weight W in pounds and
the tension in the cable if the acceleration of the 100-lb body is 10 ft/sec², the
coefficient of sliding friction is 0.5, and the pulleys are assumed to be frictionless.

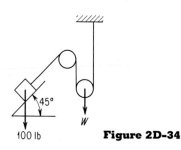

45°

100 lb W **Figure 2D-34**

2D-35 A body starts from rest, and its velocity t sec later is given by the following
table:

t, sec	v, fps
0	0
0.5	12
1	19.5
2	29.5
3	34
4	31
5	19.8
6	11
7	6.5

a. Find the distance traveled in the first 5 sec.
b. Find the acceleration at the instant when $t = 2$ sec.
If another body starts from rest under a uniform acceleration, and travels
the same distance in the first 5 sec,
c. Find its acceleration.
d. Find its velocity at the instant when $t = 5$ sec.

2D-36 A motor car has a maximum speed of 70 mph on the level, and the engine is then transmitting 60 hp to the road wheels.

a. Find the resistance to motion in pounds of weight.

b. Assuming the tractive effort at the wheels to remain constant, while the resistance to motion varies as the square of the speed, find the slope (1 in *x*) up which the car can maintain 50 mph, the weight of the car being 4000 lb.

2D-37 To measure the muzzle velocity of a bullet fired from a rifle, two cardboard disks mounted on a long axle are rotated at a constant speed, the rifle is mounted parallel to the axle, and, upon firing, the bullet passes through each of the disks in turn. In a test, the disks were 3.5 ft apart on the axle, the speed of rotation was 3000 rpm, and the bullet hole in the farther disk was displaced 18° with respect to the hole in the nearer disk. Determine the muzzle velocity.

2D-38 A 3-oz bullet is fired horizontally into an 18-lb block suspended vertically so that it can swing as a pendulum. If the center of gravity of the block is caused to rise 8 in., what was the speed of the bullet?

2D-39 Referring to Fig. 2D-39, *CD* is turning clockwise about *D* at 60 rpm. What is the tangential velocity of point *B* for the position shown?

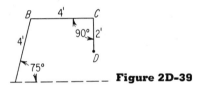

Figure 2D-39

2D-40 A steel plate 30 in. in diameter and 4 in. thick has six 3-in. holes equally spaced on a 20-in.-diameter circle (Fig. 2D-40). The disk is rotating about an axis through its center at 300 rpm. Calculate the kinetic energy stored in the disk.

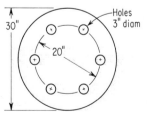

Figure 2D-40

2D-41 A rope wound around a cylindrical drum, which is free to turn in a frictionless bearing about an axis through its center, has a 16-lb weight fastened to its end (Fig. 2D-41). The drum is 1 ft in diameter and has a moment of inertia of 1.875 slug-ft². When the weight has fallen 16 ft,

a. Find the work done on the system.

b. Find the kinetic energy of the cylinder.

c. Find the kinetic energy of the weight.

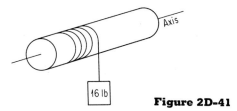

Figure 2D-41

2D-42 An elevator and load weighing 10,000 lb after a 2-sec free fall down the shaft strikes a safety device which applies a braking force of 15,000 lb.

a. What is the rate of deceleration?

b. What is the total distance traveled including the fall and deceleration to a stop?

2D-43 A drop hammer of 1 ton deadweight capacity is propelled downward by a 12-in.-diameter air cylinder. At 100 psi air pressure, what is the impact velocity if the stroke is 28 in.? What is the impact energy?

2D-44 A 10-hp 1750-rpm motor is connected directly to a 30 lb-ft torque brake. The rotating system has a total inertia ($W R^2$) of 10 lb-ft². How long will it take the system to stop if the brake is set at the instant the motor is shut off? How much energy must be dissipated?

2D-45 A train traveling at 75 mph receives a warning that it is approaching a "slow signal," and the engineer immediately reduces speed at a uniform rate of 1.8 ft/sec², so that his speed is 15 mph as he reaches the slow signal. The slow order covers a stretch of 5 miles of track being reconstructed and as soon as he reaches the end of the slow order the engineer increases his speed at such a uniform rate that the train is traveling 85 mph 90 sec thereafter.

a. What length of time elapsed between the warning and the beginning of the slow-order stretch, and what distance did the train travel during that time?

b. What distance did the train travel from the end of the slow-order stretch in attaining its speed of 85 mph?

c. How long will the train have to travel at 85 mph in order to make up the time lost by not running continuously at 75 mph?

2D-46 A 30-ton freight car with a velocity of 4 mph in coupling collides with a stationary 20-ton car. Assume 25 percent of the energy is lost in impact and failure to couple; what will be the velocities of both cars after collision?

2D-47 Find the minimum speed that a ball starting at point P must have to loop the loop on the vertical circular track without leaving it (Fig. 2D-47). (Assume frictionless contact.)

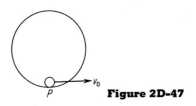

Figure 2D-47

2D-48 A mass weighing 25 lb falls a distance of 5 ft upon the top of a helical spring with a spring constant of 20 lb/in.
  a. What will the velocity of the mass be after it has compressed the spring 8 in.?
  b. Calculate the maximum compression of the spring.

2D-49 A circular cylinder, as in Fig. 2D-49, rolls on a horizontal floor. At a given instant the angular acceleration α equals 8 rad/sec², and the angular velocity ω equals 6 rad/sec. Assume that there is no slipping and that A and C are on a horizontal line. Determine the velocity and the acceleration of point A at this instant.

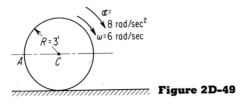

Figure 2D-49

2D-50 A boy weighing 100 lb is jumping up and down on a Pogo stick which has a 100 lb/in. spring. The boy jumps so that the lower end of the Pogo stick is 3 ft off the ground at maximum height, and at the bottom of the cycle he pushes down with 25 lb more than is required to stop his downward motion in order to overcome friction and keep up the jumping. How many inches is the spring compressed at the bottom of the cycle?

2D-51 Cylinder A weighing 100 lb is placed in a box B (Fig. 2D-51), the bottom of which is inclined 30° to the horizontal. The box is accelerated to the right at 20 ft/sec². Determine the forces F_1 and F_2.

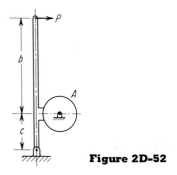

Figure 2D-51

2D-52 The wheel A in Fig. 2D-52 is a solid cylinder weighing 1000 lb, with a diameter of 8 ft and free to turn on a frictionless axle. It is desired to arrange a brake as shown, by means of which the speed of the wheel may be reduced from 120 rpm to zero in 10 sec. Determine the ratio b/c of the available force. $P = 100$ lb and the coefficient of brake friction is 0.25.

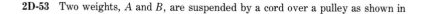

Figure 2D-52

2D-53 Two weights, A and B, are suspended by a cord over a pulley as shown in

Fig. 2D-53. Assuming that pulley and cord are weightless and frictionless and that the weights are permitted to shift from the position indicated by gravity, when A is 500 lb and B is 750 lb:

  a. Determine the acceleration of weights.
  b. Determine the velocity of A when one weight hits the floor.
  c. Determine the momentum of A at the instant when one weight hits the floor.
  d. Determine the tension in the cord.

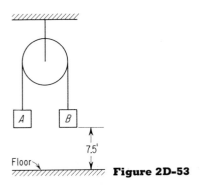

Figure 2D-53

2D-54 One end of a ladder 50 ft long is leaning against a perpendicular wall standing on a horizontal plane. Supposing the foot of the ladder to be pulled away from the wall at the rate of 3 fpm:

  a. How fast is the top of the ladder descending when the foot is 14 ft from the wall?
  b. When will the top and bottom of the ladder move at the same rate? Solve by calculus.

DYNAMICS—MULTIPLE-CHOICE PROBLEMS

For each question select the correct answer from the five given possibilities.

2DM- 1 A 16-lb weight and an 8-lb weight resting on a horizontal frictionless surface are connected by a cord, A, and are pulled along the surface with a uniform acceleration of 4 ft/sec^2 by a second cord, B, attached to the 16-lb weight. The tension in cord A is closest to:

  (a) 4 lb   (d) 1 lb
  (b) 3 lb   (e) 2 lb
  (c) 32 lb

2DM- 2 An impuse is the product of:
(*a*) force and displacement
(*b*) force and time
(*c*) force and velocity
(*d*) mass and acceleration
(*e*) mass and velocity

2DM- 3 A crane is lowering a 2-ft-diameter ball which weighs 1000 lb on the end of a weightless cable at a uniform velocity of 12 ft/sec. Assuming no friction, the total tension in the cable is:
(*a*) 750 lb   (*d*) 900 lb
(*b*) 800 lb   (*e*) 1000 lb
(*c*) 850 lb

2DM- 4 The kinetic energy of a 2400-lb automobile traveling at 100 ft/sec is closest to:
(*a*) 3.75×10^5 ft-lb
(*b*) 7.5×10^5 ft-lb
(*c*) 24×10^6 ft-lb
(*d*) 12×10^6 ft-lb
(*e*) 3.75×10^6 ft-lb

2DM- 5 A 10-in-diameter pulley is belt driven with a net torque of 250 ft-lb. The ratio of tensions in the tight to slack sides of the belt is 4:1. What is the maximum tension in the belt?
(*a*) 250 lb   (*d*) 500 lb
(*b*) 83 lb   (*e*) 333 lb
(*c*) 800 lb

2DM- 6 The second derivative of distance with respect to time is:
(*a*) speed
(*b*) velocity
(*c*) force
(*d*) distance
(*e*) acceleration

2DM- 7 Centrifugal force *is not* a function of:
(*a*) angular velocity
(*b*) radius
(*c*) mass
(*d*) moment of inertia
(*e*) any of these

2DM- 8 Kinetic energy has units of:
(*a*) ft-lb   (*d*) lb-ft/sec²
(*b*) ft-lb/sec   (*e*) Btu/hr
(*c*) lb/in²

2DM- 9 A newton is the force required to give:
(*a*) one kilogram an acceleration of 1 m/sec²
(*b*) one kilogram a velocity of 1 m/sec
(*c*) one kilogram a velocity of 1 cm/sec
(*d*) one gram an acceleration of 1 cm/sec²
(*e*) one gram a velocity of 1 cm/sec

2DM-10 A flower pot is inadvertently dropped out of a second-story window 14 ft above the pavement. What is its velocity when it strikes the ground?
(*a*) 30.0 ft/sec   (*d*) 33.6 ft/sec
(*b*) 32.2 ft/sec   (*e*) 27.8 ft/sec
(*c*) 28.1 ft/sec

3

FLUID MECHANICS

The review of fluid mechanics or hydraulics can be split into two general subdivisions—hydrostatics and hydrodynamics. Hydrostatics will be discussed first. The phases of this general subdivision which have been included in many of the past fundamentals examinations include the following: pressure due to a head of fluid, pressure gauges and manometers, the location of the center of pressure and the resultant force due to the applied fluid pressure, and Archimedes' principle of buoyancy.

3-1. HYDROSTATICS

Hydrostatics is the name applied to the study of fluids at rest; the term "fluids" includes both liquids and gases. This discussion shall be concerned primarily with the properties of liquids, taking as a guide the emphasis shown in past fundamentals examinations.

Liquids are distinguished from solids by the fact that their particles are readily displaced. A liquid, depending upon its viscosity, may offer a considerable resistance to a rapid change of shape, but this resistance will vanish once the motion has ceased. One definition of the liquid state is that "a liquid in equilibrium offers no resistance to change of shape." From this it follows that the resultant of all the forces acting upon a particle of liquid at the surface of a liquid must be perpendicular to the tangent to the surface at that point. This also follows from the fact that a liquid cannot withstand any shearing force. An example of the usefulness of this concept is shown in the determination of the shape of the surface of the fluid in an upright cylindrical container rotating about its central axis (Fig. 3-1).

First draw a sketch of a section through the center of the cylinder. To determine the required curve of the surface, then select any particle of fluid

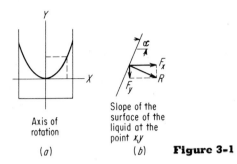

Axis of
rotation

(a)

Slope of the
surface of the
liquid at the
point x,y

(b)   **Figure 3-1**

on this surface and add the forces acting on this particular particle. F_x equals the force on the particle due to the rotation about the axis and equals the centripetal force, or

$$F_x = \omega^2 rm$$
$$F_y = \text{gravitational force} = mg$$

The slope of the surface of the liquid at the point (x, y) is perpendicular to the slope of the resultant R of the forces acting on the particle of liquid. $\tan \alpha = F_x/F_y$, but also $\tan \alpha = dy/dx$ at this particular point, so

$$dy/dx = (\omega^2 rm/mg)$$

where ω and g are constants, $r = x$, and $dy = (\omega^2/g) x \, dx$, or the equation for the curve of the surface is $y = (\omega^2/2g)x^2$ plus a constant of integration. If the axes are selected as shown in the figure, $y = 0$ for $x = 0$ and the constant of integration is zero, the final equation becomes $y = (\omega^2/2g)x^2$, which will be recognized as the equation of a parabola. The curve of the surface of the liquid is then a paraboloid of revolution.

Another example of the application of this principle is afforded by a past examination problem:

■ A container partially filled with water and having a total weight of 10 lb is sliding down a 30° inclined plane against a friction factor of $\mu = 0.30$.
  a. At what rate is the container accelerating down the inclined plane?
  b. At what angle is the water surface in the container measured from the horizontal?

In order to simplify your calculations, it is suggested that you solve for the unknowns in the triangles graphically. First, draw the figure (Fig. 3-2).

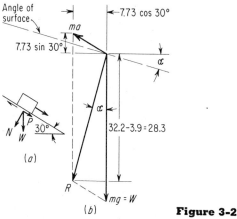

Figure 3-2

$$f = N \times \mu = W \cos 30° \times 0.30$$
$$f = 0.260W = 2.60 \text{ lb}$$
$$P = W \sin 30° = 5.00 \text{ lb}$$

The net force acting on the container to cause acceleration = 2.40 lb.

$$a = \frac{F}{m} = \frac{2.40}{10/g} = 0.240g = 7.73 \text{ ft/sec}^2$$

Taking one particle of water on the surface, we can show the forces acting upon it. If we draw the inertial force of ma at an angle of 30° and the gravitational force of mg at the correct angles, the angle of the resultant can be determined either graphically or analytically.

3-2. FLUID PRESSURE

The pressure at any point in a liquid is equal to the pressure on the surface of the liquid plus the weight of the column of liquid of unit cross-sectional area extending vertically from the point to the surface of the liquid. Furthermore, the pressure is the same at all points at the same level within a static, continuous, single-liquid system, and the pressure at a particular point is the same regardless of the direction in which the pressure is measured. Pressure is nondirectional. The pressure at a given depth in a liquid will then depend upon the specific weight of the liquid. Columns of unequal heights of different liquids may produce equal pressures. These are the bases of the manometer and the differential manometer.

3-3. MANOMETERS

An example illustrating a differential manometer is afforded by a problem which asked:

■ Carbon tetrachloride flows in pipe A in Fig. 3-3 and water flows in pipe B. The two pipes are connected by a differential gauge which contains oil (sp. gr., 0.80) and pipe A is 10 ft above pipe B. If the specific gravity of the carbon tetrachloride is 1.5 and the other dimensions are as shown, what is the difference in pressure (in feet of water) between pipes A and B, and which is the greater?

The notations P_1 and P_2 are added to the figure, and since they are at the same elevation in the same fluid system, $P_1 = P_2$.

$$P_A = P_1 + 5 \text{ ft of CCl}_4 \qquad P_B = P_2 + 2 \text{ ft of oil} + 13 \text{ ft of water}$$
$$P_1 = P_A - 5 \text{ ft of CCl}_4 \qquad P_2 = P_B - 2 \text{ ft of oil} - 13 \text{ ft of water}$$
$$P_1 = P_A - 7.5 \text{ ft of water} \qquad P_2 = P_B - 1.60 \text{ ft of water} - 13 \text{ ft of water}$$
$$P_A - 7.5 \text{ ft of water} = P_B - 14.60 \text{ ft of water}$$
$$P_B = P_A + 7.1 \text{ ft of water}$$

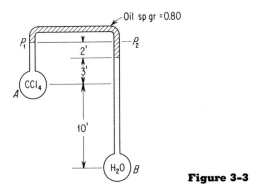

Figure 3-3

Utilizing the same reasoning, we can calculate the total force on any submerged area. The total force will equal the summation of the pressures at depths h, $h + \Delta h$, $h + 2\Delta h$, . . . times the areas at depths h, $h + \Delta h$, $h + 2\Delta h$, ... This is shown in Fig. 3-4. The force on $dA = P_h \, dA = (P_1 + wh) \times dA$.

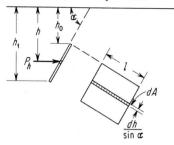
Fluid of specific weight w lb/ft^3

Figure 3-4

$$dA = 1 \times \frac{dh}{\sin \alpha}$$

$$\sum P_h \, \Delta A \rightarrow \int_{h_0}^{h_1} (P_1 + wh)\frac{dh \, l}{\sin \alpha}$$

The force acting on the surface is

$$F = \int_{h_0}^{h_1} (P_1 + wh)\frac{l \, dh}{\sin \alpha}$$

For the rectangular surface shown,

$$F = \frac{l}{\sin \alpha}\left[P_1(h_1 - h_0) + \frac{w}{2}(h_1{}^2 - h_0{}^2) \right]$$

P_1 is ordinarily only atmospheric pressure, which will act on both sides of a gate or dam subjected to hydraulic pressure, so it can be omitted from the equation. This gives:

$$F = \frac{lw}{2 \sin \alpha}(h_1{}^2 - h_0{}^2) = wA \frac{h_1 + h_0}{2}$$

since $A = l(h_1 - h_0)/\sin \alpha$, $w(h_1 + h_0)/2$ is the average pressure acting on the area.

The same result may be obtained without the use of calculus by constructing the graph of pressure vs. depth and calculating the resulting force from the relationship P psf $\times$ A sq ft $=$ F lb (Fig. 3-5). The total force represented by the trapezoid of pressure is, for a perpendicular surface,

$$F = \left[P_1 + \frac{w(h_0 + h_1)}{2} \right](h_1 - h_0) \times l$$

where l = width of the perpendicular surface. The actual pressure acts over a larger sloping surface in the ratio of $1/\sin \alpha$, so

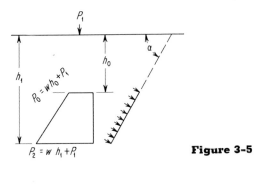

Figure 3-5

$$F = \frac{l}{\sin \alpha}\left[P + \frac{w(h_1 + h_0)}{2}\right] \times (h_1 - h_0)$$

since pressure always acts perpendicular to a surface regardless of the angle of the surface.

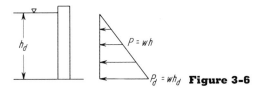

Figure 3-6

This same relationship can be arrived at for a vertical dam in another manner (Fig. 3-6). Pressure increases linearly with the depth, and the force per unit width of the dam will equal $\int p \, dh$; this is just the area of the pressure triangle shown, which in turn is equal to the product of one-half the altitude times the base or

$$(^1/_2 h_d) \times (w h_d) = w \frac{(h_d)^2}{2} \qquad \text{lb/ft of width of dam}$$

The same average pressure relationship will hold for any submerged surface that is symmetrical about its horizontal center line. This includes rectangles, circles, and ellipses.

Utilizing the above, we can also derive a more general relationship for the hydrostatic force acting on any submerged surface. First put down the equation for the force acting on a surface, but leave out the effect of atmospheric pressure. As previously noted, the atmospheric pressure force acts on both sides of the surface, so its effects are balanced. This leaves

$$F = \int_{h_0}^{h_1} wh \frac{l \ dh}{\sin \alpha}$$

but $l \ dh = dA$

so
$$F = \int_{h_0}^{h_1} w \frac{h \ dA}{\sin \alpha}$$

but
$$\frac{\int_{h_0}^{h_1} h \ dA}{A} = \bar{y}$$

This gives $F = w\bar{y}A/\sin \alpha$ where $\bar{y}$ is the vertical distance from the surface of the liquid to the centroid of the submerged area.

The total net force acting on a surface because of the pressure of the liquid may be considered as the resultant of the hydrostatic forces acting. The hydrostatic force can then be resisted by a single applied force acting against the surface on the side opposite the liquid. This single force must be located at a point about which there will be no moment due to the hydrostatic force; this point is called the "center of pressure." We can determine the center of pressure for a rectangular surface in a fashion similar to that by which we determined the force due to the hydrostatic pressure (Fig. 3-7).

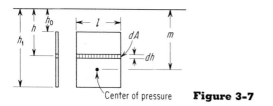

Center of pressure   **Figure 3-7**

3-4. CENTER OF PRESSURE

Locating the center of pressure is similar to locating the center of gravity of a surface. The center of pressure might be termed the center of force, since it is that point on the submerged surface about which there is no resultant moment due to the pressure forces acting on the surface. m is the distance from the surface of the liquid to the center of pressure.

$$m = \frac{\Sigma r \, \Delta F}{\Sigma \, \Delta F} = \frac{\displaystyle\int_{h_0}^{h_1} h \times hw \, dA}{\displaystyle\int_{h_0}^{h_1} hw \, dA}$$

For the rectangular surface shown, this becomes

$$m = \frac{\displaystyle\int_{h_0}^{h_1} h^2 w \times l \, dh}{\displaystyle\int_{h_0}^{h_1} hw \times l \, dh} = \frac{lw \, \frac{1}{3}(h_1^3 - h_0^3)}{lw \, \frac{1}{2}(h_1^2 - h_0^2)} = \frac{2 \, h_1^3 - h_0^3}{3 \, h_1^2 - h_0^2}$$

Since $l \, dh = dA$, a more general relationship may be deduced from the above relationship:

$$m = \frac{w \displaystyle\int_{h_0}^{h_1} h^2 \, dA}{w \displaystyle\int_{h_0}^{h_1} h \, dA}$$

where $\displaystyle\int_{h_0}^{h_1} h^2 \, dA$ = moment of inertia of the area about axis 0-0

$$\int_{h_0}^{h_1} h \, dA = \bar{y} \int dA$$

since $\bar{y} = (\int h \, dA)/\int dA$, giving the general relationship $m = I/A\bar{y}$.

where m = distance from point 0 to the center of pressure
I = moment of inertia of the area about a horizontal axis through 0
A = area
$\bar{y}$ = distance from 0 to the center of gravity (or centroid) of the area

In other words, the distance from an axis to the center of pressure of an area is equal to the moment of inertia of the area about that axis divided by the moment of the area about the same axis.

A problem illustrating this is:

■ A vertical 4- by 4-ft gate is submerged with the level top 2 ft below the water surface. The gate is exposed to hydrostatic pressure on one side. Find the magnitude and location of these hydrostatic forces.

First draw a figure and label it (Fig. 3-8). The pressure diagram will be

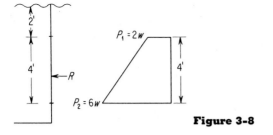

Figure 3-8

a trapezoid, as shown in the figure. The total force acting on the gate because of the pressure of the water will then equal $[(2w + 6w)/2] \times 4$ lb/ft of width $\times$ 4 ft wide; therefore,

$$F = 4w \times 4 \times 4 = 3990 \text{ lb } (w \text{ for water} = 62.4 \text{ lb/ft}^3)$$

Note that the atmospheric pressure may be disregarded since it acts equally on both sides of the gate. The location of the center of pressure will be m ft below the surface, where $m = I/A\bar{y}$. The moment of inertia of a rectangle about a transverse axis through its center is $bh^3/12$, which equals $(4 \times 4^3)/12 = 21.3$ ft^4. The moment of inertia about the surface, 4 ft from the axis of the square, would be:

$$21.3 + AS^2 = 21.3 + 16 \times 4^2 = 277.3 \text{ ft}^4$$
$$A\bar{y} = 16 \times (2 + 2) = 64$$
$$m = 277.3/64 = 4.33 \text{ ft}$$

Another problem of a similar type was:

- Flashboards placed on the crest of an ogee spillway are designed to fail when the head H is 12 ft. The boards are 4 ft wide and are supported by iron pins mounted in sockets along the crest at 10-ft intervals. Consider only hydrostatic pressure and determine the resisting moment of each pin at the crest of the spillway (Fig. 3-9).

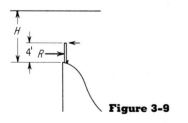

Figure 3-9

Add an arrow to represent the resultant R. The total force acting on each foot of width of the flashboards will equal

$$\frac{8 + 12}{2} \times w4 \times 1 = 2500 \text{ lb/ft}$$

(see previous example for sample pressure diagram). The location of the center of pressure $m = I/A\bar{y}$, where I is the moment of inertia of the pressurized area about an axis at the surface of the water, and $\bar{y}$ is the distance below the surface to the center of gravity of the pressurized area.

$$I/\text{ft of width} = \frac{1 \times 4^3}{12} + (1 \times 4)10^2 = 405.3 \text{ ft}^4/\text{ft}$$
$$A\bar{y}/\text{ft of width} = (1 \times 4)(2 + 8) = 40$$
$$m = 405.3/40 = 10.1 \text{ ft below the surface of the water}$$

Each pin must withstand the force acting on half of the 10-ft width of flashboard on each side of it, or each pin must withstand the hydrostatic force acting on 10 ft of flashboard. Taking moments about an axis at the top of the flashboard we find that the pins must withstand a force of

$$\frac{(2500 \text{ lb/ft} \times 10 \text{ ft})(10.1 - 8)}{4} = 13,100 \text{ lb/pin}$$

The remainder of the force of 25,000 lb/(10-ft width), or 11,900 lb/10 ft of width, must be held at the top of the flashboard.

3-5. ARCHIMEDES' PRINCIPLES OF BUOYANCY

The principles of buoyancy and flotation may be expressed as follows:

1. A body immersed in a fluid is buoyed up by a force equal to the weight of fluid displaced by the body.

2. A floating body displaces its own weight of the fluid in which it is floating.

3. The buoyant force acts upward through the center of gravity of the displaced volume.

The application of these principles can be illustrated by means of an example selected from a past fundamentals examination, which asked:

■ A rectangular caisson is being constructed for use in building a bridge pier. The caisson is 50 by 20 by 30 ft high (inside dimensions). The walls are 1 ft thick, and the weight is 18,000 lb. There is no bottom.
   a. How far below the water surface will the bottom edge be when launched?
   b. What additional weight is required to sink the caisson on the bottom of the lake, which is 25 ft deep?

c. What additional weight is required to hold the caisson on the bottom once it is seated and the water pumped out?

The volume of the walls per foot of depth would equal

$$50 \times 1 \times 2 + 22 \times 1 \times 2 = 144 \text{ ft}^3$$

The volume of water displaced for each foot the open-ended caisson sank below the surface of the water would then be 144 ft^3, and the buoyancy per foot of caisson below the surface would equal

$$144 \times 62.4 = 8980 \text{ lb/ft}$$

so the answer to part (*a*) is 18,000/8980 = 2.0 ft. The additional weight to sink it an additional 23 ft would equal

$$23 \times 8980 = 207,000 \text{ lb}$$

or, total displacement would equal $25 \times 144 = 3600$ ft^3. The maximum buoyant force would equal

$$3600 \times 62.4 = 225,000 \text{ lb}$$
$$225,000 - 18,000 = 207,000 \text{ lb} \qquad [\text{part } (b).]$$

Once the caisson was seated on the bottom and the water was pumped out, there would be no buoyant force acting upward *if* (and a very big if) the bottom of the caisson were sealed tightly so that no water could seep under it. If there were seepage, then the pressure acting upward on the bottoms of the walls would range from $25 \times 62.4 = 1560$ lb/ft^2 at the outside, where the pressure would equal the static pressure at a depth of 25 ft, to zero pressure at the inside, where only atmospheric pressure would act. The average pressure acting upward on the bottoms of the walls would then equal the average pressure between 1560 and 0, or 780 lb/ft^2. The upward force acting on the caisson would thus equal 144 ft$^2 \times 780$ lb/ft$^2 = 112,300$ lb less the 18,000 lb weight of the caisson, or 94,300 lb.

This is conservative if the space inside is kept pumped out and free of internal water. If the leakage were not removed and the water level rose inside, a greater weight would be required to keep the caisson in place.

The same principles are used to measure the specific gravities of fluids. Another question asked in a past fundamentals examination was:

■ A hydrometer bulb and stem displace 0.5 in.^{3} when immersed in fresh water. The stem has an area of 0.05 in.^{2} What is the distance on the stem between the graduations for 1.00 and 1.10 sp. gr.?

The hydrometer bulb and stem displace 0.5 in.^{3} of water, so the bulb and stem weigh $(0.5/1728) \times 62.4$ lb. They will displace this same weight of liquid when floating in the 1.1-sp.-gr. liquid, so the volume of liquid displaced will equal

$$\frac{0.5}{1728} \times 62.4 \times \frac{1728}{1.1 \times 62.4} = 0.455 \text{ in.}^3$$

The difference in the volume of fluid displaced (or the volume of bulb and stem immersed) will be $0.500 - 0.455 = 0.045$ in.^{3} Since the stem has a cross-sectional area of 0.05 in.^{2}, the stem will stick out of the liquid 0.045 in.^{3}/0.050 in.^{2} = 0.90 in. higher in the 1.10-sp.-gr. fluid than it did out of the 1.00-sp.-gr. fluid, and the distance on the stem between the graduations for 1.00 and 1.10 sp. gr. would then equal 0.90 in.

Another problem illustrating the application of Archimedes' principles was:

■ A cubical block of wood 10 cm on a side and of density 0.5 g/cm^3 floats in a container of water. Oil of density 0.8 g/cm^3 is poured on the water until the top of the oil layer is 4 cm below the top of the block.

   a. How deep is the oil layer?

   b. What is the gauge pressure at the lower face of the block in dynes/cm^2?

First draw a sketch (Fig. 3-10). The mass of the block is $10 \times 10 \times 10 \times 0.5 = 500$ g. The total volume of fluid displaced equals $(10 - 4) \times 100 = 600$ cm^3. The density of water is 1 g/cm^3, and the weight of the block must equal the weight of the fluid displaced, so

$$100y \times 0.8 \text{ g/cm}^3 + 100(6 - y) \times 1 \text{ g/cm}^3 = 500 \text{ g}$$
$$y = 5 \text{ cm}$$

or the oil layer is 5 cm deep. The gauge pressure at the lower face of the block would equal $(5 \times 0.8 \times 980)$ dynes/cm^2 due to oil $+ 1 \times 980$ dynes/cm^2 due to water; $P = 4900$ dynes/cm^2.

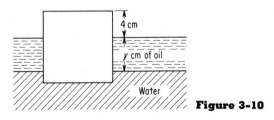

4 cm

y cm of oil

Water

Figure 3-10

The subject of bulk modulus should also be mentioned. Liquids are generally considered to be incompressible. Relative to the compressibility of gases this is true, but liquids as well as gases can be compressed; i.e., the volume of a liquid will be reduced under the application of pressure. The ratio of the change in volume divided by the original volume equals the pressure applied divided by the bulk modulus of the fluid, or $\Delta V/V = P/BM$, which equals the percentage reduction in volume. The bulk modulus of water is about 300,000 psi. This means that the volume of a given amount of water will be reduced by $5000/300,000 = 1.7$ percent if the pressure acting on the water is increased by 5000 psi.

3-6. HYDRODYNAMICS

Fluids in motion are governed in large part by the same principles that were discussed under Dynamics in Chap. 2, Mechanics. To analyze the characteristics of the flow of liquids, we must also, however, include the effects of viscosity; and analysis of the flow of gases must in addition include a consideration of the thermodynamic effects.

The bulk of the problems on this subject in past examinations have been concerned with the flow of liquids and have included examples covering flow through pipelines, open channels, orifices, and venturi tubes. Other phases covered have included the conservation of momentum, pump analyses, and a few problems concerning drag and model theory.

3-7. BERNOULLI'S EQUATION

Perhaps the most important fundamental relationship dealing with hydrodynamics is Bernoulli's equation. This equation is based on the concept of the conservation of energy; it states in algebraic form that the total energy at one point of a closed system is equal to the total energy at any other point in the system plus the energy lost between those two points. Bernoulli's equation may be written as:

$$\frac{P_1}{w} + \frac{v_1^2}{2g} + Z_1 = \frac{P_2}{w} + \frac{v_2^2}{2g} + Z_2 + h_{L_{1-2}}$$

where P_1 and P_2 = respective static pressures at points 1 and 2, psf

w = specific weight of the fluid flowing, lb/ft³

v_1 and v_2 = respective velocities, ft/sec

g = acceleration of gravity = 32.2 ft/sec²

Z_1 and Z_2 = elevations of points 1 and 2 above some selected datum plane

$h_{L_{1-2}}$ = head loss measured in feet of fluid flowing between points 1 and 2

The head loss results from the frictional resistance of the fluid because of its viscosity. An example of the use of Bernoulli's equation is the solution of the following sample examination problem:

■ From the information given on the sketch (Fig. 3-11) calculate the diameter at section B in inches. The diameter at section A is 4.00 in. Disregard friction. Kerosene is flowing at the rate of 2.5 cfs (sp. gr. kerosene = 0.80).

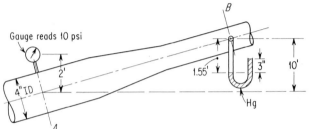

Figure 3-11

First, write out Bernoulli's equation:

$$\frac{P_A}{w} + \frac{v_A{}^2}{2g} + Z_1 = \frac{P_B}{w} + \frac{v_B{}^2}{2g} + Z_2$$

Let the elevation at point A be the zero reference plane. This gives:

$$Z_1 = 0 \quad \text{and} \quad Z_2 = 10 \text{ ft}$$

$$\frac{P_A}{w} = \frac{(10 \times 144) \text{ lb/ft}^2}{62.4 \times 0.80} + 2 = 28.9 + 2 = 30.9 \text{ ft}$$

$$\frac{P_B}{w} + 1.55 \text{ ft} = 3 \text{ in. Hg} = \left(\frac{3 \times 13.6}{12} \text{ ft of water}\right) \times \frac{1}{0.80}$$
$$= 4.25 \text{ ft of kerosene}$$

$$\frac{P_B}{w} = 4.25 - 1.55 = 2.70 \text{ ft}$$

$$v_A = \frac{Q}{A} = \frac{2.5 \text{ cfs}}{(4/12)^2 \times \pi/4 \text{ ft}^2} = 28.7 \text{ ft/sec}$$

$$\frac{v_B{}^2}{2g} = \frac{P_A}{w} - \frac{P_B}{w} + \frac{v_A{}^2}{2g} - Z_2 = 30.0 - 2.70 + \frac{28.7^2}{64.4} - 10$$

$$\frac{v_B{}^2}{2g} = 30.95 \qquad v_B = \sqrt{1995} = 44.7 \text{ ft/sec}$$

$$A_B = \frac{Q}{v_B} = \frac{2.5}{44.7} = 0.0560 \text{ ft}^2 \quad \text{or} \quad 8.06 \text{ in.}^2$$

$$D_B = \sqrt{8.06/0.785} = 3.20 \text{ in.}$$

3-8. FLOW MEASUREMENT

Many flow-measuring devices are based upon application of Bernoulli's principle. The ones we are most interested in are the orifice, the venturi tube, the nozzle, and the pitot tube. The orifice, the nozzle, and the venturi tube are quite similar in that all are reductions in flow area; the correlation between pressure and velocity upstream of these devices and at the sections of reduced area gives the velocity of flow through the throat and thus provides a measure of the flow rate.

3-9. ORIFICES

None of these three devices conforms exactly to Bernoulli's equation, however, because of fluid friction and the contraction of the jet of liquid as it flows through the section of reduced area. This leads to a correction factor called the coefficient of discharge C, which, in the case of an orifice, is the product of a velocity coefficient C_v and a coefficient of contraction C_c. For a sharp-edged orifice, $C_c = 0.62$, $C_v = 0.98$, and $C = 0.61$. This means that for a sharp-edged orifice placed in a pipe the velocity of flow through the orifice would equal $C\sqrt{2gh}$, where h would equal $\Delta P/w$. This is illustrated by means of the following selected past examination problem:

- If a 1-in. orifice has a head of 200 ft and the coefficient of discharge is 0.98, calculate the velocity of flow.

$$Q = A \times 0.98\sqrt{2g \times 200} = (^1/_{12})^2 \times \frac{\pi}{4} \times 0.98\sqrt{64.4 \times 200}$$

$$Q = 0.607 \text{ ft}^3/\text{sec} \quad \text{or} \quad 0.607 \times 60 \times 7.48 = 272 \text{ gpm}$$

In the case of an orifice in a pipe, the fluid approaching the orifice will have a velocity which must be considered in the calculation of the orifice coefficient. If we apply Bernoulli's equation to a point upstream and at the throat of the orifice, we obtain:

$$\frac{v_o^2}{2g} + \frac{P_o}{w} = \frac{v_p^2}{2g} + \frac{P_p}{w}$$

where the subscript o refers to the orifices and the subscript p refers to the pipe, since the same quantity of liquid flows through both pipe and orifice.

$$v_p = v_0 \frac{A_o}{A_p}$$

$$v_o{}^2 = \left(\frac{P_p - P_o}{w}\right) \times 2g + \left(v_o \frac{A_o}{A_p}\right)^2$$

where $\dfrac{P_p - P_o}{w} = h$ ft of fluid

$$v_o{}^2 = \left[1 - \left(\frac{A_o}{A_p}\right)^2\right] 2gh$$

$$v_o = \sqrt{\frac{2gh}{1 - (A_o/A_p)^2}}$$

To this must be added the effect of the previously discussed orifice coefficients, giving

$$v_o = c\sqrt{2gh} \times \frac{1}{\sqrt{1 - (A_o/A_p)^1}}$$

where $1/\sqrt{1 - (A_o/A_p)^2}$ is termed the "velocity-of-approach factor." If the diameter of the orifice is less than one-third the diameter of the pipe, the error introduced by ignoring the velocity-of-approach factor is less than 1 percent. As a result it is usually ignored. Unfortunately it is also often forgotten and when it is, it's lack can cause an error in calculations.

One other point that should be mentioned here is that much of the differential head across an orifice is recovered downstream. This fact is often forgotten, especially when orifices are used as flow limiters.

3-10. VENTURI METERS

Another past examination problem illustrating Bernoulli's principle was:

- A 3- by 1-in. venturi meter carries oil with a specific gravity of 0.86. A differential gauge connected between the pipe and the throat contains water with oil above the water in the manometer tubes. If the gauge shows a deflection of 57.6 in. and if the meter coefficient is 0.975, find the rate of discharge.

First, sketch the figure. Assume that the pipe and venturi meter are horizontal and that both are at the same elevation (Fig. 3-12). The difference in pressure between throat and body can be determined from the manometer reading.

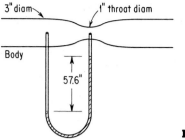

Figure 3-12

$$P_B + 57.6 \text{ in. of oil} = P_T + 57.6 \text{ in. of water}$$
$$P_B - P_T = 57.6 - 57.6 \times 0.86 = 8.06 \text{ in. of water}$$
$$\frac{P_B - P_T}{w} = \frac{8.06}{12} \times \frac{1}{0.86} = 0.782 \text{ ft of oil}$$

Bernoulli's equation gives:

$$\frac{P_B}{w} + \frac{v_B{}^2}{2g} + Z_B = \frac{P_T}{w} + \frac{v_T{}^2}{2g} + Z_T$$

but

$$Z_B = Z_T$$

and

$$v_B = \frac{A_T}{A_B} \times v_T = \frac{v_T}{9}$$

$$\frac{P_B - P_T}{w} = \frac{v_T{}^2 - v_B{}^2}{2g} = \frac{80v_T{}^2}{162g}$$

$$v_T = \sqrt{\frac{0.782 \times 162g}{80}}$$

if the meter coefficient were 100 percent. Actually

$$v_T = 0.975 \sqrt{\frac{0.782 \times 162g}{80}} = 6.97 \text{ ft/sec}$$

$$Q = (^1/_{12})^2 \times \frac{\pi}{4} \times 6.97 = 0.0380 \text{ cfs} \qquad \text{or} \qquad 17.1 \text{ gpm}$$

3-11. PITOT TUBES

A pitot tube is used to measure the velocity of a fluid by converting the energy due to the velocity (velocity head) to potential energy or pressure head. Referring to Bernoulli's equation, we get the relationship that $v = \sqrt{2gh}$, and since there is no flow, no flow coefficient is needed as in the previous example.

The relationship $v = \sqrt{2gh}$ is also known as Torricelli's theorem. In mechanics this relationship was developed to show the equivalence between change in potential energy and change in kinetic energy. The velocity is that which would be attained by a body falling freely through a height h in a vacuum at the surface of the earth. Torricelli proved that the velocity of efflux is equal to that which would have been attained in free fall from a height corresponding to that of the free surface of the liquid. He proved this almost a hundred years before Bernoulli developed his more general relationship. A further description of a pitot tube and an example of the calculations required to solve such a sample examination problem are as follows:

- A pitot tube having a coefficient of unity is inserted in the exact center of a long smooth tube of 1-in. ID in which crude oil (sp. gr. = 0.90)($\mu = 35 \times 10^{-5}$ lb-sec/ft^2) is flowing. Determine the average velocity in the tube if the velocity pressure is (a) 2.3 psi, (b) 3.9 in. water.

First make a sketch of the system under consideration (Fig. 3-13). Bernoulli's equation gives us the velocity at the center of the tube. The velocity of the fluid in contact with the pitot tube is all converted into pressure head.

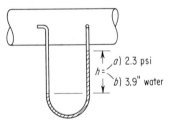

Figure 3-13

$$\frac{P_1}{w} + \frac{v_1{}^2}{2g} + Z_1 = \frac{P_2}{w} + \frac{v_2{}^2}{2g} + Z_2$$

where $Z_1 = Z_2$
$v_2 = 0$

$$\frac{v_1{}^2}{2g} = \frac{P_2 - P_1}{w}$$

For case (a)

$$\frac{P_2 - P_1}{w} = \frac{(2.3 \times 144) \text{ lb/ft}^2}{(62.4 \times 0.9) \text{ lb/ft}^3} = 5.9 \text{ ft of oil}$$

For case (b)

$$\frac{P_2 - P_1}{w} = \frac{3.9}{0.90} \times \frac{1}{12} = 0.361 \text{ ft of oil}$$

The velocity at the center of the tube would then be

$$v_a = \sqrt{5.9 \times 2g} = 19.5 \text{ ft/sec}$$
$$v_b = \sqrt{0.361 \times 2g} = 4.82 \text{ ft/sec}$$

3-12. AVERAGE VELOCITY IN A PIPE

The average velocity can be determined from the velocity at the center of the tube, which is the maximum velocity. In the case of viscous flow, $v_{avg} = v_{max}/2$, and the average velocity in the case of tubulent flow is approximately equal to $0.82v_{max}$.

Turbulent flow occurs for Reynolds numbers (see below) greater than 4000; flow is ordinarily viscous for Reynolds numbers less than 2000.

$$\text{Re} = \frac{\rho Dv}{\mu} = \frac{62.4 \times 0.90}{32.2} \frac{\text{lb-sec}^2}{\text{ft}^4} \times (\frac{1}{12})\text{ft} \times \frac{1}{35 \times 10^{-5}} \frac{\text{ft}^2}{\text{lb-sec}} \times v \text{ ft/sec}$$
$$= 415 \times v$$

For case (a)

Re = 8100—turbulent flow

For case (b)

Re = 2000—viscous flow

Average velocities will then be
For case (a)

$$v_{avg} = 0.82 \times 19.5 = 16 \text{ ft/sec, giving Re} = 6640$$

For case (b)

$$v_{avg} = 0.50 \times 4.82 = 2.41 \text{ ft/sec, giving Re} = 1000$$

3-13. REYNOLDS NUMBER

Reynolds number, abbreviated Re, is a dimensionless number equal to $\rho Dv/\mu$, where ρ = density of the fluid flowing, slugs/ft³, $\rho = w/g$; v = average fluid velocity in pipe, ft/sec; μ = absolute viscosity in lb-sec/ft², or μ = centipoises/47,800; and D = four times the hydraulic radius (D = the diameter of a circular pipe). Reynolds number also equals Dv/ν, where ν is the kinematic viscosity measured in ft²/sec (ν = centistokes divided by 93,000).

It seems advisable here to emphasize the difference between density ρ, slugs/ft³ and specific weight w, lb/ft³. Many engineers and, unfortunately, many writers, carelessly interchange these two, giving the units of lb/ft³ for ρ and even calling this density. But it is not. Density, by definition, is mass per unit volume. It is conventionally designated by the symbol for rho, ρ. In most cases, the careless interchange of terms and units creates only minor confusion because it is possible to check the units to determine what was meant regardless of what was said. There is the danger, however, particularly for the individual engineer, of making a serious mistake in an analysis because of a lack of understanding and appreciation of the difference between specific weight w and density ρ. One place where this occurs too frequently is in the calculation of a Reynolds number. As noted above, Re = $\rho\, Dv/\mu$ where ρ is density. If the units of lb/ft³ are mistakenly used here, the Reynolds number will be too large by a factor of 32.2.

3-14. HYDRAULIC RADIUS

The hydraulic radius is equal to the cross-sectional flow area divided by the wetted perimeter. For a circular pipe,

$$R_H = \frac{(\pi/4)D^2}{\pi D} = \frac{D}{4}$$

For a rectangular duct of width w and height h,

$$R_H = \frac{w \times h}{2w + 2h} = \frac{wh}{2(w + h)}$$

For a square of side length s, $R_H = s/4$. In calculating the hydraulic radius of an open trough or flume, it is important to remember the phrase "wetted perimeter." R_H for the trough shown in Fig. 3-14 of width w with water flowing at a depth h would equal $(w \times h)/(w + 2h)$, since the top is open and the "wetted perimeter" includes only the two sides and the bottom.

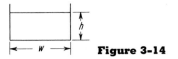

Figure 3-14

3-15. TURBULENT AND VISCOUS FLOW

The Reynolds number (Re) is a measure of the amount of turbulence in the flowing fluid; for Re less than 2000 the flow is generally considered to be laminar or viscous, for Re greater than 4000 the flow is generally considered to be turbulent. The range of Re between 2000 and 4000 is called the "critical zone," and the flow in this region is neither completely laminar nor completely turbulent. These values are average values for the general case. Specific cases can vary widely, depending upon the entrance conditions and other factors affecting the flow. Under carefully controlled laboratory conditions, turbulent flow has been obtained at Reynolds numbers of considerably less than 2000, and laminar flow has been obtained at Reynolds numbers of many times 4000. For most cases met in practice, however, the limits of 2000 and 4000 will apply.

3-16. FRICTIONAL HEAD LOSS

The head loss due to frictional flow of fluid flowing full in a pipe or duct can be determined from the relationship (Darcy's equation)

$$h_L = f\frac{Lv^2}{2gD}$$

where h_L = head, loss, ft of fluid flowing
  L = length of pipe or duct through which the fluid flows, ft
  v = fluid velocity, ft/sec
  g = acceleration of gravity
  D = four times the hydraulic radius (or the diameter of a circular pipe), ft
  f = friction factor, which is dimensionless and depends upon the value of the Reynolds number and the pipe roughness

For laminar flow, $f = 64/\text{Re}$ and f is independent of any ordinary internal pipe roughness. In the turbulent range, the value of f is somewhat dependent upon the roughness of the pipe and is usually taken from a curve of f vs. Re, such as is shown in Fig. 3-15.

■ What size pipe will be required to transport 2 ft³/sec of oil with a drop in pressure of 1 psi/1000 ft of pipe? The specific gravity is 0.85, and the absolute viscosity is 12 centipoises at a temperature of 70°F. Assume steel pipe.

$$h_L = f\frac{Lv^2}{2gD}$$

where h_L is measured in feet of fluid flowing.

$$h_L = \frac{1 \times 144 \text{ lb/ft}^2}{62.4 \times 0.85 \text{ lb/ft}^3} = 2.72 \text{ ft/1000 ft}$$

$$\frac{2gh_L}{L} = f\frac{v^2}{D} = \frac{2g \times 2.72}{1,000} = 0.175 = f\frac{v^2}{D}$$

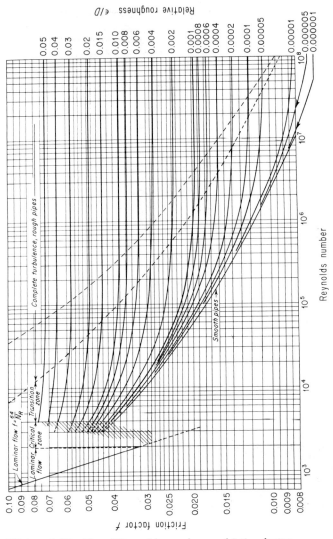

Figure 3-15 Plot of Reynolds number vs. friction factor.

Now, determine the Reynolds number.

$$\mu = \frac{0.12 \text{ poise}}{478} = 2.50 \times 10^{-4} \text{ lb-sec/ft}^2$$

$$\text{Re} = \frac{\rho D v}{\mu} = \frac{62.4 \times 0.85}{32.2} \times \frac{1}{2.50 \times 10^{-4}} \times Dv = 6580Dv$$

$$2 \text{ ft}^3/\text{sec} = Q = v \times A = v \times \frac{\pi}{4}D^2$$

giving $vD^2 = 2.55$. These three relationships reduce to

$$\text{Re} = \frac{17,500}{D} \quad \text{and} \quad f = 0.0269D^5$$

From Fig. 3-16, using the curve for steel pipe, $\epsilon/D = 0.00015$. If D is 1 ft,

$$\text{Re} = 17,500$$

and $f = 0.028$, so D should be slightly larger to satisfy the requirements exactly. The theoretical inner diameter would be approximately $12\frac{1}{8}$ in., but standard pipe is actually a little less than 12 in. ID (Schedule No. 40 pipe, 11.938 in. ID for 12-in. nominal pipe size), and this ordinarily has a $\pm\frac{1}{32}$-in. tolerance. In addition, the calculated pressure drop is not absolutely accurate, and the friction factor f may vary as much as ± 10 percent for commercial pipe. Considering these various factors, standard 12-in. pipe would be the best selection. The 12-in. ID pipe would give:

Figure 3-15 (Continued)
Table of Values of Absolute Roughness ϵ for New Pipes

	Feet
Drawn tubing, brass, lead, glass, centrifugally spun concrete, bituminous lining, transite	0.000005
Commercial steel or wrought iron	0.00015
Welded-steel pipe	0.00015
Asphalt-dipped cast iron	0.0004
Galvanized iron	0.0005
Cast iron, average	0.00085
Wood stave	0.0006–0.003
Concrete	0.001–0.01
Riveted steel	0.003–0.03

ϵ/D for use with curves of Fig. 3-15 found by dividing the value given above for ϵ by diameter of pipe in feet.

$$v = \frac{2}{0.785} = 2.55 \text{ fps}$$

$$\text{Re} = \frac{62.4 \times 0.85}{32.2} \frac{1 \times 2.55}{2.50 \times 10^{-4}} = 16,800$$

giving $f = 0.028$

$$h_L = 0.028 \frac{1000 \times (2.55)^2}{2g \times 1} = 2.82 \text{ ft/1000 ft}$$

$$2.82 \text{ ft} \times (62.4 \times 0.85) \text{ lb/ft}^3 \times \frac{1 \text{ ft}^2}{144 \text{ in.}^2} = 1.04 \text{ psi/1000 ft}$$

estimated pressure drop.

A further application of these principles is illustrated by another sample problem:

■ What quantity of water, in cubic feet per second, will flow in a 450-ft 10-in. ID pipe if the static head is 35 ft? The friction factor may be considered essentially constant, having a value of 0.025. Disregard entrance and exit losses (Fig. 3-16).

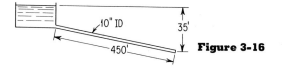

Figure 3-16

Apply Bernoulli's equation, taking as the two points the top of the reservoir and the discharge end of the pipe.

$$\frac{P_1}{w} + \frac{v_1^2}{2g} + Z_1 = \frac{P_2}{w} + \frac{v_2^2}{2g} + Z_2 + h_{L_{1-2}}$$

P_1 and P_2 are equal to atmospheric pressure, and v_1 is negligible, since the area of the reservoir is many times that of the pipe cross section. $Z_1 - Z_2 = 35.0$ ft, the difference in elevation between the two points.

$$35 = \frac{v_2^2}{2g} + h_L = 0.0155v_2^2 + h_L$$

$$h_L = f \frac{Lv^2}{2gD} = 0.025 \frac{450v^2}{2g(^{10}/_{12})} = 0.210v^2$$

$$35 = 0.0155v_2^2 + 0.210v_2^2 \qquad v = \sqrt{\frac{35}{0.226}} = 12.45 \text{ fps}$$

$$Q = v \times A = 12.45 \times (^{10}/_{12})^2 \times \frac{\pi}{4} = 6.80 \text{ ft}^3/\text{sec}$$

3-17. OPEN-CHANNEL FLOW

If the flow is in a pipe which is not running full or if it is in an open channel, the flow rate is usually calculated by means of the Manning formula:

$$v = 1.486 \frac{R^{2/3} S^{1/2}}{n}$$

where R = hydraulic radius
S = slope of the channel, ft/ft
n = Manning factor, which depends upon the material of which the channel is made and the roughness of the inside of the channel (see Table 3-1)

TABLE 3-1   Values of Roughness Coefficient, n (Manning Factor)

Surface	Average value of n
Wood, smooth, planed surface	0.011
Smooth metal	0.011
Finished concrete	0.012
Wood, unplaned	0.013
Good ashlar masonry or brick work	0.013
Unfinished concrete	0.014
Cast iron	0.015
Vitrified clay	0.015
Brick	0.016
Rubble masonry	0.017
Riveted steel	0.018
Firm gravel	0.020
Corrugated metal	0.022
Earth canals and rivers in good condition	0.025
Gravel	0.029
Earth canals with stones or weeds	0.035
Rough natural streams	0.050
Very weedy natural streams	0.100

An illustration of the use of this equation is afforded by the past examination problem:

■ A sewer made of concrete pipe (Manning coefficient 0.013) 4 ft in diameter is to carry 32.2 ft³/sec when running half full. What is the minimum slope in feet per 1000 ft required?

Sketch the pipe (Fig. 3-17). Using Manning's formula for flow in open channels, given above,

$$R = \frac{\text{cross-sectional area}}{\text{wetted perimeter}} = \frac{^1/_2(\pi/4)D^2}{^1/_2\pi D} = \frac{D}{4} = 1 \text{ ft}$$

$$\text{Required } v = \frac{Q}{A} = \frac{32.2}{^1/_2(\pi/4)D^2} = \frac{32.2}{6.28} = 5.13 \text{ fps}$$

$$5.13 = 1.486\frac{1^{2/3}S^{1/2}}{0.013}$$

$$S = \left(\frac{5.13 \times 0.013}{1.486}\right)^2 = 0.00202$$

or the slope must equal 2.02 ft/1000 ft of line.

Figure 3-17

3-18. CENTRIFUGAL PUMPS

In a centrifugal pump the fluid leaves the periphery of the impeller at a higher velocity than that at which it enters; this velocity is partly converted into pressure by the pump casing before the fluid leaves the pump through the nozzle. Thus a centrifugal pump is a device which supplies kinetic energy to a fluid (velocity head) and then converts it into potential energy (pressure head). If we apply Bernoulli's equation to a pump, we see that the head added to the fluid can be determined from the relationship

$$H = H_d - H_s + \frac{v_d^2}{2g} - \frac{v_s^2}{2g}$$

where H_d = discharge head measured at the pump discharge and referred to the pump shaft center line, and is expressed in feet of the fluid being pumped

H_s = suction head measured at the pump intake, also referred to the same datum plane of the pump shaft (Note that if the suction pressure is negative, it must be added.)

v_d and v_s = the velocity at the discharge and at the suction of the pump

The terms v_d and v_s give the difference in velocity head into and out of the pump. If the suction and discharge nozzles of the pump are the same size, these terms cancel each other.

3-19. SPECIFIC SPEED

There are a few other important points regarding centrifugal pumps; they will be discussed very briefly. The first of these is the specific speed of a pump. The specific speed

$$n_s = \frac{nQ^{1/2}}{H^{3/4}}$$

where n_s = specific speed
   n = speed of rotation, or impeller speed, rpm
   Q = flow, gpm
   H = head, ft of fluid flowing

The specific speed as a type number is constant for all similar pumps and does not change with speed for the same pump. Specific speed is a criterion of similarity for centrifugal pumps. The efficiency for a particular specific speed of a pump will remain essentially constant even though the output is varied. The efficiency vs. specific speed relationship will also hold for larger or smaller pumps of the same design (models which have been scaled up or down in size).

3-20. PUMP AFFINITY LAWS

The specific speed relationship is derived from the pump affinity laws, which are expressed by the following formulas:

$$\frac{Q_1}{Q_2} = \frac{n_1}{n_2} \qquad \frac{H_1}{H_2} = \frac{n_1{}^2}{n_2{}^2} \qquad \frac{HP_1}{HP_2} = \frac{n_1{}^3}{n_2{}^3}$$

These state that the pump capacity Q varies directly as the speed; the head, or pressure output, varies as the square of the speed; and the power varies as the cube of the speed.

3-21. NET POSITIVE SUCTION HEAD

A pump must "suck" fluid into its intake. As we know, a fluid can withstand little tensile stress, so the absolute pressure at the intake must be great enough to prevent cavitation (or "pulling apart") of the fluid. This pressure requirement, which will be different for different pumps, is called the "net positive suction head," or the NPSH. It is defined as the pressure of the fluid in feet at the inlet to the pump referred to the center line of the pump, minus the vapor pressure of the fluid at the fluid temperature in feet, plus the velocity head of the fluid at this point.

An example of a type of pump problem the examinee may expect is the following:

■ Several water piping systems are under consideration for use with a pump which has the operating characteristics shown in Fig. 3-18. Curves showing the operating characteristics of the piping systems under consideration are given in the lower figure. It is desired to use the piping system which will permit the pump to operate at its highest efficiency. (1 gal of water weighs 8.33 lb.)

a. Select the piping system which will allow the most efficient pump operation. Show calculations to substantiate your choice.

b. How many gallons per minute will be pumped by the pump through the piping system selected?

c. What is the efficiency of the pump when operating at the chosen flow rate?

d. What size motor would you recommend so that the pump may operate over its full capacity range without overloading?

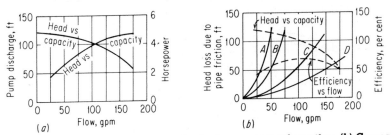

Figure 3-18 (*a*) Curves represent actual pump test information. (*b*) Curves show head loss for piping systems A, B, C, and D.

The first step is to plot the curve of pump efficiency vs. flow. The efficiency will equal the ratio

$$\frac{\text{horsepower out of pump}}{\text{horsepower input to pump}}$$

The horsepower out of the pump equals the water horsepower, which can be determined from the relationship

$$\text{hp} = \frac{(\text{gpm} \times 8.33)\ \text{lb/min} \times (\text{head})\ \text{ft}}{33,000\ \text{ft-lb/min}} = 2.525 \times 10^{-4}\ (\text{gpm}) \times (\text{head})$$

$$\text{Efficiency} = \frac{2.525 \times 10^{-4}(\text{gpm}) \times (\text{head})}{\text{hp input}}$$

Taking different points from the pump data curve gives the following:

Gallons per minute	Feet of head	Water horsepower	Horsepower input	Efficiency, %
25	119	0.75	1.8	42
50	115	1.45	2.5	58
75	110	2.08	3.2	65
100	102	2.58	3.8	68
125	93	2.94	4.3	69
150	78	2.95	4.5	66
175	55	2.43	4.6	53

The resulting curve of efficiency vs. flow can be plotted on the curve showing the head losses for piping systems A, B, C, and D. When this is done (the efficiency curve and the head vs. capacity curve are shown dotted in the given figure), it can be seen that piping system C will permit the pump to operate more efficiently than any of the other systems. The operating point will be at 130 gpm and 90 ft of head. The pump efficiency at this point is 68 percent. The power requirement equals 4.34 hp. It would be best to use a 5-hp motor to power the pump, since the operating point requires 4.34 hp and if someone opened a valve to drop the back pressure then the flow would increase and the power requirement would rise to almost 5 hp.

Another example selected from a past examination is as follows:

- A pump for the water system shown in Fig. 3-19 is to be selected from the characteristic curves shown. The pump is to service both tanks, pumping into one tank at a time. All piping is 6-in. diameter, new steel (standard strength, schedule No. 40) with $f = 0.017$. The pump desired is to deliver more than 500 gpm to each tank.
  - a. Under these conditions, which pump would be the most efficient for the system?
  - b. What maximum horsepower *out* of the motor is necessary for operation of the system?
  Use the given graph for solution.

Designating the points as R, F, and S for reservoir, first tank, and second tank, we can write Bernoulli's equation for each case.

$$\frac{P_R}{w} + \frac{v_R^2}{2g} + Z_R + h_P = \frac{P_F}{w} + \frac{v_F^2}{2g} + Z_F + h_{L(R-F)}$$

where h_P = head supplied by pump

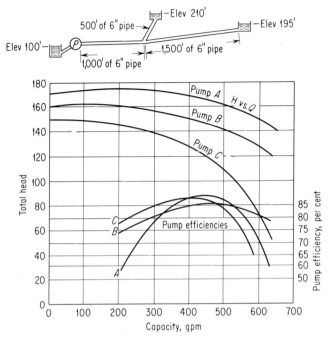

Figure 3-19

This gives

$$h_P = \frac{v_F{}^2}{2g} + 110 + h_{L(R-F)}$$

where v_F = velocity through pipe

The inner diameter of 6-in. schedule No. 40 pipe is given in pipe data tables as 6.065 in. $D = 6.065/12 = 0.505$ ft.

$$500 \text{ gpm} = \frac{500}{60 \times 7.48} = 1.115 \text{ ft}^3/\text{sec}$$

$$\text{Cross-sectional area} = \left(\frac{6.065}{12}\right)^2 \times \frac{\pi}{4} = 0.200 \text{ ft}^2$$

$$v_F = \frac{Q}{A} = \frac{1.115}{0.200} = 5.58 \text{ fps}$$

$$h_{L(R-F)} = f\frac{L}{D}\frac{v_F{}^2}{2g} = 0.017 \times \frac{1500}{0.505} \times \frac{v_F{}^2}{2g} = 0.784 v_F{}^2$$

$$h_p = \frac{v_F{}^2}{2g} + 110 + 0.784 v_F{}^2 = 0.483 + 110 + 24.4 = 134.9 \text{ ft}$$

For the second tank,

$$h_p = \frac{v_s^2}{2g} + 95 + h_{L(R-S)}$$

$$h_{L(R-S)} = 0.017 \times \frac{2500}{0.505} \times \frac{v^2}{2g} = 1.31v_s^2$$

$$h_p = \frac{v_s^2}{2g} + 95 + 1.31v_s^2 = 0.483 + 95 + 40.8 = 136.3 \text{ ft}$$

The pump would then have to supply at least 136.3 ft of head at a flow of 500 gpm. This rules pump C out because the characteristic curve for this pump shows that it produces only about 110 ft of head at a flow rate of 500 gpm. Either pump A or pump B could handle the operation. Pump A is more efficient at a flow rate of 500 gpm, but to operate at this flow rate it would have to pump against a pressure of slightly over 160 ft, as shown on its characteristic curve.

Some explanation may be in order here regarding the pump characteristic curve. This is a curve of the actual operational characteristics of a pump. Such a curve is constructed from data obtained during a test of the pump. It shows the output characteristics of a pump of a particular size and design. Take pump A for example. The characteristic curve could have been obtained by putting a throttling valve and a flowmeter on the discharge, a pressure gage between the pump outlet and the throttling valve, and then operating the pump at the desired speed. The valve was probably closed at the start of the test. At this condition the pressure gage would have read the 172 ft shown as the shutoff pressure on the curve. Next, the valve might have been opened all the way and then closed slightly until the pressure gage read 150 ft. The flowmeter would show a flow rate of 600 gpm at that condition. The valve could be closed a bit more until the discharge pressure rose to 162 ft. At that pressure the flow rate shown by the flowmeter would be 500 gpm. Similarly, at a discharge pressure of 170 ft, the flow rate would be 400 gpm; at 170 ft, 300 gpm; and so on. The data obtained from such a test would plot to form the characteristic curve of pump A.

Note that the discharge pressure is measured in feet of fluid flowing rather than pressure. This is another characteristic of centrifugal pump operation. The pressure produced by the pump is a function not only of the pump, but also of the liquid being pumped. Remember that a centrifugal pump converts centrifugal force to pressure head. It would, then, seem reasonable to assume that the output pressure would be proportional to the density of the liquid being pumped, and it is. Pump A, for example, would pump against a discharge pressure of 170 ft at a flow rate of 400 gpm. If water were being pumped, the 170 ft of head would correspond to a gage pressure of $170 \times 0.433 = 73.6$ psi. $(0.433 = 62.4/144$ and has the units of psi per foot of water). If oil with a specific gravity of 0.84 were being pumped, rather than the water, then the gage would register a pressure of only 73.6 × 0.84 = 61.8 psi, at the flow rate of 400 gpm. On the other hand, if the pump were used to pump carbon tetrachloride, (sp. gr. approximately 1.6), the discharge pressure at a flow rate of 400 gpm would rise to 73.6 × 1.6 = 118 psi.

So we can see that if pump A were used in this system, the quantity of liquid handled by the pump would have to increase until the pressure required to force the larger flow through the system matched the new flow rate on the characteristic curve for pump A.

The system pressure drop to the second tank equals

$$h_p = \frac{v_s{}^2}{2g} + 95 + 1.31v^2 = 95 + 1.32v^2$$

The velocity will be directly proportional to the flow rate in gpm, so the head loss for a 600 gpm flow rate would equal

$$95 + (600/500)^2 \times 41.3 = 154.4 \text{ ft}$$

This is almost on the curve for pump A. At this point, 600 gpm flow rate, the efficiency of pump A drops to 72 percent. This is the point at which pump A would operate if it were used with the given system.

Pump B would also pump more than 500 gpm since the discharge pressure of this pump at the 500 gpm rate is 140 ft and the system backpressure is only 136.4 ft. It would pump only a little more, however, until the flow rate and system backpressure matched a point on the curve for pump B. The operating point for the second tank would shift to approximately 515 gpm and 139 ft of head. At this point, the efficiency would equal 84 percent. This is an appreciably better efficiency than could be obtained with pump A *in this particular system.*

The pump would pump approximately 525 gpm at 137 ft of head into the first tank, again with an efficiency of 84 percent.

The maximum horsepower out of the motor would be

$$\frac{525 \text{ gpm} \times 8.34 \text{ lb/gal} \times 137 \text{ ft}}{33{,}000 \times 0.84 \text{ eff}} = 21.7 \text{ hp}$$

The graph of pump characteristics gives "total head" as the item listed on the ordinate. Total head would equal pressure head plus velocity head. The velocity head is the term $v_s{}^2/2g$. This has been included in the calculations, so the analysis is complete. However, we might check the magnitude of this factor for the 525 gpm flow. For this rate of flow

$$v = \frac{525}{7.48 \times 60 \times 0.200} = 5.85 \text{ ft/sec}$$

$$\frac{v^2}{2g} = 0.531 \text{ ft}$$

so the velocity head is small enough to have safely been ignored. This is usually the case in pump calculations, but the magnitude of the velocity head should always be checked before it is disregarded. There are occasional cases where it is important.

3-22. FLUID MOMENTUM

Another phase of fluid mechanics treated in past fundamentals examinations is the subject of fluid momentum and the force exerted by a fluid when it is subjected to a change in momentum.

We know from our study of mechanics that the impulse equals the change in momentum, or $F \Delta t = \Delta(Mv)$, which becomes $F \Delta t = M \Delta v$ for a mass which remains constant. This can also be recognized as an expression of Newton's second law, $F = ma$, or $F = m(dv/dt)$.

The mass M, for the case of fluid flow, can be written as

$$M = \rho Q \Delta t$$

and when this relationship is substituted in the impulse-momentum relationship, an equation is obtained for the calculation of the force exerted on (and by) the fluid.

$$F \Delta t = \rho Q \Delta t \Delta v$$

$$F = \rho Q \Delta v = \frac{wQ}{g} \Delta v$$

where Δv = change in velocity, or $\Delta v = \bar{v}_2 - \bar{v}_1$

where the lines over the v_2 and the v_1 indicate that these velocities are vector quantities and the subtraction is a vectorial subtraction. This is important and should be kept in mind. The velocities are vector quantities and must be treated as such. The resultant force is also a vector quantity and possesses both magnitude and direction. These principles are illustrated by means of sample problems, as follows:

■ A horizontal bend in a pipeline reduces the pipe from a 30-in. diameter to an 18-in. diameter while bending through an angle of 135° from its original direction. The flow rate is 10,000 gpm, and the direction of flow is from the 30-in. to the 18-in. pipe. If the pressure at the entrance is 60 psig, what must be the magnitude of the resultant force necessary to keep the bend in place? Consider the bend to be adequately supported in a vertical direction.

First draw the figure (Fig. 3-20), and then calculate the velocities and pressures in and out of the bend. Then calculate the changes in momentum and obtain the forces acting on the elbow.

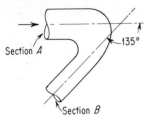

Section A

$135°$

Section B **Figure 3-20**

$$10,000 \text{ gpm} \times \frac{1}{60 \times 7.48} = 22.3 \text{ ft}^3/\text{sec}$$
$$\text{Area}_A = 4.91 \text{ ft}^2$$
$$\text{Area}_B = 1.77 \text{ ft}^2$$
$$v_A = \frac{22.3}{4.91} = 4.54 \text{ ft/sec}$$
$$v_B = \frac{22.3}{1.77} = 12.6 \text{ ft/sec}$$

Determine the pressure at section B by means of Bernoulli's equation:

$$\frac{P_A}{w} + \frac{v_A{}^2}{2g} = \frac{P_B}{w} + \frac{v_B{}^2}{2g}$$
$$P_A - P_B = \frac{w}{2g}(v_B{}^2 - v_A{}^2)$$
$$P_A - P_B = \frac{62.4}{64.4}(159 - 20.6) = 134 \text{ psf, or } 0.931 \text{ psi}$$

The pressure at section B is $P_B = 59$ psi.

First determine the momentum forces acting on the elbow. 10,000 gpm, assuming water flow, corresponds to 1388 lb/sec weight flow or 43.1 slugs/sec mass flow $= m/t$. The momentum force acting on the elbow would equal the mass flow rate times the change in velocity.

$$F_M = (m/t) \, \Delta v$$

or, in the X direction

$$F_{Mx} = (m/t)(v_{1x} - v_{2x})$$

in the Y direction

$$F_{My} = (m/t)(v_{1y} - v_{2y})$$

For this case,

$$v_{1x} = 4.54 \text{ ft/sec}$$
$$v_{2x} = -12.6 \times \cos 45° = -8.91 \text{ ft/sec}$$
$$v_{1y} = 0$$
$$v_{2y} = -12.6 \times \sin 45° = -8.91 \text{ ft/sec}$$
$$F_{Mx} = 43.1 \ (4.54 + 8.91) = 580 \text{ lb}$$
$$F_{My} = 43.1 \ (0 + 8.91) = 384 \text{ lb}$$

Thus the fluid exerts a force on the elbow to the right equal to 580 lb and a force upward of 384 lb. To withstand these forces, the elbow must exert equal and opposite forces on the fluid. The resultant force which must be exerted on the elbow to balance the fluid momentum forces would then equal 580 lb to the left and 384 lb downward. The resulting force which must be applied externally to the elbow to withstand the fluid momentum forces would equal $\sqrt{580^2 + 384^2} = 695.6$ lb. It would act to the left at an angle equal to $\tan^{-1} (384/580) = 33.5°$ down from the horizontal (see Fig. 3-21).

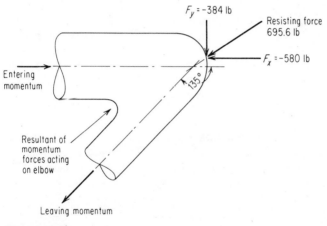

Figure 3-21

Alternatively, it might be assumed that all of the momentum force carried by the fluid entering the region is dissipated and that all of the momentum carried by the fluid leaving the region is created. Then the net momentum force acting on the elbow in this example would equal the momentum force entering minus the momentum force leaving the elbow, bearing in mind that momentum is a vector quantity.

For the example considered:

$$\text{Entering } F_M = 43.1 \text{ slugs/sec} \times 4.54 \text{ ft/sec} = 196 \text{ lb}$$
$$\text{Leaving } F_M = 43.1 \text{ slugs/sec} \times 12.6 \text{ ft/sec} = 543 \text{ lb}$$

The entering F_M is in the X direction acting toward the right, so

$$\text{Entering } F_{Mx} = 196 \text{ lb}$$
$$\text{Entering } F_{My} = 0$$

The leaving F_M is downward to the left at an angle of 45° with the horizontal, so

$$\text{Leaving } F_{Mx} = -543 \cos 45° = -384 \text{ lb}$$
$$\text{Leaving } F_{My} = -543 \sin 45° = -384 \text{ lb}$$

The sum of the momentum forces acting on the elbow are:

$$\text{Total } F_{Mx} = 196 - (-384) = 580 \text{ lb}$$
$$\text{Total } F_{My} = 0 - (-384) = 384 \text{ lb}$$

The resultant of the momentum forces acting on the elbow equals 696 lb acting upward to the right at an angle of 33.5° with the horizontal as previously determined.

To determine the total force acting on the elbow, it is necessary to add the pressure forces to the momentum forces.

$$\text{Pressure force at } A = \left(30^2 \times \frac{\pi}{4}\right) \times 60 = 42,400 \text{ lb} = F_A$$
$$\text{Pressure force at } B = \left(18^2 \times \frac{\pi}{4}\right) \times 59 = 15,000 \text{ lb}$$

The force at B can be resolved into forces in the x and y directions.

$$F_{B-x} = 15,000 \times 0.707 = 10,600 \text{ } lb$$
$$F_{B-y} = 15,000 \times 0.707 = 10,600 \text{ lb}$$

The total forces add to (see Fig. 3-22):

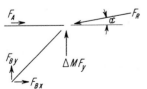

Figure 3-22

$$F_x = F_A + F_{B-x} + \Delta M_{Fx} = 53,581 \text{ lb}$$
$$F_y = F_{B-y} + \Delta MF_y = 10,985 \text{ lb}$$
$$F_R = \sqrt{F_x^2 + F_y^2} = 54,800 \text{ lb}$$
$$\alpha = \tan^{-1}\left(\frac{F_y}{F_x}\right) = \tan^{-1} 0.205 = 11.6°$$

Remember that a force is a vector quantity and that the direction of the action of the force must also be given.

3-23. JET PROPULSION

The principle of momentum force, $F = (m/dt) \, \Delta v$, has been utilized in rocketry and jet propulsion. It has also been utilized in helping to control unruly crowds. For example:

■ A firehose with a 1-in.-diameter nozzle is supplied with water at 100 psi pressure. What force would be exerted by the jet if it impinged on a flat (or slightly rounded) surface?

The velocity of the jet can be determined by calculating the change from potential energy (100 psi) to kinetic energy (velocity head). Assume that the nozzle is 98 percent efficient. Then $v = 0.98\sqrt{2gh} = 7.86\sqrt{h}$. The head of water equivalent to 100 psi equals

$$\frac{(100 \times 144) \text{ lb/ft}^2}{62.4 \text{ lb/ft}^3} = 231 \text{ ft}$$

and the jet velocity would equal 119.5 ft/sec.
The quantity of flow would equal

$$119.5 \text{ ft/sec} \times (0.785/144) \text{ ft}^2 = 0.651 \text{ ft}^3/\text{sec}$$

Mass flow would then equal

$$0.651 \times 62.4/32.2 = 1.26 \text{ slugs/sec}$$

The velocity of the jet in the axial direction would change from 119.5 ft/sec to zero after the jet had impinged on the flat surface, so $\Delta v = 119.5$ ft/sec.
Then,

$$\text{Jet force} = 1.26 \times 119.5 = 151 \text{ lb}$$

The reaction force acting on the nozzle would equal essentially the same. That is why it takes two (or more) firefighters to hold the nozzle on a firehose.

A different type of application of the principles of jet propulsion is afforded by the following problem:

■ A jet-propelled airplane is flying through standard air (68°F, 14.7 psia) at a speed of 500 mph. The hot gases leave the jet engine with a velocity of 1900 ft/sec relative to the 10-in.-diameter discharge nozzle. The temperature of the jet equals 1200°F. The hot gases can be assumed to have the same properties as air at that temperature. Assume that the pressure at the discharge end of the nozzle equals one atmosphere. What is the magnitude of the thrust on the airplane? What is the equivalent horsepower of the jet engine?

The volume flow of the discharge gases would equal

$$1900 \text{ ft/sec} \times (10^2 \times \pi/4 \times 1/144) \text{ ft}^2 = 1036 \text{ ft}^3/\text{sec}$$

The specific weight of the gases can be estimated with the aid of the gas law:

$$w = \frac{P}{RT} = \frac{14.7 \times 144}{53.3 \times 1660} = 0.0239 \text{ lb/ft}^3$$

The weight rate of flow would then equal $1036 \times 0.0239 = 24.76$ lb/sec, and the mass flow rate would equal $24.76/32.2 = 0.769$ slugs/sec. The reactive force on the jet nozzle would equal

$$(m/dt) \, \Delta v = 0.764 \times 1900 = 1461 \text{ lb}$$

Horsepower equals ft-lb/sec divided by 550. In this case, 24.76 lb of gases are being discharged from the jet engine at a velocity of 1900 ft/sec. Since the bulk of the discharge gases are made up of air, the jet engine can be assumed to approximate a pump which takes in air at a velocity of 500 mph (733 ft/sec) and discharges it at a velocity of 1900 ft/sec. The change in energy through the "pump" which must be supplied by the jet engine would equal $1/2 m(v_2^2 - v_1^2) = 1,536,400$ ft-lb/slug. Since the flow through the engine is 0.769 slug/sec the total energy supplied to the gases equals 1,181,500 ft-lb/sec, which is equivalent to 2,148 horsepower.

Another way of looking at the problem is to determine the amount of "head" added to each pound of gas as it flows through the engine. The velocity head at entrance would equal

$$\frac{(v_{\text{in}})^2}{2g} = \frac{733^2}{64.4} = 8343 \text{ ft}$$

The velocity head at the exit would equal

$$\frac{(v_{out})^2}{2g} = \frac{1900^2}{64.4} = 56,056 \text{ ft}$$

The energy added to the fluid as it passed through the jet engine would then equal 47,713 ft.

$$47,713 \text{ ft} \times 24.76 \text{ lb/sec} = 1,181,370 \text{ ft-lb/sec, or } 2148 \text{ hp}$$

3-24. COMPRESSIBLE FLOW

There have been only a few references to compressible fluid flow in past fundamentals examinations and most of these could be treated satisfactorily by the application of the principles discussed in the section on incompressible fluid flow. A general relationship to keep in mind is that the error introduced by considering (and treating) a compressible fluid as an incompressible fluid is approximately equal to one-fourth of the square of the Mach number of the fluid flowing for velocities appreciably less than the sonic velocity. In other words,

$$\text{Error} = (\text{approximately})\frac{1}{4}\left(\frac{v}{c}\right)^2$$

where c = velocity of sound at the existing conditions

The quantity v/c is the ratio of the actual velocity to the local velocity of sound and is known as the Mach number. The sonic velocity can be determined from the relationship

$$c = \sqrt{kgRT}$$

where k = ratio of the specific heats at constant pressure divided by the specific heat at constant volume

g = acceleration of gravity

R = gas constant and = (approximately) $\dfrac{1544}{\text{molecular weight}}$

T = absolute temperature of the gas and = °F + 460

The value of k may be taken as 1.66 for monatomic gases such as helium, 1.40 for diatomic gases such as air, and 1.30 for gases such as carbon dioxide, ammonia, methane, and acetylene. Illustrative examples are afforded by a pair of representative past examination problems.

■ A blower-type fan delivers air at the rate of 2200 ft³/min against a static pressure of 2 in. of water. How much air horsepower is being delivered if the air has a density of 0.075 lb/ft³? Velocity pressure may be disregarded.

The very low pressure of compression of 2 in. of water (0.0722 psi) indicates that the work required to compress the air is negligible so, as discussed previously, the air may be treated as an incompressible fluid. Disregarding the velocity head,

$$\text{hp} = \frac{\text{ft-lb/min}}{33,000} = \frac{(2200 \times 0.075) \text{ lb/min} \times {}^2\!/_{12} \times 62.4/0.075 \text{ ft of air}}{33,000}$$

$$\text{hp} = 0.694$$

where

$$\frac{2 \text{ in. water}}{12 \text{ in./ft}} \times \frac{62.4 \text{ lb/ft}^3 \text{ water}}{0.075 \text{ lb/ft}^3 \text{ air}} = \frac{1}{6} \text{ ft water} \times 833 \frac{\text{ft air}}{\text{ft water}}$$

$$= 138.6 \text{ ft of air, equivalent to 2 in. of water}$$

The problem could also be worked by

$$\text{hp} = \frac{\text{ft}^3/\text{min} \times \text{lb/ft}^2}{33,000}$$

$$\text{hp} = \frac{2200 \text{ ft}^3/\text{min} \times ({}^2\!/_{12} \times 0.433 \text{ psi/ft water} \times 144) \text{ lb/ft}^2}{33,000} = 0.694$$

Another past examination problem was as follows:

■ An air-conditioning duct is designed to carry 2,300 ft³ of air per minute (cfm) with a pressure loss of 0.4 in. of water using standard air weighing 0.075 lb/ft³. During construction it was found that several right-angle turns had to be added to avoid other piping. Upon completion the duct was tested and found to have a pressure loss of 0.6 in. of water at 2300 ft³/min, using air having a "density" of 0.074 lb/ft³. What is the percentage increase in horsepower required?

The pressure loss in the system will equal the friction loss

$$h_L = f\frac{Lv^2}{2gD}$$

and the losses in the elbows. The losses in the elbows can also be measured in "velocity heads." A velocity head equals $v^2/2g$.

$$\text{hp}_{\text{design}} = \frac{2300 \text{ ft}^3/\text{min} \times 0.4/12 \times 0.433 \times 144 \text{ lb/ft}^2}{33,000} = 0.145 \text{ hp}$$

$$\text{hp}_{\text{measured}} = \frac{2300 \text{ ft}^3/\text{min} \times 0.6/12 \times 0.433 \times 144 \text{ lb/ft}^2}{33,000} = 0.217 \text{ hp}$$

or a 50 percent increase in horsepower will be required.

The use of 0.075 lb/ft³ air and then 0.074 lb/ft³ air does not enter into this method of calculation, since the volume and the pressure were the only factors considered. If the horsepower had been determined by multiplying pounds per minute of air flow times feet of fluid pressure rise, the specific weight terms would have canceled out, since the pressure differential was measured by a different medium, not with the fluid (air) being pumped.

$$\text{hp} = \frac{(2300 \text{ ft}^3/\text{min} \times 0.075 \text{ lb/ft}^3) \times 0.4/12 \text{ ft water} \times 62.4/0.075}{33,000}$$

$$\text{hp}_{\text{design}} = \frac{2300 \times 0.075 \times 0.4/12 \times 62.4/0.075}{33,000} = 0.145 \text{ hp}$$

$$\text{hp}_{\text{measured}} = \frac{2300 \times 0.074 \times 0.6/12 \times 62.4/0.074}{33,000} = 0.217 \text{ hp}$$

3-23. LIFT AND DRAG COEFFICIENTS

The subject of aerodynamics has also been touched upon in past fundamentals examinations and will be mentioned briefly. Two of the most important characteristics of an airfoil are the drag coefficient

$$C_D = \frac{D}{(\rho v^2/2) \times A}$$

and the lift coefficient

$$C_L = \frac{L}{(\rho v^2/2) \times A}$$

where D and L = drag and lift, respectively, lb
ρ = mass density of the medium (air)
v = velocity of the airfoil (or of the air past the airfoil)
A = projected area

An example of an aerodynamics problem is as follows:

■ The graph in Fig. 3-23 shows test values for typical lift and drag coefficients for a Clark Y airfoil.

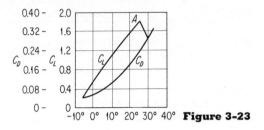

Figure 3-23

a. Explain why the peak point of the lift coefficient curve (point A) is so important.

b. When the Clark Y airfoil section was tested at an angle of approach of $\alpha = 5°$, a total drag of 50 lb was measured. What is the magnitude of the total lift under these conditions?

a. Point A is the point of maximum lift, or "stall," and determines the stalling angle above which lift no longer continues to increase with the angle of attack. At this point (also called the "burble point") the flow separates from the upper side of the wing and forms a turbulent wake, which increases the profile drag. This increased drag is accompanied by a large drop in lift and lift coefficient because of increased pressure on the upper side of the wing.

b. At 5°, $C_D = 0.08$ and $C_L = 0.8$. The drag force,

$$F_D = C_D A \frac{\rho v^2}{2}$$

and the lift force, $F_L = C_L A[(\rho v^2)/2]$. Since $A[(\rho v^2)/2]$ will be the same for the same test point, $(F_D/F_L) = (C_D/C_L)$, or

$$F_L = \frac{C_L}{C_D} \times F_D \qquad F_L = \frac{0.8}{0.08} \times 50 = 500 \text{ lb}$$

SAMPLE PROBLEMS

3- 1 Find the force on the end of a triangular upright trough having an apex angle of 90° when filled with water to a height of 3 ft.

3- 2 From the information shown on the schematic diagram in Fig. 3-P-2, calculate the velocity of the fluid at section B in feet per second.

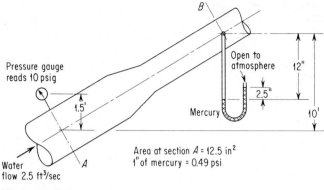

Pressure gauge
reads 10 psig

Open to
atmosphere   12"

2.5"

Mercury

10'

1.5'

Area at section A = 12.5 in^2
1" of mercury = 0.49 psi

Water
flow 2.5 ft^3/sec

B

A

Figure 3-P-2

3- 3   Water flows over the triangular weir shown in Fig. 3-P-3 with a head of 1 ft, a
coefficient of discharge C of 0.60, and the temperature of the water is 140°F.
Compute the flow over this weir in gallons per hour.

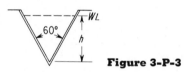

60°

h

WL

Figure 3-P-3

3- 4   A piece of glass weighs 125 g in air, 75 g in water, and 92 g in gasoline. What is
the specific gravity of the gasoline?

3- 5   A 12- by 12-in. timber 12 ft long floats level in fresh water with 4.3 in. above
the water surface. One end is just touching a rock which prevents that end
from sinking any deeper. How far out from the supported end can a 150-lb man
walk before the free end submerges?

3- 6   *a.* An 8-in. pipeline, 3500 ft long, conveys water from a pump to a reservoir
whose water surface is 450 ft above the pump which is pumping at the rate
of 3 ft^3/sec. Use constants for rough cast-iron pipe, disregard velocity head
and minor losses, and determine the gauge pressure in pounds per square
inch at the *discharge end* of the *pump*.

   b. Water is flowing, horizontally, from a reservoir through a circular orifice
under a head of 50 ft. The coefficient of discharge is 0.7, and the diameter of
the orifice is 4 in. How many cubic feet per second are being discharged
from the orifice?

3- 7 A swimming pool is being cleaned by a rubber hose 2 in. in diameter with a brush fixture on the end through which the water is drawn by a vacuum pump. The hose is 50 ft long, leading to the pump 12 ft above the bottom of the pool. If the water in the pool is 8 ft deep, what suction (in pounds per square inch) would be required at the pump to draw 600 gpm through the hose? (Assume a loss coefficient at the brush at $0.5V^2/2g$, where V is the velocity in the hose. Assume kinematic viscosity of water $= 1.2 \times 10^{-5}$ ft²/sec.)

3- 8 Water flows from a supply tank through 80 ft of welded steel pipe 6 in. in diameter to a hydraulic mining nozzle 2 in. in diameter. The coefficient of the nozzle may be assumed to be 0.90, and the temperature of the water may be assumed to be 75°F with a kinematic viscosity of 0.00001 ft²/sec. If 1.00 ft³/sec. of water is flowing, how far must the nozzle lie below the elevation of the water in the tank?

3- 9 Water falling from a height of 100 ft at the rate of 2000 ft³/min drives a water turbine connected to an electric generator at 120 rpm. If the total resisting torque due to friction is 400 lb at 1-ft radius and the water leaves the turbine blades with a velocity of 12 ft/sec, find the horsepower developed by the generator.

3-10 A venturi meter with a throat 2 in. in diameter is placed in a pipeline 4 in. in diameter carrying fuel oil. If the differential pressure between the upstream section and the throat is 3.5 psi, how much oil is flowing in gallons per minute? The discharge coefficient of the venturi meter is 0.97. The specific gravity of the oil is 0.94, and the specific weight is 58.7 lb/ft³.

3-11 A rectangular sluice gate, 4 ft wide and 6 ft deep, hangs in a vertical plane. It is hinged along the top (4-ft) edge. If there are 21 ft of water above the top of the gate, what horizontal force applied at the bottom of the gate will be necessary to open it?

3-12 A pump on the diagram shown in Fig. 3-P-12 pumps water to an elevation of x feet above the center of the pump. When the pump is running, the pressure gauge A registers 39.2 psi. When the pump is shut off, the pressure gauge A

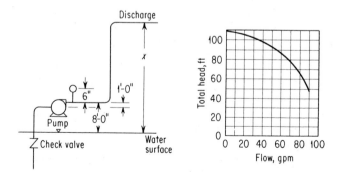

Figure 3-P-12

reads 10 psi because of the water in the line. A characteristic curve, as furnished by the manufacturer for the pump, is shown.

  a. What is the total head, including the suction head, against which the pump must operate if the friction in the inlet line can be considered negligible?

  b. What is the head loss due to friction in the line from the gauge to discharge?

  c. How many gallons per minute are being pumped when the pressure gauge registers 39.2 psig?

  d. Calculate the efficiency of the pump with 1.25-hp input and 39.2-psig discharge pressure.

3-13 A gravity dam composed of concrete is 45 ft high. It is 3 ft wide on top and the base is 30 ft wide. The water side of the dam is vertical and the other side slopes uniformly from top to bottom. The water behind the dam is 42 ft deep and there is no water pressure under the dam.

  a. Compute the position of the resultant at the base of the dam.

  b. Compute the pressure in pounds per square foot at the toe and at the heel of the dam.

  c. Compute the position of the resultant with the reservoir empty.

  d. Would the dam be safe against sliding?

3-14 A smooth nozzle on a fire hose is 1¹/₈ in. in diameter and discharges 250 gpm. What is the force with which this stream of water strikes a flat surface normal to the stream if placed only a short distance away?

3-15 A pipe 12 in. in diameter carries water by gravity from a reservoir. At a point 500 ft from the reservoir and 28 ft below its surface, a pressure gauge reads 10.5 psi; at a point 8500 ft from the reservoir and 28.5 ft below its surface, a pressure gauge reads 61.5 psi. Compute the discharge in gallons per minute.

3-16 The instrument readings for a pump test are shown on the schematic diagram in Fig. 3-P-16. Find the following:

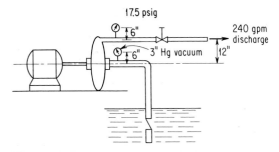

Figure 3-P-16 3460 rpm at 7.89 lb-ft torque; 220 volts, 21.2 amp at 88 percent power factor single phase; weight of water = 8.33 lb/gal; weight of mercury = 0.49 lb/cu in.

    a. Kilowatt input to motor
    b. Horsepower output of motor
    c. Water horsepower output of pump
    d. Efficiency of pump under conditions listed

Assume the pipes leading to the gauges are filled with water and the difference between the inlet and outlet velocity head is negligible.

3-17 A horizontal 36-in. pipe curving through an angle of 60° contains water with an average velocity of 7.1 ft/sec and pressure of 50 psi. Find the force exerted on the bend by this discharge (Fig. 3-P-17).

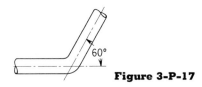

Figure 3-P-17

3-18 The hinged top of a 4-ft square gate is submerged 10 ft below the surface of the water in the tank (Fig. 3-P-18).
    a. What is the distance x to the center of pressure on the gate?
    b. Find the force F required to balance the water pressure on the gate.

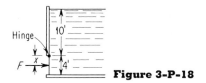

Figure 3-P-18

3-19 A circular concrete culvert 8 ft in diameter and having a slope of 1 ft in 1000 ft is carrying water 6 ft in depth. Compute the discharge in cubic feet per second, disregarding loss of head at entrance. Assume that Manning's coefficient equals 0.013.

3-20 A jet of water is entirely reversed in direction as it hits a vane. If the vane is fixed, the velocity of the jet is 100 ft/sec, the weight flowing equals 200 lb/sec, and friction is such that the water leaves the vane at 0.9 × the velocity with which it strikes it. Determine:

    a. The force exerted on the vane
    b. The horsepower of the jet and the power exerted on the vane

3-21   *a*. Find the hydraulic radius of the section shown in Fig. 3-P-21 when the water is flowing at a depth of 5.0 ft from the bottom of the V notch.
    b. What quantity of water, in cubic feet per second, will flow through this section at the above-mentioned depth, if the slope is 1.0 ft/100 ft, and the coefficient of roughness is 0.013? Manning's equation for velocity is $V = (1.486H^{2/3}S^{1/2})/n$.

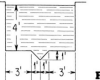

Figure 3-P-21

3-22 The graph in Fig. 3-P-22 shows test values for typical lift-drag coefficients on a Clark Y airfoil.

  a. Explain the reasons why the peak point of the lift coefficient curve (point A) is so important.

  b. When the Clark Y airfoil section was tested at an angle of approach of $\alpha = 5°$, a total drag of 50 lb was measured. What is the magnitude of the total lift under these conditions?

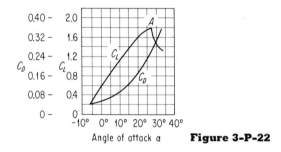

Angle of attack α

Figure 3-P-22

3-23 A penstock 10 ft in diameter carrying 250 ft³/sec of water bends down through a 45° angle. What is the force F_x at the bend? (See Fig. 3-P-23.)

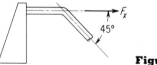

Figure 3-P-23

3-24 A hydrometer weighs 50 g. The area of the cross section of the stem is 1 cm².

  a. What is the distance on the stem between the 0.8 gradation and the 0.9 gradation?

  b. Will these numbers be placed above or below the 1.0 on the stem?

3-25 In the venturi meter shown in Fig. 3-P-25, the liquid being measured is water and the coefficient of the meter is 0.98. Compute the rate of flow in cubic feet per second.

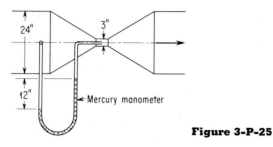

Figure 3-P-25

3-26 Using Fig. 3-P-26, determine the following:

    a. Total loss of head, in feet, from *A* to *B* due to fluid friction in the pipe and fittings

    b. Head loss due to pipe friction

    c. Head loss due to valves and fittings

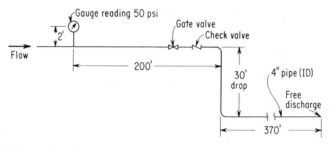

Figure 3-P-26

    In Fig. 3-P-26, flow at *A* = 1.25 ft³/sec; actual inside diameter of pipe = 4.02 in.; actual inside area of pipe = 12.7 in.²; friction factor for 4-in. pipe = 0.024.

3-27 A manhole of 30-in. diameter in the side of a penstock is closed with a flat plate. The static head at the center of the manhole is 25 ft (Fig. 3-P-27).

    a. What is the total pressure in pounds on the cover plate?

    b. Where is the center of pressure with reference to the horizontal center line of the manhole?

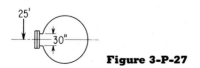

Figure 3-P-27

3-28 A standard 10-in. galvanized iron pipe 15 ft long, weighing 32 lb/ft, is closed at one end by a pipe cap with a weight attached so that the pipe will float upright. The cap and weight total 25 lb and displace $1/2$ cu ft of water. How much of the pipe will show above the surface of seawater of 1.025 sp. gr.?

3-29 A tank in the form of a frustum of a cone, with bases horizontal and axis vertical, is 10 ft in height. The top diameter is 8 ft; bottom diameter is 3 ft. What time is required to empty the tank through a 3-in. standard square orifice for which the discharge coefficient is 0.61?

3-30 An oil pipeline has a 10-in. ID, is 100 miles long, and is made of welded steel. The discharge end is 500 ft above the intake. The rate of flow is 4 ft/sec. What is the total pressure in pounds per square inch gauge at intake when the whole length is full of gasoline at 60°F and 0.68 sp. gr.?

3-31 A wood flume of cross section as in Fig. 3-P-31 is to carry 200 ft³/sec of water. What slope in feet per 1000 feet is required?

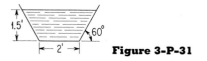

Figure 3-P-31

3-32 A centrifugal motor-driven water pump is connected to a pipeline as shown in Fig. 3-P-32.
 a. What is the average velocity in the pipe?
 b. If the input to the pump motor is 1105 watts, what is the combined efficiency of the motor and pump?
 c. What is the friction factor for the 300-ft length of pipe?

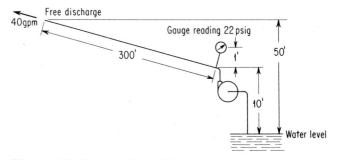

Figure 3-P-32 Pipe ID = 2.067 in.; pipe flow area = 3.36 sq in.; 7.48 gal/cu ft.

3-33 A 1-in.-diameter cable is used to anchor a buoy near the entrance to a harbor where the ocean current reaches a velocity of 4 mph. A model is made of the buoy and cable to one-tenth scale for tests in the laboratory. Using fresh water in the test, results show that for a velocity of 5 ft/sec the drag on 1 ft of model cable is 2 lb. The density of sea water is 63.8 lb/ft³. If the relation between the model and prototype can be expressed as $F/(\rho A V^2)$ constant, where F = drag force, ρ = density of fluid, A = area (projected area perpendicular to flow), V = velocity of fluid relative to model or prototype, what is the drag per foot of prototype cable as installed in the harbor entrance?

3-34 A liquid is forced through a pipe 1 meter long with an internal diameter of 0.6 cm by a pressure difference of 1.2×10^5 dynes/cm². If 570 cm³ of liquid was discharged in 5 min, what was the viscosity of the liquid in centipoises?

3-35 Disregarding friction, find the gauge pressures and speeds at positions 1, 2, and 3 in Fig. 3-P-35 representing flow of water from a tank with an internal gauge pressure of 15 psi through a discharge pipe. h = 100 ft, y = 50 ft, A_1 = 1 ft², A_2 = ³/₄ ft², A_3 = 12 ft².

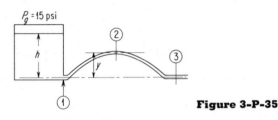

Figure 3-P-35

3-36 Turpentine is stored in a tank on the second floor of a plant making paints. A pipeline is to be installed to a mixer on the first floor feeding the turpentine by gravity through an equivalent length of 55 ft of straight steel pipe. The bottom of the tank is 15 ft above the point of discharge to the mixer. Calculate the minimum pipe diameter which will ensure a flow of 10 gpm to the mixer. Minimum temperature may be taken at 60°F, at which temperature the viscosity of turpentine is 1.6 centipoises and the specific gravity is 1.025.

3-37 A cubical block of wood 10 cm on a side and of density 0.5 g/cm³ floats in a container of water. Oil of density 0.8 g/cm³ is poured on the water until the top of the oil layer is 4 cm below the top of the block.

a. How deep is the oil layer?

b. What is the gauge pressure at the lower face of the block in dynes per square centimeter? (See Fig. 3-P-37.)

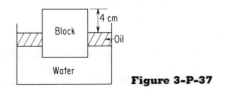

Figure 3-P-37

3-38 Estimate the total time required to empty a tank which is a paraboloid of revolution with every horizontal section a circle whose radius equals the square root of the height above the bottom, through which is cut a 1-in.-round sharp-edged orifice. The depth of water at the start of time is 10 ft.

3-39 A 12-in. ID smooth cast-iron pipe, 4000 ft in length, conveys water from a pump to a reservoir whose water surface is 250 ft above the pump, which is pumping at a rate of 3 ft³/sec. Determine the gauge pressure, in pounds per square inch above atmospheric, at the pump discharge. Assuming the efficiency of the pump to be 89 percent, determine the horsepower input.

3-40 A pitot tube having a coefficient of unity is inserted in the exact center of a long, smooth tube of 1-in. ID in which crude oil (sp. gr. = 0.9; $\mu = 35 \times 10^{-5}$ lb-sec/ft²) is flowing. Determine the average velocity in the tube if the velocity pressure is (a) 2.3 psi; (b) 3.9 in. of water.

3-41 Water flows through 3000 linear ft of 36-in.-diameter pipe which branches into 2000 linear ft of 18-in.-diameter pipe and 2400 linear ft of 24-in.-diameter pipe. These rejoin, and the water continues through 1500 ft of 30-in.-diameter pipe. All pipes are horizontal, and the friction factors are 0.016 for the 36-in. pipe, 0.017 for the 24-in. and 30-in. pipes, and 0.019 for the 18-in. pipe. Find the pressure drop in pounds per square inch between the beginning and the end of the system if the steady flow is 60 ft³/sec in the 36-in. pipe. Disregard minor losses.

3-42 A piece of lead (sp. gr. 11.3) is attached to 40 cm³ of cork (sp. gr. 0.25). When fully submerged the combination will just float. What is the weight of the lead?

3-43 A weighted timber 1 ft square and 6 ft long floats upright in fresh water with 2 ft of its length exposed. What will be the length projecting above the surface if the water is covered with a layer of oil 1 ft thick? The specific gravity of the oil is 0.8.

3-44 The outlet and inlet of a venturi meter are each 4 in. in diameter, and the throat is 3 in. in diameter. The inlet velocity is 10 ft/sec. The inlet has a static head of 10 ft of water. If there is no loss due to friction between the inlet and the throat of the meter, what will be the head in the throat?

3-45 A conduit having a cross section of an equilateral triangle of sides b has water flowing through it at a depth $b/2$. Find the "hydraulic radius" (Fig. 3-P-45).

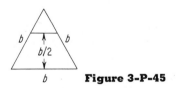

Figure 3-P-45

3-46 A mass of copper, suspected of being hollow, weighs 523 g in air and 447.5 g in water. If the specific gravity of copper is 8.92, what is the volume of the cavity, if any?

3-47 A 60° pipe bend reduces from a diameter of 8 in. to a diameter of 6 in. The static pressure at the 8-in. diameter is 20 psi. If the flow is 3.0 ft³/sec and the fluid is water, find:

a. Dynamic force on the bend
b. Static force on the bend
c. Total force on the bend

The centers of the 6- and 8-in. apertures lie in the same horizontal plane (Fig. 3-P-47).

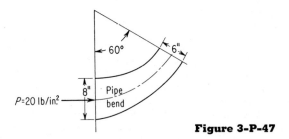

Figure 3-P-47

3-48 A gravity dam has a trapezoidal section 18 ft high, 2 ft wide on top, and 8 ft wide at the base. The water side is vertical. The depth of the water is 15 ft. Assume that the concrete weighs 141 lb/ft³ and that the foundation is sealed so that no water gets under the dam. What is the vertical component of stress on the foundation at the upstream edge of the base? Is this safe?

3-49 For the condition shown in Fig. 3-P-49, find force F needed to open the gate. The width of gate is 4 ft and the height is 6 ft.

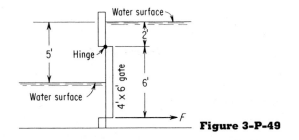

Figure 3-P-49

3-50 For the arrangement shown in Fig. 3-P-50, calculate the pressure difference (in pounds per square inch) between A and B. (Specific gravity of mercury is 13.6 and of turpentine is 0.8.)

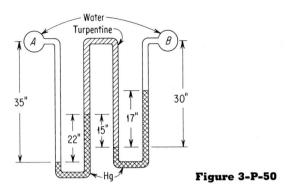

Figure 3-P-50

3-51 A 36-in. water main, carrying 60 ft³/sec, branches at point X into two pipes, one 24 in. in diameter and 1500 ft long, and the other 12 in. in diameter and 3000 ft long. They come together at Y and continue as a 36-in. pipe. If the factor f is equal to 0.022 in all pipes, find the rate of flow in each of the branches.

3-52 The water system shown in Fig. 3-P-52 is made up of fairly smooth cast-iron pipe.
a. What quantity of water is flowing past C when 2 ft³/sec is flowing past A?
b. What is the pressure drop between A and B? Disregard minor losses.

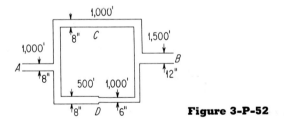

Figure 3-P-52

MULTIPLE-CHOICE PROBLEMS

For each question select the correct answer from the five given possibilities.

3M- 1 If absolute viscosity has the units of lb-sec/ft², then kinematic viscosity has units of:
    (a) centipoise   (d) ft³/sec
    (b) Btu/sec   (e) ft²/sec
    (c) ft²/sec-lb

3M- 2 The energy of a fluid flowing at any section in a pipeline is a function of:
  (*a*) the pressure only
  (*b*) the pressure, height above a chosen datum, velocity of flow, density of fluid, viscosity of fluid, and temperature of fluid
  (*c*) the height above a chosen datum, density, internal energy, pressure, and velocity of flow
  (*d*) the velocity of flow only
  (*e*) none of these

3M- 3 If a body weighs 100 lb in air, and 25 lb in fresh water, its volume in cubic feet is:
  (*a*) 0.75   (*d*) 1.3
  (*b*) 1.1   (*e*) 1.5
  (*c*) 1.2

3M- 4 Friction head in a pipe carrying water varies:
  (*a*) inversely with gravity squared
  (*b*) directly with diameter
  (*c*) inversely with diameter
  (*d*) directly with velocity
  (*e*) inversely with the coefficient of friction (f)

3M- 5 A fluid flows at a constant velocity in a pipe. The fluid completely fills the pipe, and the Reynolds number is such that the flow is just subcritical and laminar. If all other parameters remain unchanged and the viscosity of the fluid is decreased a significant amount, one would generally expect the flow to:
  (*a*) not change   (*d*) increase
  (*b*) become turbulent   (*e*) temporarily increase
  (*c*) become more laminar

3M- 6 A barge loaded with rocks floats in a canal lock with both the upstream and downstream gates closed. If the rocks are dumped into the canal lock water, with both the gates still in the closed position, the water level in the lock will theoretically:
  (*a*) rise
  (*b*) rise and then return to original level
  (*c*) fall
  (*d*) fall and then return to original level
  (*e*) remain the same

3M- 7 The locus of the elevations to which water will rise in a piezometer tube is termed:
  (*a*) the phreatic line
  (*b*) the energy gradient
  (*c*) the hydraulic gradient
  (*d*) friction head
  (*e*) critical depth

3M- 8   If a liquid has laminar flow through a pipe, which of the following may be true?
(*a*) The Reynolds number is 35,000 or higher.
(*b*) A great deal of energy is lost due to turbulence.
(*c*) The roughness of the pipe will have no effect on the friction factor
(*d*) The velocity at the center of the pipe will be the same as at the pipe wall.
(*e*) None of these.

3M- 9   Cavitation is the result of:

(*a*) static pressure in a fluid becoming less than fluid vapor pressure
(*b*) rivets under impact load
(*c*) exposure of concrete to salt water
(*d*) heat treatment of a low-carbon steel
(*e*) improper welding technique.

3M-10   A cylinder of cork is floating upright in a container partially filled with water. A vacuum is applied to the container such that the air within the vessel is partially removed. The cork will:
(*a*) rise somewhat in the water
(*b*) sink somewhat in the water
(*c*) remain stationary
(*d*) turn over on its side
(*e*) sink to the bottom of the container

4
THERMODYNAMICS

The coverage in past examinations has included questions concerning applications of the general energy equation, different forms of the equation of state, thermodynamic cycles, entropy and enthalpy, heat transfer, the thermodynamics of air, Mollier charts, and other fundamental principles of this branch of engineering.

4-1. GENERAL ENERGY EQUATION

The first, and perhaps most widely applicable, of these subjects which we shall review is the general energy equation. Just as Bernoulli's equation is universally used in fluid mechanics, so, with the necessary modifications, it is widely applied in thermodynamics. Bernoulli's equation, we recall, merely states that the energy contained in a flowing fluid at point A must be equal to the energy in that same fluid at a later or downstream point B plus the energy lost or minus that gained between the points A and B. With incompressible fluids the energy contained in the fluid at any point is essentially equal to the sum of the kinetic energy and the potential energy due to its elevation and pressure. The difference in the internal energy of the incompressible fluid between points A and B was so slight as to be inconsequential. This is not the case, however, with a compressible fluid; account must be taken of the change in internal energy, and additional terms must be added to Bernoulli's equation to make it applicable to compressible fluids. When these terms are added, Bernoulli's equation becomes, for 1 lb of fluid:

$$\frac{P_1}{w} + \frac{v_1^2}{2g} + \text{internal energy}_1 + \text{heat added} + Z_1 = \frac{P_2}{w} + \frac{v_2^2}{2g}$$
$$+ \text{internal energy}_2 + Z_2 + W$$

where W = work done by the gas between points 1 and 2

173

When dealing with gases the change in energy due to the difference in elevation between the two points considered is seldom large enough to affect the accuracy of the result, so Z_1 and Z_2 are usually omitted. The energy is usually given in Btu per pound rather than in foot-pounds per pound as for incompressible fluids, so all the terms in the equation are given in Btu. The specific weight w is generally replaced by its reciprocal for gases $(1/w) = V$, specific volume. The internal energy is denoted by the symbol u, and the heat added to the substance by q, both in Btu per pound. The work done by the substance, W, is also measured in Btu per pound of gas. The equation given then becomes:

$$\frac{P_1 V_1}{J} + \frac{v_1{}^2}{2gJ} + u_1 + q = \frac{P_2 V_2}{J} + \frac{v_2{}^2}{2gJ} + u_2 + W$$

where $J = 778$ ft-lb/Btu, the mechanical equivalent of heat

The internal-energy term and the PV/J term are usually combined, since $u + (PV/J) = h$, the enthalpy. The general energy equation then becomes:

$$\frac{v_1{}^2}{2gJ} + h_1 + Q = \frac{v_2{}^2}{2gJ} + h_2 + W$$

where the value of h depends upon the pressure and the temperature and may be looked up in a table giving the properties of that particular gas or on a Mollier chart.

4-2. FIRST LAW OF THERMODYNAMICS

The general energy equation is actually an expression of the first law of thermodynamics, which can be stated: "Energy can be neither created nor destroyed, but only converted from one form to another." Or the statement may be glamorized to the following: "If any system is carried through a cycle (the end state being precisely the same as the initial state), then the summation of the work delivered to the surroundings is proportional to the summation of the heat taken from the surroundings." In essence, the first law of thermodynamics is only a restatement, with amplification, of what was called the law of conservation of energy in mechanics and which was later restated in fluid mechanics as Bernoulli's equation.

Examples of the application of the general energy equation are afforded by many past examination problems. One of these asked:

■ In thermodynamics the following terms are most generally employed in connection with the steady flow of thermodynamic fluids: Q, h, v^2, Z, and W.

a. Write the general energy equation between two points (1) and (2) in a system.

b. Solve this equation for Q.

c. Write this equation for velocity in a perfect nozzle, assuming the point (1) is the entrance of the nozzle and the point (2) is the exit of the nozzle.

Part (a) is what we have just been discussing, and the general energy equation is written:

$$\frac{v_1{}^2}{2gJ} + h_1 + Z_1 + Q = \frac{v_2{}^2}{2gJ} + h_2 + Z_2 + W$$

Part (b) consists merely of rearranging the terms to give

$$Q = \frac{v_2{}^2 - v_1{}^2}{2gJ} + (h_2 - h_1) + (Z_2 - Z_1) + W$$

In a perfect nozzle—part (c)—no heat is added or lost and no work is done by the fluid. The nozzle is a device for converting some of the internal energy into kinetic energy. The general energy equation then reduces to

$$\frac{v_1{}^2}{2gJ} + h_1 = \frac{v_2{}^2}{2gJ} + h_2$$

since $Z_1 = Z_2$, $Q = 0$, and $W = 0$.

This is probably a converging nozzle with a relatively large entrance area, in which case v_1 will be small and may be disregarded, giving $v_2{}^2/2gJ = h_1 - h_2$, limited, however, by the sonic velocity at the throat or discharge section of a converging nozzle, which is the smallest section of the nozzle. In a converging-diverging nozzle the gas will expand in the diverging section and the velocity will be supersonic, provided the critical pressure ratio is exceeded.

Another example of the application of the general energy equation is in the determination of the quality of steam in a line by means of a throttling calorimeter. An example of this is afforded by a past examination problem:

■ The pressure and temperature in a throttling calorimeter are 14.7 psia and 240°F, respectively. If the pressure of the steam in the main is 150 psia, what is its quality?

The problem asks for the *quality* of the steam. This implies that part of the steam in the line will be condensed (in the liquid state) and the mixture in the line will contain both steam (gas) and droplets of water (condensate). In other words, the steam is said to be *wet*.

The quality of a mixture is the percent by weight that is gas. It might be noted here that steam ("dry" steam, that is) is a gas just as are air, nitrogen, carbon dioxide, and all other gases. And it conforms to the same laws as do the other gases. However, steam is not a "perfect" gas (neither is any other gas). Since steam is so widely used and so much equipment has been designed and built to extract energy from, and add energy to, steam, it has been found desirable to determine what the true properties are at different state points instead of estimating those properties with the perfect-gas relationship.

In most cases, reasonably good values of steam properties can be obtained with the aid of the perfect-gas relationships. And these are adquate for preliminary calculations. However, when a large sum of money is to be invested in a steam installation, it is highly desirable to have available more accurate values than can be obtained from the perfect-gas relationships. That is the reason for the development of the steam tables.

The concept of steam quality can be explained with the aid of a temperature-entropy diagram (see Fig. 4-1). At point A the water is 100 percent liquid. At point B the water is 100 percent gas. At some intermediate point, C, it is a mixture of gas and liquid. Note that the entire mixture is at the same temperature; i.e. the line AB is an isotherm.

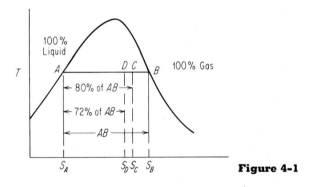

Figure 4-1

The change in entropy times temperature, $T\Delta S$, is equal to the change in enthalpy less $V\,dP$ (see Sec. 4-8). For this case, evaporation of water, the pressure and temperature are constant from A to B. The change in heat content over the span is then directly proportional to the change in entropy. Thus the increase in heat content of the steam from A to B is directly proportional to the percentage of the water that is in the gaseous state. If, for example, the temperature is 250°F, then, from the steam tables, the heat content of the 100 percent liquid water at point A equals 218.5 Btu/lb. The heat content of the saturated steam at point B is 1164.0 Btu/lb. The difference between the two heat contents is 945.5 Btu/lb, the heat of vaporization. Similarly, $T\Delta S$ between saturated liquid and saturated vapor equals 710°R × 1.332 = 945.7 Btu/lb.

If point C is 80 percent of the distance from A to B, it represents a condition of 80 percent quality. That is, 80 percent of the water at that point will be in the form of a gas. The heat content of the mixture would be 218.5 + 0.80 × 945.5 = 974.9 Btu/lb. If, on the other hand, the heat content of a wet-steam mixture at some other point D equaled 901.2 Btu/lb, the quality of the steam at point D would equal (901.2 − 218.5)/945.5 = 72 percent. The two points between which the general energy equation should be written are (1) in the main where the quality of the steam is to be found, and (2) in the calorimeter.

The steam flows into the calorimeter through a throttling orifice so that no heat is added or subtracted and no work is done in the process. The elevations of the two points are the same, and the velocity at point (1) is essentially zero, as is the velocity in the calorimeter. The general energy equation then reduces to $h_1 = h_2$, and the enthalpy at point (2) can be found in a steam table. At point (2) the enthalpy of the steam (14.7 psia and 240°F) is 1164.2 Btu/lb. This means that the enthalpy at point (1) must also equal 1164.2 Btu/lb.

The pressure in the main is 150 psia. The enthalpy of the saturated liquid at 150 psia is 330.51 Btu/lb, that of the saturated vapor is 1194.1 Btu/lb, and the heat of vaporization is 863.6 Btu/lb. Since the enthalpy of the steam in the main is only 1164.2 Btu/lb, it must be wet steam. If we take 1164.2 − 330.5 = 833.7 Btu/lb, this means that only 833.7 Btu/lb is available for evaporating the saturated liquid into steam, whereas 863.6 Btu/lb is required for total evaporation of the saturated liquid.

Then, 833.7/863.6 = 96.5 percent of the liquid has been evaporated, and the quality is 96.5 percent.

A slightly different application of the general energy law occurs in determining the power output of a turbine. An example of this application is afforded by a past examination problem which asks:

■ A steam turbine receives 3600 lb of steam per hour at 110 ft/sec velocity and 1525 Btu/lb enthalpy. The steam leaves at 810 ft/sec and 1300 Btu/lb. What is the horsepower output?

Nothing is said about the efficiency of the energy transfer from the steam to the turbine, so assume it to be 100 percent; i.e., all the energy given up by the steam appears as work done by the turbine. Applying the general energy equation between points (1) entering the turbine and (2) leaving the turbine, we obtain ($Z_1 = Z_2$ and Q is assumed to be zero):

$$\frac{110^2}{2gJ} + 1525 = \frac{810^2}{2gJ} + 1300 + W \qquad \text{for each pound of steam}$$
$$W = (1525 - 1300) + (0.24 - 13.1) = 212 \text{ Btu/lb of steam}$$
$$\frac{3{,}600}{60}\text{lb/min} \times \frac{1}{33{,}000 \text{ ft-lb/min hp}} \times 212 \text{ Btu/lb} \times 778 \text{ ft-lb/Btu} = 300 \text{ hp}$$

A slightly different application is the case of an impulse turbine where the heat energy in the steam is converted into velocity through a nozzle and only kinetic energy is extracted from the steam by the turbine blades. A past problem exemplifying this case was:

- In an impulse turbine 5 lb of steam per second impinges upon the buckets at an initial velocity of 4000 ft/sec and the steam leaves the buckets at a velocity of 1000 ft/sec. Disregarding friction, what horsepower would be developed and what would be the efficiency?

The general energy equation applies here, but $h_1 = h_2$, $Q = 0$, and $Z_1 = Z_2$, giving:

$$\frac{v_1{}^2}{2gJ} = \frac{v_2{}^2}{2gJ} + W$$
$$W = \frac{v_1{}^2 - v_2{}^2}{2g} = \frac{4{,}000^2 - 1{,}000^2}{2g} = 233{,}000 \text{ ft-lb/lb of steam}$$
$$\text{hp} = \frac{5 \text{ lb/sec} \times 233{,}000 \text{ ft-lb/lb}}{550 \text{ ft-lb/sec hp}} = 2120 \text{ hp developed}$$

When the steam leaves the turbine it still has a velocity of 1000 ft/sec and possesses $1000^2/2g = 15{,}500$ ft-lb/lb of mechanical energy. When the steam entered the turbine it had $4000^2/2g = 248{,}500$ ft-lb/lb mechanical energy. The turbine then extracted 233,000 ft-lb/lb of mechanical energy from the steam, and the efficiency of the turbine is then $233{,}000/248{,}500 = 93.8$ percent.

4-3. SECOND LAW OF THERMODYNAMICS

There are many different ways of stating the second law. One of these is the first enunciation as given by Lord Kelvin as, "It is impossible to transfer heat from a colder system to a warmer system without other simultaneous changes occurring in the two systems or in their environment." The second enunciation is, "It is impossible to take heat from a system and to convert it into work without other simultaneous changes occurring in the system or in its environment." Clausius stated the second law as, "Heat cannot, of itself, pass from a colder to a hotter system." The second law can also be stated, "It is impossible to construct an engine which will work in a complete cycle and produce no other effect except the raising of a weight and the cooling of a heat reservoir." The proof of these different statements of the second law is only that they have never been disproved.

4-4. THIRD LAW OF THERMODYNAMICS

To complete the discussion of the "laws" of thermodynamics, the third law will be given here. The third law is also known as Nernst's postulate. It can be stated as, "The entropy remains unchanged in any isothermal process taking place in a condensed system in the vicinity of the absolute zero." A slightly different form is "The entropy of a substance becomes zero in a state at the absolute zero of temperature."

4-5. EQUATION OF STATE

The equation of state may be given, for 1 lb of a gas, as

$$PV = RT$$

Where, in English units, P = pressure, lb/ft^2

V = specific volume, ft^3/lb

T = temperature, °Rankine or °Fahrenheit absolute

R = specific gas constant, ft/°Rankine or ft-lb/(lb)(°Rankine)

The equation of state is written correctly in a number of different ways which are all equivalent. It is sometimes written $PV = WRT$, where P, R, and T are as before, W = pounds of gas, and V = total volume in cubic feet or

$$V \text{ ft}^3/\text{lb} \times W \text{ lb} = \text{Vol, ft}^3$$

If we let $W = M$ lb, where M is equally numerically to the molecular weight, we have

$$PMV = MRT$$

where V = specific volume, ft^3/lb

Rearranging terms gives

$$MR = \frac{P(MV)}{T}$$

From Avogadro's law all perfect gases have the same number of molecules in any given volume at a given temperature and pressure. It follows from Avogadro's law, that a pound mole (a weight of gas equal in pounds to the molecular weight of the gas) of any perfect gas will occupy the same volume as that of any other perfect gas for the same conditions of temperature and pressure. At 32°F and atmospheric pressure this volume is called a *mole volume* and equals 359 ft³ (in the cgs system the mole volume equals 22.4 liters, which is called the "gram-molecular volume" and contains 6.024×10^{23} molecules).

Referring to the equation

$$MR = \frac{P(MV)}{T}$$

we can see that MV (lb $\times$ ft³/lb) will be a constant number for all perfect gases at a particular pressure and temperature. At standard conditions $MV = 359$ ft³. We know that PV/T is a constant (equal to R) for a particular gas. Then $[P(MV)]/T$ equals a constant for any perfect gas and MR is a constant for that particular gas. MR is called the universal gas constant or the molar gas constant and is equal to (very nearly) 1544. With this we can determine R for a given gas from the relationship $R = 1544/$mol. wt.

The equation of state will always hold for a perfect gas and will very nearly hold for all common gases in moderate pressure and temperature ranges.

For special cases we have modifications of the equation of state. For constant pressure we have that, since $PV = RT$,

$$\frac{V_1}{T_1} = \frac{V_2}{T_2} = \text{constant}$$

which is called "Charles' law,"

$$\frac{V_1}{V_2} = \frac{T_1}{T_2} \qquad (P = \text{constant})$$

For constant temperature we have that $P_1 V_1 = RT = P_2 V_2$, or

$$P_1 V_1 = P_2 V_2 \qquad (T = \text{constant})$$

This is called "Boyle's law."

The equation of state will be examined in more detail using past examination problems as examples. One of these examination questions asked:

- When air at atmospheric pressure is pumped into a tire, its volume is compressed to one-third its original value and the temperature rises from 60 to 100°F. What is the gauge pressure of the air in the tire?

The equation of state gives us

$$\frac{P_1 V_1}{T_1} = \frac{P_2 V_2}{T_2} = R$$

where R is very nearly constant over the range considered. Then

$$P_2 = P_1 \times \frac{T_2}{T_1} \times \frac{V_1}{V_2}$$

The initial pressure of the air is 14.7 psia.

$$T_1 = 60°F = 60 + 460 = 520°R$$
$$T_2 = 100°F = 100 + 460 = 560°R$$

and the final volume is one-third the initial volume, or $V_2 = \frac{1}{3}V_1$.

$$P_2 = 14.7 \times \frac{560}{520} \times \frac{1}{1/3} = 47.5 \text{ psia}$$

The gauge pressure of the air in the tire would then equal

$$47.5 - 14.7 = 32.8 \text{ psig}$$

Another example is afforded by the question:

- A completely airtight cylinder 300 cm long is divided into two parts by a freely moving, airtight, heat-insulating piston. When the temperatures in the compartments are equal and equal 27°C, the piston is located 100 cm from one end of the cylinder. How far will the piston move if the gas in the smaller part of the cylinder is heated to 74°C, the temperature of the larger section remaining constant? Assume perfect gas laws hold, and disregard the thickness of the piston.

For this problem it is best to draw the figure first (Fig. 4-2).

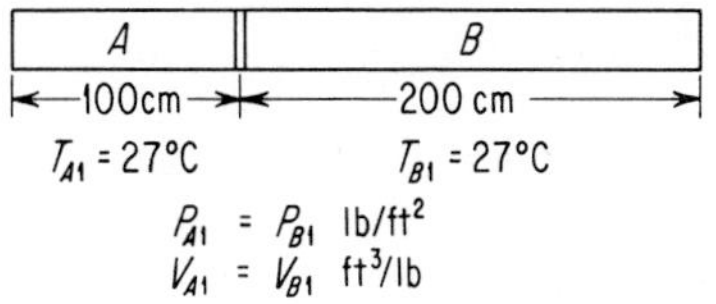

$T_{A1} = 27°C \qquad T_{B1} = 27°C$
$P_{A1} = P_{B1} \text{ lb/ft}^2$
$V_{A1} = V_{B1} \text{ ft}^3/\text{lb}$

Figure 4-2

This problem may be done in two steps, or the answer may be found in only one step. Let us take it step by step first.

Assume the piston is held stationary while the heat is added, and calculate the pressure increase in volume A. This would be a constant-volume process.

$$P_{A2} = P_{A1} \times \frac{T_2}{T_1} = P_{A1} \times \frac{74 + 273}{27 + 273} = 1.16 P_{A1}$$

Then let the piston move until the pressures in chambers A and B are equalized, holding the temperatures constant.

$$P_{A2}V'_{A2} = P_{A3}V'_{A3} \qquad \text{and} \qquad P_{B1}V'_{B1} = P_{B3}V'_{B3}$$

plus the equation $\qquad V'_{B1} - V'_{B3} = V'_{A3} - V'_{A2}$
or $\qquad\qquad\quad V'_{A2} + V'_{B1} = V'_{A3} + V'_{B3}$

since the total volume remains constant. Here we are using the relationship $PV' = WRT$, where V' is measured in cubic centimeters.

$$V'_{A2} = 100 \times A \qquad V'_{B1} = 200 \times A$$

where A = the cross-sectional area

The same relationships will hold regardless of what the cross-sectional area might be. To simplify the calculation, let us use $A = 1$.

$$P_{A2} = 1.16 P_{A1} = 1.16 P_{B1}$$
$$V'_{A1} = 100 = V'_{A2} \qquad \text{and} \qquad V'_{B1} = 200$$
$$1.16 P_{A1} \times 100 = P_{A3}V'_{A3}$$
$$P_{B1}V'_{B1} = P_{A1} \times 200 = P_{B3}V'_{B3}$$
$$P_{A3} = P_{B3} \qquad \text{and} \qquad V'_{A3} + V'_{B3} = 300$$

Solving these equations simultaneously gives

$$V'_{A3} = 110 \qquad \text{and} \qquad V'_{B3} = 190$$

The piston will then move 10 cm to the right in the figure as drawn.

To determine the new volume in one step we utilize the same reasoning, regarding cross-sectional area as before, and we have:

$$\frac{P_{A1}V'_{A1}}{T_{A1}} = \frac{P_{A2}V'_{A2}}{T_{A2}} = W_A R \qquad \text{constant}$$

and $\qquad \dfrac{P_{B1}V'_{B1}}{T_{B1}} = \dfrac{P_{B2}V'_{B2}}{T_{B2}} = W_B R \qquad \text{constant}$

where $P_{A1} = P_{B1}$, $P_{A2} = P_{B2}$, $V'_{A1} = 100$, $V'_{B1} = 200$, $T_{B1} = T_{B2}$, and $V'_{A2} + V'_{B2} = 300$.

$$\frac{100 P_{A1}}{27 + 273} = \frac{P_{A2}V'_{A2}}{74 + 273} \qquad \text{or} \qquad P_{A1} = 0.00864 V'_{A2} P_{A2}$$
$$P_{B1} \times 200 = P_{B2} \times V'_{B2} \qquad P_{A1} = 0.005 P_{A2} V'_{B2}$$

Combining these two gives $V'_{B2} = 1.728 V'_{A2}$, which, combined with the relationship $V'_{A2} + V'_{B2} = 300$, gives $V'_{A2} = 110$, as before.

As another example, consider the following past examination problem:

■ A tank contains 10 ft³ of oxygen at 500 psia and 85°F. Part of the oxygen was used, and later the pressure was observed to be 300 psia and the temperature, 73°F.

  a. How many pounds of oxygen were used?

  b. What volume would the used oxygen occupy at standard atmospheric pressure and 70°F?

At the beginning we have

$$P_1V_1 = W_1RT_1$$
$$R = \frac{1544}{\text{mol. wt}} = \frac{1544}{32} = 48.2$$
$$P_1 = 500 \times 144 \qquad V_1 = 10 \text{ ft}^3 \qquad T_1 = 460 + 85$$

Then

$$W_1 = \frac{(500 \times 144) \times 10}{48.2 \times 545} = 27.4 \text{ lb} \qquad \text{oxygen at start}$$

At the later time,

$$W_2 = \frac{P_2V_2}{RT_2} = \frac{300 \times 144 \times 10}{48.2 \times 533} = 16.8 \text{ lb} \qquad \text{oxygen remaining}$$

The amount of oxygen used equaled $27.4 - 16.8 = 10.6$ lb. The specific volume of oxygen may be calculated from the equation of state, $PV = RT$.

$$V = \frac{RT}{P} = \frac{48.2(460 + 70)}{14.7 \times 144} = 12.07 \text{ ft}^3/\text{lb}$$

The 10.6 lb of oxygen used would then occupy

$$10.6 \times 12.07 = 128 \text{ ft}^3 \text{ of volume}$$

A further example of the usefulness of the equation of state is afforded by yet another past examination problem:

■ A hydrocarbon has the following composition: $C = 82.66$ percent, $H = 17.34$ percent. The density of the vapor is 0.2308 g/l at 30°C and 75 mm. Determine its molecular weight and its molecular formula.

The equation of state gives $PV = RT$, or $w = P/RT$, where $w = 1/V = $ g/cm³. Knowing w, P, and T, we can solve for R, where $R = P/wT$, $p = $ 75 mm Hg, and mercury weighs 13.6 g/cm³.

Therefore, $P = 13.6 \times 7.5$ cm $\times$ 980 dynes/g $= 10^5$ dynes/cm^2

$\quad\quad w = 2.308 \times 10^{-4}$ g/cm^2

$\quad\quad T = 273 + 30 = 303$K

$$R = \frac{10^5}{2.308 \times 10^{-4} \times 303} = 1.43 \times 10^6 \frac{\text{dyne-cm}}{\text{(g)(K)}} = 1.43 \times 10^6 \frac{\text{ergs}}{\text{(g)(K)}}$$

The universal gas constant in the cgs system equals 8.312×10^7 ergs/(mole)(K). So

$$\frac{8.312 \times 10^7}{1.43 \times 10^6} = 58.1 \text{ g/mole}$$

or the molecular weight of the gas is 58.1.

The equation of the gas is C_nH_m, and the molecular weight is $n \times 12 + m \times 1.008$, where 12 is the atomic weight of carbon and 1.008 is the atomic weight of hydrogen.

$$n \times 12 = 0.8266 \times 58.1 = 48$$
and $\quad\quad m \times 1.008 = 0.1734 \times 58.1 = 10.1$

The required answers are, then: the molecular weight = 58.1, and the molecular formula is C_4H_{10}.

Another method of approach is to determine the weight of 22.4 l of the gas at standard conditions. The weight of 1 mole also equals the molecular weight. Standard conditions are 1 atm of pressure (760 mm) and 0°C.

$$P = wRT \quad\quad R = \text{constant} = \frac{P}{wT}$$

giving $\quad\quad \dfrac{P_1}{w_1 T_1} = \dfrac{P_2}{w_2 T_2} \quad\quad w_2 = w_1 = \dfrac{P_2}{P_1} \times \dfrac{T_1}{T_2}$

$$w_2 = 0.2308 \times \frac{760}{75} \times \frac{273 \times 30}{273} = 2.60 \text{ g/l}$$

at standard conditions.

$$2.60 \text{ g/l} \times 22.4 \text{ l} = 58.2 \text{ g/mole}$$

or the molecular weight is 58.2 and the molecular formula may be determined as before.

4-6. WORK AND THERMODYNAMIC PROCESSES

The equation of state can be used to find the pressure, volume, or temperature of a gas at any point at which two of the three properties are known. It tells nothing, however, about the way in which any of these properties varies between the two points. The work required to change a gas from one state point to another or the work which can be obtained from a gas as it expands from one state point to another depends upon the path which the gas follows as it changes from one point to another. This can be easily shown by an examination of a graph showing the relation between the pressure and volume of a gas as it changes from one state point to another (Fig. 4-3).

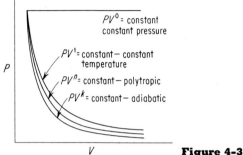

Figure 4-3

We know from our study of mechanics that Work $= \int F\ ds$. For a gas, Force $=$ pressure $\times$ area, and $ds = (dV)/A$, so we have

$$F\ ds = P \times A \times \frac{dV}{A} = P\ dV$$

and the work done by an expanding gas equals $\int P\ dV$. We recognize this as equivalent to the area under the P-V diagram. This means that the process by which the gas changes from one state point to another is of extreme importance when it comes to determining the work required by or obtainable from a gas.

The more important processes are:

Constant temperature, isothermal	$PV =$ constant
Constant pressure, isobaric or isopiestic	$V/T =$ constant, $P =$ constant
Constant volume, isometric	$P/T =$ constant, $V =$ constant
Adiabatic	$PV^k =$ constant
Polytropic	$PV^n =$ constant

where $k = c_p/c_v$, the ratio of specific heats, and n is an exponent, ordinarily somewhat less than k.

The different processes are shown in Fig. 4-3, and the differences in the work required, or developed, are apparent. The work required to go from point (1) to point (2) can be calculated from the relationship Work $= \int_{1}^{2} P\ dV$. For constant pressure we have:

$$W = P \int_1^2 dV = PV_2 - PV_1$$

For the constant temperature, polytropic, and adiabatic processes, we have a similar condition in that the relationship $PV^n = $ constant applies to all three cases, with n equaling 1, n, and k, respectively. Here we have, taking logarithms of both sides,

$$\ln P + n \ln V = \ln C$$

and differentiation gives

$$\frac{dP}{P} + n\frac{dV}{V} = 0$$

or

$$dV = -\frac{V}{n}\frac{dP}{P}$$

then

$$W = \int_1^2 -\frac{1}{n}V \, dP$$

but

$$V = \left(\frac{C}{P}\right)^{1/n}$$

where $PV^n = C$

$$W = -\frac{1}{n}\int_1^2 C^{1/n}P^{-1/n} \, dP = -\frac{C^{1/n}}{n}\frac{1}{1 - 1/n}P^{(1-1/n)} \Big|_1^2$$

$$= -\frac{P^{1/n}V}{n - 1}P^{(1-1/n)} \Big|_1^2$$

which gives

$$W = \frac{P_2V_2 - P_1V_1}{1 - n}$$

for work done by the gas. W will be negative if work is done on the gas. This equation can be used to calculate the work done by (or on) the gas for $n = k$ and $n = n$, the polytropic exponent. For an isothermal process where $PV = $ constant and $n = 1$, this relationship gives the indeterminate fraction, $0/0$, so the relationship for work for an isothermal process must be found in another way.

$$PV = RT \text{ constant} \qquad P = \frac{RT}{V}$$

$$W = \int P \, dV = \int \frac{RT}{V}dV = RT \ln\frac{V_2}{V_1} = RT \ln\frac{P_1}{P_2}$$

$$\text{Work} = P_1V_1 \ln\frac{P_1}{P_2}$$

In the case of constant volume, $dV = 0$ and no work is done by the gas. There will, however, be a change in the enthalpy of the gas for a change in the pressure when the volume remains constant. This follows from the fact that $h = u + (PV)/J$ or $h = u + (RT)/J$ and also that $du = c_v\,dT$, where u, the internal energy of the gas, is a function of temperature alone. There will, then, be a change in both the internal-energy term and also the PV/J term for a constant-volume process.

Utilizing the same relationships, it can be seen that an isothermal process is a constant-enthalpy process and also a constant-energy process for a perfect gas.

With the relationships given, the work done by a gas (positive or negative) to go from one state point to another can be calculated.

The work may also be determined by the application of the general energy equation; often this will be simpler than calculating the integral of $P\,dV$. We have

$$W = \frac{v_1{}^2 - v_2{}^2}{2gJ} + h_1 - h_2 + Z_1 - Z_2 + Q$$

In most cases this can be reduced to $W = h_1 - h_2 + Q$, and it is then only necessary to determine Δh and Q to calculate W.

The enthalpy h is equal to the internal energy u plus the quantity PV/J. For a perfect gas and a reversible adiabatic process, the following relationships will hold:

$$du + P\,dV = 0$$
$$du = c_v\,dt \quad \text{or} \quad \Delta u = c_v(T_2 - T_1)$$
$$\Delta h = wc_p\,\Delta T \quad \text{or} \quad \Delta h = c_p\,\Delta T$$
$$R = c_p - c_v \quad \text{for the specific gas, numerically } R = (c_p - c_v)J$$
$$W = c_v(T_1 - T_2) = \text{work done by gas on a piston}$$
$$h_1 - h_2 = \frac{k}{k-1}(P_1V_1 - P_2V_2)$$

also, for an isentropic process, $dh - V\,dP = 0$. And since

$$h_2 - h_1 = \int_1^2 dh$$

we have
$$\frac{k}{k-1}(P_1V_1 - P_2V_2) = \int_1^2 V\,dP$$

For a reversible polytropic process ($Pv^n = $ constant), it can be shown that

$$Q = U_2 - U_1 + \int_1^2 P\,dV = \left(\frac{R}{n-1} - c_v\right)(T_1 - T_2) = \frac{c_p - nc_v}{n-1}(T_1 - T_2)$$

The derivation of these relationships is somewhat too lengthy and involved to be included here; the interested reader is referred to any good text on thermodynamics for a more complete discussion.

For illustration let us look at a past examination problem:

■ A centrifugal compressor pumps 100 lb of air per minute from 14.7 psia and 60°F to 50 psia and 270°F by an irreversible process. The temperature rise of 50 lb/min of circulating water about the casing is 12°F. What horsepower is required? Disregard any changes in kinetic energy.

The general energy equation will always apply, and if we write it for the two points of the process, we can calculate the work required, provided we can determine the value of the individual components.

$$\frac{v_1{}^2}{2gJ} + h_1 + Z_1 + Q = \frac{v_2{}^2}{2gJ} + h_2 + Z_2 + W$$

We have been told to disregard changes in kinetic energy, and we shall assume that the difference in elevation between the inlet and outlet of the compressor is small enough to be ignored without affecting the accuracy of the answer. This gives us the equation for the work:

$$W = h_1 - h_2 + Q$$

Q, the heat lost by the air during the compression process, can easily be determined from the data given for the cooling water.

$$Q = -50 \text{ lb/min} \times 12°F \times 1 \text{ Btu/(lb)(°F)} = -600 \text{ Btu/min}$$

From the relationship $\Delta h = c_p \, \Delta T$, which holds for a steady-flow process, we find that

$$h_2 - h_1 = 0.24(210) = 50.4 \text{ Btu/lb} = -(h_1 - h_2)$$
$$W = -(100 \text{ lb/min} \times 50.4 \text{ Btu/lb}) - 600 = -5640 \text{ Btu/min}$$
$$\frac{5640 \text{ Btu/min} \times 778 \text{ ft-lb/Btu}}{33,000 \text{ ft-lb/min-hp}} = 133 \text{ hp}$$

which is the required answer, considering the gas as a "perfect" gas.

The process is given as an irreversible one so the work cannot be correctly calculated from the relationship

$$\text{Work} = \int P \, dv$$

This follows from the fact that the process is not reversible and, therefore, is not frictionless and that the amount of work done against friction cannot be calculated from the reversible (or frictionless) work relationships. The path of the process can, however, be described by means of the relationship PV^n = constant. Let us calculate what the work requirement would have been had the compression been a reversible polytropic process. This process can be represented by a sketch in the P-V plane as shown in Fig. 4-4.

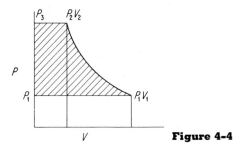

Figure 4-4

A centrifugal compressor is essentially a centrifugal pump which converts velocity head to pressure head and then pushes the compressed gas at constant pressure to wherever it goes. The curve of the process would be first a constant-pressure intake of air at pressure P_1 from zero volume to V_1. The gas would then be compressed polytropically, in accordance with the relationship PV^n = constant, to the state point represented by P_2V_2. Finally the compressed gas would be pushed out at a constant pressure to the state point represented by $P_3(P_3 = P_2)$. We can assume a clearance of zero since it is a centrifugal compressor. The shaded area of the diagram in Fig. 4-4 indicates the work done on the gas and equals the area under that portion of the curve. The shaded area equals

$$\int_1^2 P \ dV + P_2V_2 - P_1V_1$$

where P_2V_2 is the work represented by the constant-pressure discharge and P_1V_1 is the work represented by the constant-pressure intake at atmospheric pressure.

$$\int_1^2 P \ dV = \int_1^2 CV^{-n} \ dV$$

where $PV^n = C$, giving $P = CV^{-n}$.

$$\int_1^2 CV^{-n} \ dV = \frac{C}{1-n}V^{(1-n)}\Big|_1^2 = \frac{PV^n}{1-n}(V_2^{1-n} - V_1^{1-n}) = \frac{P_2V_2 - P_1V_1}{1-n}$$

which is negative, indicating, according to our convention, that work is being done on the gas. This is the work done on the gas during the polytropic compression.

The constant-pressure process work

$$\int_2^3 P \, dV = P(V_3 - V_2) = -P_2 V_2$$

This is also negative according to our convention, indicating work done on the gas.

The work done during the intake stroke equals $\int_0^1 P \, dV = P_1 V_1$, which is positive, indicating work done by the gas. This gives

$$\text{Work} = \int_0^1 P \, dV + \int_1^2 P \, dV + \int_2^3 P \, dV = P_1 V_1 + \frac{P_2 V_2 - P_1 V_1}{1 - n} - P_2 V_2$$

which combines to give $n/(1 - n)(P_2 V_2 - P_1 V_1)$ work done by the gas.

This could also have been obtained from the relationship

$$W = \int_1^2 V \, dP$$

which also would give the shaded area.

Since $PV^n = \text{constant}$, we find by differentiating that

$$nV^{n-1} P \, dV - V^n \, dP = 0$$

or $V \, dP = -nP \, dV$. This gives $\int V \, dP = -n \int P \, dV$, indicating that the work required to go from $P_1 V_1$ to $P_2 V_2$ equals

$$\left(\frac{n}{1 - n} \right) (P_2 V_2 - P_1 V_1)$$

as before.

This also represents the difference between flow work and nonflow work. The nonflow work is represented by the shaded area in Fig. 4-5. In this process a given volume of gas is compressed from V_1 to V_2 (or expands from V_2 to V_1). There is no flow of gas here; the quantity of the gas is constant, no gas being added and none removed. Thus we have the term "nonflow work," with the work being represented by $\int_1^2 P \, dV$.

In the case of Fig. 4-4 there is actually a flow of gas. It is taken in at a pressure of P_1 to a volume of V_1, compressed to P_2 and V_2, and then discharged at a pressure P_2. There is an actual flow of gas through the compressor—thus the term "flow work" with the work equal to

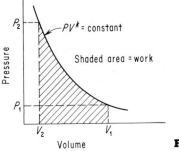

Volume

Figure 4-5

$\int_{1}^{2} V\ dP.$ Since $PV^n = $ constant, $P_1 = 14.7$ psia, $P_2 = 50$ psia, $P_1V_1{}^n = P_2V_2{}^n$, and

$$\ln P_1 + n \ln V_1 = \ln P_2 + n \ln V_2$$
$$n = \frac{\ln P_1 - \ln P_2}{\ln V_2 - \ln V_1} = \frac{\ln (P_1/P_2)}{\ln (V_2/V_1)}$$

$V_1 = RT_1/P_1 = 53.3 \times (460 + 60)/(14.7 \times 144) = 13.10$ ft³/lb

$V_2 = 5.40$ cu ft/lb $\qquad n = \dfrac{\ln (14.7/50)}{\ln (5.42/13.1)} = \dfrac{-1.225}{-0.884} = 1.386$

which gives $W = -40{,}035$ ft-lb/lb, or 40,035 ft-lb/lb work done on the gas for a reversible (frictionless) polytropic cycle. This amount of work would require 121.3 hp for the 100 lb/min of air handled by the compressor. The frictional work would equal the difference, or 11.5 hp.

A somewhat similar example is provided by another past examination problem:

- Derive an equation in terms of P_1, V_1, P_2, and V_2 for the work required to compress a gas adiabatically from P_1, V_1 to P_2, V_2.

This is similar to the analysis used in the previous example. We know the work will equal the area shown on the P-V diagram (Fig. 4-5).

$$\text{Work} \int_{1}^{2} P\ dV$$

As before

$$\int_{1}^{2} P\ dV = \frac{P_2V_2 - P_1V_1}{1 - k}$$

which is negative, indicating work done on the gas.

$$\text{Work} = \frac{P_2V_2 - P_1V_1}{k - 1} \qquad \text{ft-lb/lb of gas}$$

to compress the gas adiabatically from P_1V_1 to P_2V_2, where $P =$ lb/ft² and $V =$ ft³/lb.

Another type of problem that is sometimes encountered in examinations deals with the flow of gas through a nozzle. As an example:

■ A satellite contains a gas generator which produces gas at a pressure of 100 psig and 150°F. The gas has an equivalent molecular weight of 22 and flows out of an ideal nozzle into space. The specific heat of the gas at constant pressure equals 0.26. The design atmospheric pressure is 4 psia. The throat of the nozzle is 1 in. in diameter. Determine: (a) weight rate of flow, (b) exit diameter of nozzle, (c) exit temperature, (d) exit velocity, (e) exit specific volume, and (f) thrust.

The problem gives the initial gas pressure as 100 psig and the local atmospheric pressure as 4 psia. The chamber pressure then equals 104 psia. Similarly the initial temperature is given as 150°F, which equals 610°R. The gas constant equals 1544/MW, giving $R = 1544/22 = 70.18$ ft/°R.

The velocity in the throat of the nozzle would be sonic because of the large expansion ratio, 104/4 = 26. Then throat velocity is

$$\sqrt{kgRT} = \sqrt{1.3 \times 32.2 \times 70.2 \times 610} = 1339 \text{ ft/sec}$$

The ratio of specific heats is assumed to equal 1.3 since it is assumed that the generated gas is formed of molecules made up of three or more atoms. To a good approximation the ratio of specific heats will equal 1.66 for monatomic gases, 1.40 for diatomic gases, and 1.30 for gases which consist of molecules containing three or more atoms. Steam, for example, has a $k = 1.28$.

The throat area equals 0.7854 in.², or 0.00545 ft².

The specific volume of the gas at throat conditions would equal

$$V_1 = \frac{RT_1}{P_1} = 70.18 \times \frac{610}{104 \times 144} = 2.86 \text{ ft}^3/\text{lb}.$$

The corresponding specific weight $w = 0.350$ lb/ft³

The weight rate of flow would equal $v \times A \times w$, or

$$W = 1339 \times 0.00545 \times 0.350 = 2.55 \text{ lb/sec} \tag{a}$$

To determine the exit diameter of the nozzle, it is necessary to calculate the specific weight of the gas at the exit conditions and the velocity of the gas at exit. Then, knowing the weight flow rate, the required area can be determined.

$$A = \frac{W}{v \times w} = \frac{WV}{v}$$

The gas does no work and no heat is transferred as it flows out of the nozzle, so the expansion would be adiabatic, or

$$PV^{1.3} = \text{constant}$$
$$V_2 = V_1 \times \left(\frac{P_1}{P_2}\right)^{1/k} = 2.86\left(\frac{104}{4}\right)^{0.769} = 35.06 \text{ ft}^3/\text{lb}$$
$$T_2 = T_1 \times \left(\frac{P_2}{P_1}\right)^{(k-1)/k} = 610\left(\frac{4}{104}\right)^{0.231} = 288°\text{R, or } -172°\text{F}$$

The accuracy of the calculations can be checked using the universal gas law at the exit state point.

$$PV = RT$$
$$4 \times 144 \times 35.06 = 70.18 \times T$$

giving $T = 288°$R.

The velocity of the gas would increase as it expanded and the temperature lowered. The decrease in heat energy would be matched by an increase in kinetic energy. From the general energy equation (Sec. 4-1)

$$\frac{v_1^2}{2gJ} + h_1 = \frac{v_2^2}{2gJ} + h_2$$
$$h_1 - h_2 = c_p(T_1 - T_2) = 0.26 \times 322 = 83.72 \text{ Btu/lb}$$

Then the change in the kinetic energy of the gas would equal

$$83.72 \times 778 = 65,134 \text{ ft-lb/lb}$$

The kinetic energy at the throat $\text{KE}_1 = \frac{1}{2}mv_1^2$

$$\text{KE}_1 = \left(\frac{1}{2g}\right) \times 1339^2 = 27,840 \text{ ft-lb/lb}$$
$$\text{KE}_2 = 27,840 + 65,134 = 92,974 \text{ ft-lb/lb}$$
$$v_2 = \sqrt{2g \times 92,974} = 2447 \text{ ft/sec}$$
$$A = \frac{W}{v \times w} = \frac{WV}{v} = \frac{2.55 \times 35.06}{2447} = 0.0365 \text{ ft}^2, \text{ or } 5.26 \text{ in.}^2$$

The exit diameter of the nozzle would equal 2.59 in.

The thrust would equal the force due to the change in momentum or

$$F = \left(\frac{m}{t}\right) \Delta v = (2.55/g) \text{ slugs/sec} \times 2447 \text{ ft/sec} = 194 \text{ lb}$$

Note that if the discharge cone had not been added to the nozzle, the discharge velocity would have equaled only the throat velocity of 1339 ft/sec. The thrust would then have equaled $(2.55/32.2) \times 1339 = 106$ lb. By adding the discharge cone to the nozzle, the thrust was increased by 83 percent.

4-7. CYCLES

If we put together a series of processes so that the beginning and ending points are identical, the sequence is termed a cycle. One of the more important cycles is the Carnot cycle, which consists of a constant-temperature process, a constant-entropy process, a second constant-temperature process, and a second constant-entropy process, as shown in the diagram (Fig. 4-6).

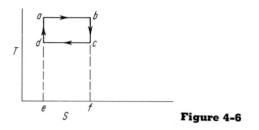

Figure 4-6

4-8. ENTROPY

This brings up the question which is so bothersome to many engineers: "What is entropy?"

Entropy cannot be pictured; it has no physical shape or substance. It is actually a mathematical concept and is the name given to the quantity $\int dQ/T$, which appears so frequently in thermodynamic analyses that it seemed desirable to give it a name. There is no absolute value of entropy; we deal with differences of entropy between state points. We can only equate differences of entropy for reversible processes, and for a reversible process

$$S_2 - S_1 = \int_1^2 \frac{dQ}{T}$$

from which can be obtained other expressions which can be used to solve for differences of entropy for a perfect gas.

$$ds = \frac{dQ}{T} = \frac{du + P \, dV}{T}$$

$$ds = c_v\left(\frac{dT}{T}\right) + R\left(\frac{dV}{V}\right)$$

which can be integrated to give

$$s_2 - s_1 = c_v \ln\left(\frac{T_2}{T_1}\right) + R \ln\left(\frac{V_2}{V_1}\right)$$

$$s_2 - s_1 = c_p \ln\left(\frac{V_2}{V_1}\right) + c_v \ln\left(\frac{P_2}{P_1}\right) = c_p \ln\left(\frac{T_2}{T_1}\right) - R \ln\left(\frac{P_2}{P_1}\right)$$

Also, the increase of entropy during a polytropic process can be shown to equal

$$s_2 - s_1 = \frac{c_p - nC_v}{n - 1}\ln\left(\frac{T_1}{T_2}\right) = c_v\left(\frac{k - n}{1 - n}\right)\ln\left(\frac{T_2}{T_1}\right)$$

One past examination problem asked:

■ What is the difference between entropy and enthalpy?

We recall that enthalpy $h = u + pV$, or

$$dh = du + P \, dV + V \, dP$$

whereas the change in entropy $ds = (du + P \, dV)/T$; these equations combine to give $dh = T \, ds + V \, dP$.

Another question in a past examination asked:

■ A volume of gas having an initial entropy of 3000 Btu/°R is heated at constant temperature of 1000°F until the entropy is 4500 Btu/°R. How much heat is added?

We know that $S_2 - S_1 = \int_1^2 dQ/T$, and if T is a constant we can integrate easily, obtaining $S_2 - S_1 = (1/T)(Q_2 - Q_1)$. Then we have $(S_2 - S_1)T = \Delta Q$, or

$$\Delta Q = (4500 - 3000)(1000 + 460) = 2{,}190{,}000 \text{ Btu}$$

added.

Another examination question was:

■ A gas has a constant-pressure specific heat of

$$c_p = 6.6 + 7.2 \times 10^{-4} \times T \text{ Btu/(mole)(°R)}$$

Determine the entropy of this gas at 2000°R and 5 atm pressure if the base of zero entropy is taken at 520°R and 1 atm.

We note that c_p is given in Btu/(mole)(°R). If we refer to the list of relationships given previously, we see that the relationship $ds = c_p(dT/T) - R(dP/P)$, from

$$ s_2 - s_1 = c_p \ln\left(\frac{T_2}{T_1}\right) - R \ln\left(\frac{P_2}{P_1}\right) $$

would appear to apply. We are not, however, given any value for R. We recall that

$$ R = \frac{1544}{\text{mol. wt}} \left(\frac{\text{ft-lb}}{\text{lb-°R}}\right) $$

The molecular weight is also the pounds per pound-mole, so we have that $R = 1544[\text{ft-lb/(mole)(°R)}]$. We also know that the mechanical equivalent of heat is 788 ft-lb/Btu, so we can modify this value of R to give $R = 1.985[\text{Btu/(mole)(°R)}]$. Now all that remains is to substitute in the relationship for ds and grind out the answer.

$$ ds = (6.6 + 7.2 \times 10^{-4} \times T)\frac{dT}{T}\frac{\text{Btu}}{\text{(mole)(°R)}} - 1.985\frac{dP}{P}\frac{\text{Btu}}{\text{(mole)(°R)}} $$

$$ \int ds = \int 6.6\frac{dT}{T} + \int 7.2 \times 10^{-4}dT - \int 1.985\frac{dP}{P} $$

$$ s_2 - s_1 = 6.6 \ln\left(\frac{T_2}{T_1}\right) + 7.2 \times 10^{-4}(T_2 - T_1) - 1.985 \ln\left(\frac{P_2}{P_1}\right) $$

$$ s_2 - s_1 = 8.888 + 1.065 - 3.19 = 6.76\frac{\text{Btu}}{\text{(mole)(°R)}} $$

The difference in entropy between points 1 and 2 is 6.76 Btu/(mole) (°R). Since $s_1 = 0$, the entropy of the gas is, then, 6.76 Btu/(mole)(°R) at 2000°R and 5 atm pressure.

Another example:

■ One pound of saturated steam at 400°F expands isothermally to 60 psia. Determine: (1) change of entropy; (2) heat transferred; (3) change of enthalpy; (4) change of internal energy; (5) work.

From a steam table the pressure corresponding to a saturated-steam temperature of 400°F is 247.3 psia. Also $h = 1201$ Btu/lb for saturated vapor, $s = 1.5272$ Btu/(°F)(lb), and $B = 1.863$ ft³/lb.

Since the expansion is isothermal, the final product will be superheated steam at 400°F and 60 psia. The steam table gives $h = 1233.6$ Btu/lb, $s = 1.7135$ Btu/(°F)(lb), and $V = 8.357$ for this condition.

The change of entropy will then be

$$1.7135 - 1.5272 = 0.1863 \text{ Btu/(lb)(°F)}$$

For a perfect gas, $\Delta h = c_p(T_2 - T_1)$, which is zero for an isothermal process. Also, $\Delta u = c_v(T_2 - T_1) = 0$ for an isothermal process for a perfect gas. Steam is not, however, a perfect gas, and the change of enthalpy is $1233.6 - 1201.0 = 32.6$ Btu/lb. From the general energy equation:

$$\frac{v_1^2}{2gJ} + h_1 + Z_1 + Q = \frac{v_2^2}{2gJ} + h_2 + Z_2 + W$$

We see that

$$Q = W + h_2 - h_1$$

since $v_1 = v_2 = 0$ and $Z_1 = Z_2$.

The work $= \displaystyle\int_1^2 P\ dV$, which equals $PV \ln (P_1/P_2)$, as determined previously. Unfortunately,

$$P_1V_1 = 247.3 \times 144 \times 1.863 = 66{,}200 \text{ ft}$$

and $P_2V_2 = 60 \times 144 \times 8.357 = 72{,}200$ ft, so PV is not a constant and the work calculated by this method will not be quite correct. Averaging the two values, this relationship gives

$$W = 69{,}200 \ln \frac{247.3}{60} = 98{,}000 \text{ ft-lb/lb} \qquad \text{work done by the gas}$$

For a reversible isothermal process we have that $Q = T\,\Delta s$. This is not a reversible process, but applying this relationship gives $Q = (400 + 460) \times 0.1863 = 160.2$ Btu/lb heat added to the gas.

Checking, we have $Q - (h_2 - h_1) = 160.2 - 32.6 = 127.6$ Btu/lb, or 99,300 ft-lb/lb, which is close to the average value for W of 98,000 ft-lb/lb.

We recall that the work done by a gas depends upon the path it follows, and since PV is not a constant, we do not know the equation of the path accurately. Also the relationship $Q = T\,\Delta s$ Btu/lb is for a reversible process, which this is not. For an ideal gas $\Delta Q = \Delta u + \int P\ dV$. Since $h = u + (PV/J)$, Δh is the same, in the case of an isothermal process, as Δu. In this actual case it is not. In this case,

$$u_1 = h_1 - \frac{PV_1}{J} = 1201.0 - \frac{66{,}200}{778} = 1115.7 \text{ Btu/lb}$$

and $$u_2 = 1233.6 - \frac{72{,}200}{778} = 1140.8 \text{ Btu/lb}$$

The work done will equal the heat added, $T \, \Delta S = 160.2$ Btu/lb, less the increase in internal energy of 25.1 Btu/lb, or 135.1 Btu/lb.

The value of 135.1 Btu/lb of work can also be obtained by iteration—calculating the values of $P \, \Delta V$ over small increments of pressure from 247.3 to 60 psia and adding the increments to obtain the total over the complete range of pressure. By the iteration process the actual process can be closely approximated even though the actual P-V path cannot be expressed mathematically.

The answers are, then:

1. $\Delta s = 0.1863$ Btu/(lb)(°F) increase
2. $\Delta Q = 160.2$ Btu/lb heat added to gas
3. $\Delta h = 32.6$ Btu/lb increase
4. $\Delta u = 25.1$ Btu/lb increase
5. Work = 135.1 Btu/lb work done by gas

4-9. CARNOT CYCLE

The ideal Carnot cycle consists of four reversible processes, two isothermals and two adiabatics. Referring to Fig. 4-6:

1. a to b, heat enters the engine at a constant temperature, reversible isothermal expansion
2. b to c, reversible adiabatic expansion until temperature drops
3. c to d, heat removed from the engine at constant temperature, reversible isothermal compression
4. d to a, reversible adiabatic compression until temperature rises to the initial temperature

This cycle is a work cycle, with the engine doing work during the adiabatic expansion. A Carnot cycle can also be run in reverse to act as a refrigerator. As an example, a past examination problem asked:

■ If a Carnot refrigerator which extracts heat from a cold-storage plant at −20°F and gives it to the atmosphere at 100°F has a 100-"ton"-per-day capacity, what horsepower is needed to drive the refrigerator?

A "ton" of refrigeration capacity is the amount equal to freeze 2000 lb of water in 24 hr. Since the heat of fusion of water is 144 Btu/lb, this means 288,000 Btu/24-hr day, 12,000 Btu/hr, or 200 Btu/min.

When the Carnot heat engine cycle is reversed, a refrigeration cycle is obtained. The work supplied to the refrigerator is represented by the area *abcd* in Fig. 4-6. The heat extracted is represented by the area *dcfe*. This means that work in the amount of $(T_a - T_d) \Delta S$ is required for the operation of the refrigerator, and that heat in the amount of $T_d \Delta S$ is extracted from the cold body. From this we can see that for the normal operating temperature range, more heat is extracted from the cold body than is required to operate the Carnot refrigerating machine. In this particular case we have $(460 - 20) \Delta S = 440 \Delta S$ Btu extracted from the cold source, while only $(560 - 440) \Delta S$ Btu is required to operate the engine. This means that for every Btu of energy used to run the refrigerator,

$$\frac{440 \Delta S}{(560 - 440) \Delta S} = 3.67 \text{ Btu}$$

is removed from the cold source. This value is called the coefficient of performance (c.o.p.), which is defined as the ratio of the amount of refrigeration effected to the work supplied. The c.o.p. may be greater than 1 or less than 1. Usually it is greater than 1. It might be noted in passing that the thermodynamic advantage as evidenced by a c.o.p. is the reason for the development of the heat pump. In the case of the heat pump all the energy will be transferred to the hot body and the c.o.p. of a heat pump is equal to the c.o.p. of the same machine used as a refrigerator plus 1. In the illustration the c.o.p. for this machine used as a heat pump would be

$$\frac{560 \Delta S}{(560 - 440) \Delta S} = 4.67 \qquad \text{or} \qquad 3.67 + 1.00$$

Since 1 Btu of refrigeration in the example would require only 1/3.67 Btu of work, 100 tons of refrigeration would require

$$\frac{100 \times 200}{3.67} = 5450 \text{ Btu/min} \qquad \frac{5450 \times 778}{33,000} = 128.5 \text{ hp}$$

The horsepower required to drive the refrigerator is 128.5.

It might be noted here that the Carnot efficiency, which equals the work done divided by the heat supplied, is the maximum efficiency possible for any thermal engine operating between two temperatures. No other cycle may exceed this efficiency and it is not necessary for a different cycle efficiency to equal it. Other reversible cycles are as efficient as a Carnot cycle, irreversible cycles are not so efficient as a Carnot cycle. No other cycle can be more efficient than a Carnot cycle.

Another interesting past examination problem is:

- A constant-temperature heat source at $t = 1000°F$ supplies heat to a Carnot engine which receives it at 600°F. The engine rejects heat at

100°F to a cold body (or sink) which is at a constant temperature of 60°F. Ten percent of the energy available to the Carnot engine is converted into friction and lost. 1000 Btu is supplied by the hot body.
a. What is the efficiency of the cycle?
b. What is the total increase in entropy due to irreversible processes?
c. What is the increase in unavailable energy?

The maximum cycle efficiency possible between the two reservoir temperatures would be $(1460 - 520)/1460 = 64.4$ percent. If the engine receives the heat at 600°F and discharges it at 100°F, the maximum possible engine efficiency would be

$$\frac{(1060 - 560)}{1060} = 47.2 \text{ percent}$$

The actual efficiency of the cycle will be somewhat less than 47.2 percent, however, because 10 percent of the heat supplied (10 percent × 1000, or 100 Btu) is lost in overcoming friction. This will give us a modified type of Carnot cycle as shown in Fig. 4-7. The actual engine efficiency will equal the work done by the engine divided by the heat supplied to the engine. Since

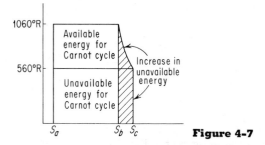

Figure 4-7

the heat is supplied to the engine at a constant temperature of 600°F (1060°R), the difference in entropy at the start would equal heat divided by the temperature, or $\Delta S = Q/T = 1000/1060 = 0.944$ Btu/°F. The heat discharged at the cold temperature by an ideal Carnot engine would be $Q = T \times \Delta S = 560 \times 0.944 = 528$ Btu. The engine loses another 100 Btu in overcoming friction so the total heat lost would equal 628 Btu. The heat utilized in doing useful work would then be $1000 - 628 = 372$ Btu. The cycle efficiency would be 372/1000, or 37.2 percent. The difference in entropy would increase from 0.944 Btu/°F at the higher temperature to

$$\frac{628}{560} = 1.12 \text{ Btu/°F}$$

at the lower temperature, giving an increase in the difference in entropy of 0.18 Btu/°F, or a 19 percent increase in the difference in entropy. The unavailable energy for the ideal cycle would equal 528 Btu as previously calculated (see also Fig. 4-7). The real engine would also dissipate 100 Btu in overcoming friction, giving an actual amount of 628 Btu unavailable energy, or a 100 Btu increase over the ideal Carnot cycle. This would be 100/528 = 19 percent increase, just as for the increase in the entropy difference from high temperature to low temperature.

Another problem asked:

■ A Carnot engine utilizes a perfect gas as the medium. At the beginning of the isothermal expansion the pressure is 120 psia and the volume is 1 ft³. At the end of the adiabatic expansion the pressure is 20 psia and the volume is 3 ft³. What is the efficiency?

The temperatures are not given but may be determined from the relationship $PV = WRT$, since this is a perfect gas and the weight of the gas W will remain constant.

$$T_1 = \frac{P_1 V_1}{WR} = \frac{120 \times 144 \times 1}{WR} = \frac{17,280}{WR}$$

$$T_2 = \frac{20 \times 144 \times 3}{WR} = \frac{8640}{WR}$$

$$\text{Eff} = \frac{T_1 - T_2}{T_1} = \frac{17,280/WR - 8640/WR}{17,280/WR} = \frac{8640}{17,280} = 50 \text{ percent}$$

The cycle efficiency equals 50 percent.

Another type of refrigeration problem is as follows:

■ A mechanical refrigerator has a capacity of 200 tons and operates on a vapor-compression cycle. The refrigerant is R-12 (formerly known as Freon 12). The cooling coil is maintained at 10°F and the vapor leaves the compressor at 200°F and 180 psia. During compression 3 Btu/lb of refrigerant are transferred to the cylinder walls of the compressor. The condensed liquid enters the control valve at a temperature of 90°F. Calculate the: (a) coefficient of performance; (b) horsepower required; (c) pounds of refrigerant circulated per minute.

The refrigerator will operate on a modified Carnot cycle (see Fig. 4-8). From a *to* b the refrigerant throttles at constant enthalpy by expanding to a lower pressure. Part of the liquid flashes into vapor and the mixture cools to the evaporator temperature. In this case that is 10°F. From b to c the remaining cold liquid boils at constant pressure and temperature and ab-

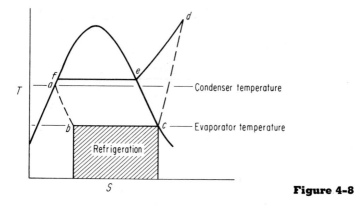

T

Condenser temperature

Evaporator temperature

Refrigeration

S

Figure 4-8

sorbs heat. From c to d is the compression portion of the cycle in which the gas which has boiled off is compressed to some point in the superheat region. For this case that is 200°F and 180 psia. The superheated vapor enters the condenser at these conditions and heat is discharged to the atmosphere. The refrigerant cools to the condensation temperature at point e. Heat is extracted at constant pressure and temperature from e to f at which point the refrigerant is a saturated liquid. The liquid is cooled further to point a, still at the pressure of 180 psia. The cooled liquid then enters the expansion valve.

First look up the properties of R-12 at the three state points: a, c, and d.

Point a (saturated liquid at 90°F):

$$\text{enthalpy } h_f = 28.713 \text{ Btu/lb}$$

Point c (saturated vapor at 10°F):

$$\text{Enthalpy } h = 78.335 \text{ Btu/lb}$$
$$\text{Entropy } s_g = 0.16798 \text{ Btu/(lb)(°F)}$$

Point d (superheated vapor 200°F, 180 psia):

$$\text{Enthalpy } h = 103.291 \text{ Btu/lb}$$
$$\text{Entropy } s = 0.18556 \text{ Btu/(lb)(°F)}$$

The compression of the vapor from point c to point d is not isentropic. This is apparent from the fact that the entropy of the vapor is higher at point d than it is at point c. In addition, it is stated that heat in the amount of 3 Btu/lb is transferred to the cylinder walls during the compression of the gas.

The work done by the compressor from point c to point d would equal the increase in the enthalpy of the fluid plus the heat lost during the compression. The engine must supply both the heat retained and the heat lost.

$$h_d - h_c = 103.291 - 78.335 = 24.956 \text{ Btu/lb}$$

To this must be added the heat lost, so the work done would equal 24.956 + 3 = 27.956 Btu/lb.

The heat extracted from the cold box by the refrigerant would equal the difference in enthalpies between points b and c. The expansion from a to b is a constant-enthalpy process so

$$h_a = h_b$$
$$\text{Heat extracted} = h_c - h_a = 78.335 - 28.713 = 49.622 \text{ Btu/lb}$$

The coefficient of performance equals the ratio of the heat extracted divided by the work done.

$$\text{c.o.p.} = \frac{\text{refrigeration}}{\text{net Work}} = \frac{49.622}{24.956} = 1.775 \qquad (a)$$

The ideal coefficient of performance of a refrigeration machine operating between a high temperature of T_1 and a low temperature of T_2 equals

$$\text{c.o.p.} = \frac{T_2}{T_1 - T_2}$$

For this case the ideal c.o.p. would equal

$$\frac{470}{660 - 470} = 2.774$$

For a refrigeration capacity of 200 tons the refrigeration system would have to extract heat in the amount of

$$200 \times 200 = 40,000 \text{ Btu/min}$$

A ton of refrigeration is that amount of refrigeration required to freeze one ton of water in 24 hr. The heat of fusion of water is 144 Btu/lb. So a ton of refrigeration equals

$$\frac{2000 \times 144}{24 \times 60} = 200 \text{ Btu/min}$$

One pound of refrigerant extracts 49.622 Btu; so the required rate of flow of refrigerant would equal

$$\frac{40,000}{49.622} = 806 \text{ lb/min}$$

The work required to compress one pound of refrigerant vapor equals 27.956 Btu/lb. The required input power would then equal $806 \times 27.956 = 22{,}532.5$ Btu/min.

One horsepower equals 33,000 ft-lb/min, or

$$\frac{33{,}000}{778} = 42.42 \text{ Btu/min}$$

So the power required to operate the refrigeration unit would equal

$$\frac{22{,}532.5}{42.42} = 531 \text{ hp} \qquad\qquad (b)$$

The horsepower per ton of refrigeration equals $4.71/\text{c.o.p.}$, which gives, for this case, a requirement of

$$\frac{4.71}{1.775} = 2.65 \text{ per ton, or a total of 531 horsepower}$$

Three other cycles touched upon in past examinations are the Ranking, Diesel, and Otto cycles.

4-10. RANKINE CYCLE

The Rankine cycle is a more practical cycle for a vapor-liquid system than is the Carnot cycle. It has a lower thermal efficiency than the Carnot cycle but a higher work ratio. This is the theoretical cycle for a steam power plant. We can discuss this cycle by reviewing a past examination problem:

- Draw a Rankine vapor cycle on the pressure-volume diagram shown in Fig. 4-9. Indicate the thermodynamic process involved in each of the following parts of the diagram:

  a. Expansion through turbine
  b. Condensation in condenser
  c. Pumping of condensate into boiler
  d. Change of liquid to saturated vapor in boiler

The Rankine vapor cycle has been sketched over the saturated-steam line in Fig. 4-9. The expansion through the turbine is represented by the line a, b, c. This is a reversible adiabatic expansion of superheated steam, denoted by a, into wet steam at point c. The transition point from dry to wet steam occurs at point b, where the cycle line crosses the saturated-steam line.

From c to d the wet steam condenses into water at a constant temperature and a constant pressure, reversibly.

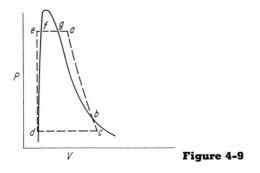

Figure 4-9

The condensate is then pumped into the boiler along a reversible adiabatic line *de*. This is practically a constant-volume process.

The liquid is then changed to superheated steam along the constant-pressure line *e*, *f*, *g*, *a*. From *e* to *f* the water is heated at a constant pressure to the boiling temperature; this is a reversible process. From *f* to *g* the water is vaporized into steam reversibly and isobarically. From *g* to *a* there is a reversible isobaric heating of the steam into superheated steam at a higher temperature.

4-11. DIESEL CYCLE

The Diesel cycle is the cycle which is the basis of the operation of the diesel engine. Another past examination problem was:

- Sketch an ideal Diesel cycle diagram (pressure vs. volume). Label the suction stroke, compression stroke, expansion stroke, and exhaust stroke.

The Diesel cycle is shown in Fig. 4-10. The intake stroke is shown by the line *ab*, a constant-pressure process. The air is next compressed

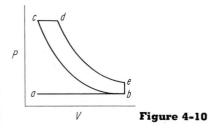

Figure 4-10

adiabatically from b to c. Heat is then added (fuel injected) at constant pressure from c to d, and the gas expands adiabatically from d to e. The exhaust valve opens at e and exhaust takes place from e to b and during the exhaust stroke, b, a. This is a four-cycle engine as shown, with suction stroke ab, compression bc, expansion c, d, e, and exhaust stroke e, b, a.

4-12. OTTO CYCLE

The Otto cycle, shown in Fig. 4-11, is the cycle on which the operation of the ordinary gasoline engine is based. The intake stroke is from a to b. The mixture is compressed adiabatically from b to c, at which point ignition takes place and heat is added, raising the temperature and pressure to d. An adiabatic expansion takes place from d to e, at which point the exhaust valve opens and combustion gases flow out. The path eb represents a constant-volume drop in temperature and an extraction of heat. The exhaust stroke is represented by the line ba.

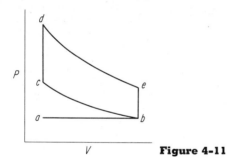

Figure 4-11

The cycle efficiency of an Otto cycle can be shown to equal $1 - (1/r^{k-1})$, where r is the compression ratio.

4-13. INDICATOR DIAGRAMS

Indicator cards are P-V diagrams obtained during engine operation. The area under a P-V diagram equals the work done during a cycle, so the net area (work is required to draw air into the cylinder during the intake stroke) is a measure of the work done on a piston during one cycle. The area of the P-V diagram is frequently divided by the length of the stroke to give average pressure; this is known as the "mean effective pressure" (MEP). The indicated horsepower is then:

$$\frac{\text{MEP(psf)} \times \text{volume(cu ft)} \times \text{power(strokes/min)}}{33,000 \text{ ft-lb/min hp}}$$

Sometimes the mean effective pressure is converted to the output of the engine, making allowances for frictional losses in the engine and power transmission system. It is then termed the "brake mean effective pressure," or the BMEP. This is usually calculated from dynamometer readings taken at the output shaft.

4-14. SPECIFIC HEAT AND LATENT HEAT

Problems concerning specific heat and latent heat merely require a balance of the heat added, lost, and retained. A good illustration is afforded by a past examination problem:

■ A tank holds 1000 lb of water at a temperature of 90°F. A piece of ice at 32°F was added and completely melted in the water, reducing its temperature to 40°F. Assuming no heat losses, find (a) how many pounds of ice were added; (b) how many pounds of steam at 212°F must be added to bring the tank contents to 140°F? Given data:
Latent heat of ice = 80 cal/g
Latent heat of steam at 212°F = 540 cal/g

First convert the values for the latent heats to Btu/lb. Since there are 252 cal/Btu and 454 g/lb, the latent heat of fusion of ice will be

$$\frac{80 \text{ cal}}{252 \text{ cal/Btu}} \times \frac{1}{\text{g}} \times \frac{454 \text{ g}}{\text{lb}} = 144 \text{ Btu/lb}$$

The latent heat of steam will be $(540/252) \times 454 = 973$ Btu/lb.

The ice was added at 32°F; it melted and rose to 40°F, so that each pound of ice absorbed $(144 + 8)$ Btu. The 1000 lb of water would give up $1000(90 - 40) = 50,000$ Btu in cooling. Total ice required = $50,000/152 = 329$ lb. The total amount of water at 40°F would then be 1329 lb. The heat required to raise its temperature to 140°F would equal $1329(140 - 40) = 132,900$ Btu. Each pound of steam would give up $973 + (212 - 140) = 1045$ Btu. The steam required would then be $132,900/1045 = 127$ lb.

4-15. PSYCHROMETRY

The principal point to remember here is that psychrometry is an application of Dalton's law of partial pressures and the Gibbs-Dalton law. As an illustration let us refer to a past examination problem:

- Atmospheric air has a dry-bulb temperature of 85°F, and a wet-bulb temperature of 64°F. The barometric pressure is 14.10 psia. Calculate (without the use of the psychrometric chart):
  a. Specific humidity in grains per pound of dry air
  b. Relative humidity
  c. Dew-point temperature
  d. Enthalpy of mixture in Btu per pound of dry air

As the air passes over the wet bulb, it absorbs moisture, thus lowering the temperature of the air and of the water, and increasing the amount of the moisture in the air. The final temperature of the air, when it has absorbed all the moisture it can hold, is the wet-bulb temperature. It is also the dew-point temperature of the air after it has absorbed the additional moisture, which is higher than the dew point of the air originally. This is also called the temperature of adiabatic saturation. The derivation of the relationship used to calculate the specific humidity is somewhat complex and will not be gone through here. The process can be described by means of a temperature-entropy diagram for steam, as shown in Fig. 4-12. Here the vapor in the air is initially at temperature T_c, and the vapor is in the superheat region, as denoted by point 1. If the air, and the vapor in the air, were cooled along the constant-pressure line 1-2, condensation would start to occur at T_a, where the constant-pressure line crosses the saturated-steam line. This is the dew point of the original air-vapor mixture. As the air absorbs moisture, the saturation temperature will rise, and the wet-bulb temperature will be between T_a and T_c at point 3 on the saturated-steam line. From this we can see that the lowest air temperature which could be obtained through evaporation of water, evaporative cooling, would be the wet-bulb temperature.

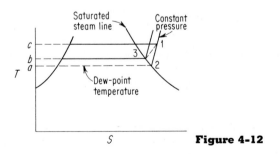

Figure 4-12

An approximate relationship derived by Carrier and most often used to find humidity is:

$$P_{wv} = P_{swb} - \frac{P - P_{swb}}{2{,}830 - 1.44t_{wb}}(t - t_{wb})$$

where P_{wv} = pressure of water vapor
P_{swb} = saturation pressure at wet-bulb temperature
P = atmospheric pressure
t_{wb} = wet-bulb temperature, °F
t = dry-bulb temperature, °F

For the example problem, $P_{swb} = 0.2951$ psia; $P = 14.10$ psia; saturation pressure at 85°F = 0.5959 psia.

$$P_{wv} = 0.2951 - \frac{14.10 - 0.2951}{2{,}830 - 1.44(64)}(85 - 64) = 0.1891 \text{ psia}$$

From Dalton's law of partial pressures we know that the pressure of the air will then be $14.10 - 0.189 = 13.91$ psia.

$$w_{\text{air}} = \frac{13.91 \times 144}{53.3 \times 545} = 0.0692 \text{ lb/ft}^3$$

$$w_{\text{water vapor}} = \frac{0.1891 \times 144}{85.8 \times 545} = 0.000584 \text{ lb/ft}^3$$

Since there are 7000 grains/lb, a cubic foot of water vapor would weigh 4.09 grains. One pound of dry air would occupy

$$\frac{1}{0.0692} = 14.45 \text{ ft}^3$$

so there would be $4.09 \times 14.45 = 59.1$ grains of moisture per pound of dry air.

The relative humidity will equal the ratio of the actual water vapor pressure to the water vapor pressure at saturation.

$$\text{Relative humidity} = \frac{0.1891}{0.5959} = 31.7 \text{ percent}$$

The dew-point temperature would be the saturation temperature for the amount of moisture in the air; following down the constant-pressure line in Fig. 4-12 from point 1 to point 2 gives us that the dew point (saturation temperature for $p = 0.1891$ psia) equals 51.8°F.

The enthalpy of the mixture will equal the sum of the enthalpies of the gases making up the mixture. The steam at 85°F is superheated steam which has a saturation temperature equal to the dew point. The enthalpy of saturated steam at 51.8°F is 1084.4 Btu/lb. The steam is then superheated 33.2°F. The specific heat of steam at low temperature and pressure may be taken as 0.446 Btu/(lb)(°F), giving an enthalpy for the vapor of

$$1084.4 + 33.2 \times 0.446 = 1099.2 \text{ Btu/lb}$$

The enthalpy of saturated steam at 85°F is 1098.8 Btu/lb, which would have served just as well for this purpose.

The enthalpy of the air at low pressures is $c_p(t - t_0)$, and since the point of zero enthalpy for air is 0°F, the enthalpy of the air is $0.240(85) = 20.4$ Btu/lb.

Since we have $59.1/7000 = 0.00845$ lb of steam per pound of air, the total enthalpy per pound of dry air is:

$$20.4 + 0.00845 \times 1099.2 = 29.68 \text{ Btu/lb dry air}$$

4-16. HEAT TRANSFER

The subject of heat transfer has also received some attention in past examinations. The emphasis has been primarily on conduction. If we look at Fig. 4-13 we see that as heat flows from t_1 to t_2 it is hindered first by the film factor on the outside of the first layer of the wall, then by the first layer, the second layer, the third layer, and lastly by the film factor on the inside of the inner layer. These may all be combined to give one equivalent heat-transfer factor for the assembly. The relationship for the combined heat-transfer factor is:

$$\frac{1}{U} = \frac{1}{h_1} + \frac{L_1}{k_1} + \frac{L_2}{k_2} + \frac{L_3}{k_3} + \frac{1}{h_2}$$

where U = overall heat-transfer coefficient
h_1 = first film factor
k_1 = thermal conductivity of the material of first layer
L_1 = thickness of first layer
k_2 = thermal conductivity of second layer
L_2 = thickness of second layer
k_3 = thermal conductivity of third layer
L_3 = thickness of third layer
h_2 = film factor at the other face of the wall

Since k is ordinarily given in Btu/(hr)(ft²)(°F)/(ft), the units of U will be Btu/(hr)(ft²)(°F), which must be multiplied by area (ft²) times temperature differential (°F). The rate of heat transfer through the wall is then:

$$Q = UA(t_1 - t_2) \qquad \text{Btu/hr}$$

This relationship can be easily derived with the aid of Fig. 4-13. Here the flow of the heat is from t_1 to t_2. The temperature drop through the first film is from t_1 to t_a. The temperature drops from t_a to t_b through layer A,

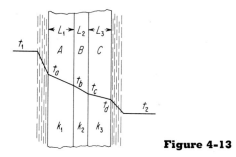

Figure 4-13

from t_b to t_c through layer B, and so on. Since the system has reached a steady state, the same amount of heat will flow through each film and through each layer. We shall call the rate of heat flow Q Btu/(hr)(ft²); that is, Q Btu passes through each square foot of the wall per hour. We can set up relationships for the rate of heat flow through each layer of the resistance.

$$Q = (t_1 - t_a)h_1 \qquad \text{giving } t_1 - t_a = Q/h_1$$
$$Q = (t_a - t_b)k_1/L_1 \qquad \text{giving } t_a - t_b = QL_1/k_1$$
$$Q = (t_b - t_c)k_2/L_2 \qquad \text{giving } t_b - t_c = QL_2/k_2$$
$$Q = (t_c - t_d)k_3/L_3 \qquad \text{giving } t_c - t_d = QL_3/k_3$$
$$Q = (t_d - t_2)h_1 \qquad \text{giving } t_d - t_2 = Q/h_2$$

Adding together the relationships on the right we obtain

$$t_1 - t_2 = Q\left(\frac{1}{h_1} + \frac{L_1}{k_1} + \frac{L_2}{k_2} + \frac{L_3}{k_3} + \frac{1}{h_2}\right)$$

which reduces to $Q = U(t_1 - t_2)$ per square foot or

$$Q \text{ Btu/hr} = UA(t_1 - t_2)$$

As an illustration, let us look at a past examination problem which asked:

■ Given the air temperature in a room is 75°F and outside temperature 30°F. The inside temperature of the window pane is 45°F and the outside temperature 40°F. The window pane is ³/₁₆ in. thick and has a conductivity $k = 1.0$ Btu/(hr)(°F)(sq ft surface)(ft thick). The resistances to heat flow consist of the air film on the inside surface, the outside surface air film, and the window pane itself. Calculate the flow of heat per square foot of window pane, and the film coefficients for the inside and outside surfaces of the pane.

This problem must be calculated in steps, and one must remember that the heat which flows through the inner film equals the heat which flows through the pane and the heat which flows through the outer film. This gives us three equations. A sketch of the temperature variation is shown in Fig. 4-14.

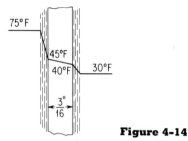

Figure 4-14

The heat which flows out is Q. The heat-transfer coefficient through this film equals h_1, which gives us the relationship

$$Q = h_1(75 - 45) \text{ Btu/hr}$$

for 1 sq ft area.

This same amount of heat must flow through the pane, and the inverse of the coefficient of heat transfer through the pane will equal

$$\frac{1}{U} = \frac{L}{k} = \frac{^3/_{16} \times ^1/_{12}}{1.0} = 0.01563$$

giving $U = 64$ Btu/(hr)(ft²)(°F).

$$Q = U(t_1 - t_2) = 64(45 - 40) = 320 \text{ Btu/(hr)(ft²)}$$

Similarly, the flow through the outer film will be $Q = 10h_2$. This gives us:

$$Q = 320 \text{ Btu/hr} \qquad \text{for each sq ft of window pane}$$
$$Q = 30h_1 \qquad h_1 = 10.7 \text{ Btu/(hr)(ft²)(°F)}$$
$$Q = 10h_2 \qquad h_2 = 32 \text{ Btu/(hr)(ft²)(°F)}$$

Flow of heat through pipe insulation is a bit different because the area of the insulation and of heat flow increases as the radius increases. This gives rise to a different relationship for the heat transfer, namely:

$$Q = \frac{2\pi r_m l(t_1 - t_2)}{r_m/(r_o h_o) + (r_o - r_i)/k_m + r_m/(r_i h_i)} \qquad \text{Btu/hr}$$

where r_m = the logarithmic mean radius = $(r_o - r_i)/(\ln r_o/r_i)$
  r_i = inner radius (Fig. 4-15)
  r_o = outer radius
  l = length of pipe
  k_m = mean thermal conductivity

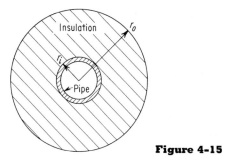

Figure 4-15

As can be seen, this equation ignores the insulating effect of the metal pipe which can safely be disregarded in most cases.

As an example, let us look at a past examination problem:

■ A $1^{1}/_{2}$-in. pipe is lagged with 85 percent magnesia insulation $1^{1}/_{4}$ in. thick. The pipe temperature is 230°F, and the outside temperature of the lagging is 85°F.
  a. Calculate the heat loss per hour for 200 ft of pipe
  b. What is the average area of heat transfer per foot of pipe?
  c. Calculate the coefficient of heat transfer based on the (1) inside area, (2) outside area, (3) average area.

Consideration of the film coefficients has been eliminated through the specification of the temperatures at the inner and outer edges of the insulation.

If we assume that the outer diameter (OD) of the pipe is $1^{1}/_{2}$ in., then $r_i = 0.75$ in. and $r_o = 2.00$ in.

$$r_m = \frac{1.25}{\ln 2.67} = 1.27$$

$$Q = \frac{2\pi \times (1.27/12) \times 200(230 - 85)}{1.25/(12 \times 0.041)} = 7580 \text{ Btu/hr}$$

The average area would be $2\pi(r_o + r_i)/2 \times {}^{1}/_{12} = 0.72$ ft²/ft. Similarly the inside area would be 0.393 ft²/ft, and the outside area would be 1.05 ft²/ft.

The coefficient of heat transfer based on the inside area would be 7580/(200 × 0.393) = 96.5 Btu/(hr)(ft²); on the outside area, 36.1 Btu/(ft²)(hr); and on the average area, 52.7 Btu/(ft²)(hr).

SAMPLE PROBLEMS

4- 1 An engine operates on the Rankine cycle with complete expansion (Fig. 4-P-1). Steam, the working medium, is admitted to the engine at a pressure of 200 psia and a temperature of 600°F; it is exhausted at a pressure of 4 in. Hg.

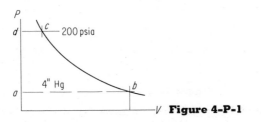

Figure 4-P-1

a. Determine the ideal thermal efficiency of this cycle.

b. Determine the temperature of the exhaust steam leaving the engine.

c. Determine the theoretical horsepower which could be developed if the cycle consumes 1 lb of steam per second.

d. Specify the volume in cubic feet per hour of the exhaust steam leaving the engine.

e. Sketch the T-S diagram.

4- 2 Five pounds of a gas mixture is 30 percent CO_2 and 70 percent O_2 by weight. It is compressed from (1) a pressure of 15 psia to (2) a pressure of 150 psia by a process characterized by the equation $P_1V_1{}^n = P_2V_2{}^n$, where the value of n is 1.0. The temperature of the mixture is 60°F at the beginning of the process. Determine:

a. Volume V_1 of the mixture at the beginning of compression

b. Temperature t_2 of the mixture at the end of compression in°F

c. Work W in foot-pounds required to compress the 5 lb of mixture

d. Change in internal energy ΔE

The equation for work $W = \int_1^2 P\ dV$, and internal energy $E = c_v T$.

4- 3 Saturated steam is used in a mixing tank to heat and mix 200 gal of solution. The solution is to be heated from 62 to 150°F by introducing steam directly into the solution. The specific heat of the solution is about 0.85 Btu/(lb)(°F) at 62°F and remains essentially constant over the temperature range stated. The specific weight of the solution is 8.40 lb/gal.

How many pounds of saturated steam at a gauge pressure of 20 psi will be required for heating if all the steam is condensed in the solution and the heat loss to the surroundings is assumed to be negligible? (See Fig. 4-P-3.)

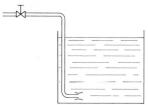

Figure 4-P-3

		Specific volume ft³/lb		Enthalpy or heat content, Btu/lb		Entropy, Btu/(°F)(lb)	
Absolute pressure, psi	Temperature, °F	Saturated liquid	Saturated vapor	Saturated liquid	Saturated vapor	Saturated liquid	Saturated vapor
5.3	165	0.0164	69.0	132	1131	0.238	1.839
14.7	212	0.0167	26.8	180	1150	0.312	1.756
20.0	228	0.0168	21.0	196	1156	0.335	1.732
34.7	259	0.0171	11.9	227	1167	0.381	1.687

4- 4 The specific heat of a gas at constant pressure is 0.2025 Btu/(lb)(°F) and at constant volume is 0.1575 Btu/(lb)(°F). What is the final volume of 10.0 ft³ of gas at an initial pressure of 25 in. Hg and a final pressure of 5 atm? Assume the process to be adiabatic. (Mercury weighs 0.49 lb/in.³)

4- 5 A system consisting of 1 lb of water initially at 70°F is heated at a constant pressure of 30 psia while confined inside a cylinder equipped with a piston. If the addition of 1200 Btu of heat raises the temperature to 400°F, what is the change in the internal energy of the system?

4- 6 A masonry wall has a 4-in.-thick brick facing wall bonded to an 8-in.-thick concrete backing. On a day when the room temperature is 68°F and the outside temperature is 12°F, the inner-surface temperature of the concrete is 57°F and the outer-surface temperature of the brick is 19°F. The thermal conductivity of the brick is 0.36 and of the concrete 0.68 Btu/(hr)(ft)(°F). Determine:

a. Overall heat-transfer coefficient
b. Convection coefficients of the vertical concrete and brick walls

4- 7 A small steam generating plant burns oil as a fuel. The oil has a specific gravity of 1.008 and costs $2.50 per barrel. The heating value of the oil is 18,250 Btu/lb. The boiler operates at 75 percent thermal efficiency when taking in feedwater at 180°F and delivering dry saturated steam at 100 psig. What is the fuel cost per 1000 lb of steam on this basis? 1 bbl = 42 gal.

Absolute pressure, psi	Temperature, °F	Specific volume, ft³/lb		Enthalpy or heat content, Btu/lb		Entropy, Btu/(lb)(°F)	
		Saturated liquid	Saturated vapor	Saturated liquid	Saturated vapor	Saturated liquid	Saturated vapor
14.7	212.0	0.0167	26.82	180.0	1150.2	0.312	1.756
85.3	316.5	0.01758	5.150	286.5	1183.6	0.459	1.615
100.0	327.8	0.01771	4.426	298.3	1186.6	0.474	1.602
114.7	337.8	0.01782	3.889	308.8	1189.0	0.487	1.591

4- 8   The internal energy of a gas is given by $u = 0.08T + 0.002T^2$. If 1 lb of this gas expands without heat flow but with a temperature drop from 300 to 100°F, how much work is done?

4- 9   The volumetric analysis for a gas mixture shows that it consists of 70 percent nitrogen, 20 percent carbon dioxide, and 10 percent carbon monoxide. If the pressure and temperature for the mixture are 20 psia and 100°F, respectively, compute:

   a. Partial pressures
   b. Weight analysis

4-10   The relation between mass and energy as derived by Einstein is $E = MC^2$, where E = energy, M = mass, and C = speed of light. If 1 lb of uranium is caused to fission and there is a loss of one-tenth of 1 percent of the mass in the process, how many kilowatthours of energy are liberated?

4-11   A steam turbine receives 3600 lb of steam per hour at 110-ft/sec velocity and 1525 Btu/lb. The steam leaves at 810 ft/sec and 1300 Btu/lb. What is the horsepower output?

4-12   Atmosphere air has a dry-bulb temperature of 85°F and a wet-bulb temperature of 64°F. The barometric pressure is 14.0 psia. Determine:

   a. Specific humidity in grains per pound of dry air
   b. Relative humidity
   c. Dew-point temperature

4-13   Air in the cylinder of a diesel engine is at 86°F and 20 psi in compression. If it is further compressed adiabatically to one-eighteenth of its original volume, find:

   a. Final temperature and pressure
   b. Work done in compression if the displacement volume of the cylinder is ½ ft³

   Assume $\gamma = 1.4$ for air.

4-14   An indicator card of a four-cycle automobile engine taken with a 420-lb spring is 3 in. long and has an area of 1.25 in.² If the engine has 8 cylinders, 3.6-in. diameter by 3-in. stroke, and travels at 2800 rpm, what is the developed horsepower?

4-15   A wall of 0.8-ft thickness is to be constructed from material which has an average thermal conductivity of 0.75 Btu/(hr)(ft)(°F). The wall is to be insulated with material having an average thermal conductivity of 0.2 Btu/(hr)(ft)(°F) so that the heat loss per square foot will not exceed 580 Btu/hr. Calculate the thickness of insulation required, if the surface temperatures of the walls are 2400 and 80°F, respectively.

4-16   There was 3 in. of ice on a pond when the atmospheric temperature was 2°F. Calculate the rate in inches per hour at which the ice was increasing in thickness.

Heat of fusion of ice = 144 Btu/lb
Thermal conductivity of ice = 12 Btu [in./(hr)(ft^2)(°F)]

4-17   A copper bar 15 cm long and of 6.0 cm^2 cross section is placed with one end in a steam bath and the other in a mixture of ice and water at atmospheric pressure. The sides of the bar are thermally insulated. How much ice melts in 2 min? How much steam condenses in this time? The thermal conductivity of copper is 0.90 cal/(cm)(sec)(°C).

4-18   A steam turbine receives 3600 lb of steam per hour at 1100 ft/sec velocity and 1525 Btu/lb enthalpy. The steam leaves at 810 ft/sec and 1300 Btu/lb. Determine horsepower output.

4-19   One pound of nitrogen with initial volume of 6 ft^3 and pressure of 300 psia expands according to the law PV^n = constant, to a final volume of 30 ft^3 and a pressure of 40 psia. Determine:

a. Value of n
b. Work done by gas in foot-pounds
c. Heat received or rejected in Btu's

4-20   Three moles of nitrogen, γ = 1.4 C_v = 4.6 (g-cal/mole), are at atmospheric pressure and 20°C. The gas is then heated at constant volume to 40°C. It then undergoes adiabatic expansion and returns to 20°C. Finally it is compressed isothermally to its original state.

a. Draw the P-V diagram for this cycle, indicating the successive states $P_1V_1T_1$, $P_2V_2T_2$, $P_3V_3T_3$.
b. Find P, V, and T of each state in terms of $P_1V_1T_1$.
c. Find the joules of work done in adiabatic expansion.
d. What is the value of the initial volume V_1?

4-21   An automobile cooling system of 26 qt of water and antifreeze has a freezing point of −12°F. How much of the above mixture must be drained out in order to add sufficient antifreeze to protect the engine at −15°F? The table printed on the antifreeze can indicates that 2^1/$_2$ qt will protect a 13-qt cooling system to −12°F, and 4^1/$_2$ qt to −15°F.

4-22   A steam turbine operating with steady-flow conditions produces 1350 hp. All losses except exhaust amount to 144,000 Btu/hr. The steam flow is 150 lb/min. Inlet condition is 450 psia and 740°F, and exhaust occurs at a pressure of 0.8 psia. Determine the condition of the exhaust steam (quality if wet, or temperature if superheated).

4-23 A gas has a constant-pressure specific heat of $C_p = 6.6 - 7.2 \times 10^{-4}T$ [Btu/(mole)(°R)]. Determine the entropy of this gas at 2000°R and 5 atm pressure if the base of zero entropy is taken as 520°R and 1 atm.

4-24 A balloon weighs 425 lb and has a volume of 35,000 cu ft when filled with hydrogen gas weighing 0.0056 lb/ft³. What load in pounds will the balloon support in air weighing 0.081 lb/ft³?

4-25 A gas company buys gas at 90 psig and 75°F and sells it at 3.80 in. of water pressure and 28°F. Disregard the losses in distribution and determine how many cubic feet are sold for each cubic foot purchased.

4-26 An indicator card of a four-cycle automobile engine taken with a 425-lb spring is 2.80 in. long and has an area of 1.20 in.² If the engine has six cylinders, 3.5 in. in diameter, and the stroke is 3 in. traveling at 3000 rpm, determine the developed horsepower of this engine.

4-27 The Btu's required to heat 1 lb of a certain substance is given by the equation $Q = 0.32t - 0.00008t^2$, where Q is expressed in Btu/lb and t is in Fahrenheit. What is the instantaneous specific heat of the substance at 150°F?

4-28 *a.* How much heat is supplied and rejected in Btu/hr if a 100-hp heat engine operates on the ideal Carnot cycle between the temperatures of 540 and 140°F?
  b. What is the thermal efficiency?

4-29 An air-speed indicator of a plane is operated on the same principle as the pitot tube. It is calibrated to read knots. If the indicator reads 200 knots at an altitude where the density of air is one-half the density at which it was calibrated, what is the true speed of the plane?

4-30 A barometer tube 88 cm long contains some air above the mercury. It gives a reading of 68 cm when upright. When the tube is tilted to an angle of 45°, the length of the mercury column becomes 80 cm. What is the reading of an accurate barometer?

4-31 One mole of a monatomic ideal gas at 300° abs is subjected to three consecutive changes: (1) the gas is heated at constant volume until its temperature is 900° abs; (2) the gas is then allowed to expand isothermally until its pressure drops to its initial value; (3) the gas is then cooled at constant pressure until it returns to its original state.
  a. Sketch a curve in the *P-V* diagram showing the above changes.
  b. Calculate the total work done by the gas in the above changes, the total amount of heat transferred to or from the gas, and the total change in internal energy of the gas. (Express your answer in calories.)

$$C_v = 3 \text{ cal/(mole)(K)} \qquad R = 2 \text{ cal/(mole)(K)}$$

4-32 Find the physical state of a definite quantity of water in a vessel of known size.
  a. 10 lb of water is injected into an evacuated vessel which is maintained at 350°F. The volume of the vessel is 10 ft³. What will be the resulting pressure in the vessel? What percentage of the water will be evaporated?
  b. Same as (*a*) except that only 1 lb of water is injected.

4-33 Steam expands behind a piston doing 50,000 ft-lb of work. If 12 Btu of heat is radiated to the surrounding atmosphere during the expansion, what is the change of internal energy?

4-34 Two pounds of dry and saturated steam is confined in a tank. A gauge on the tank indicates a pressure of 145 psi. Barometric pressure is 29 in. Hg. What is the temperature of the steam in the tank? Calculate its volume, enthalpy, entropy, and internal energy.

4-35 Fourteen hundred cubic feet of air at 85°F dry-bulb temperature and 50 percent relative humidity is mixed with 200 cfm of air at 54°F dry-bulb temperature and 40 percent relative humidity. On the basis of psychrometric chart values, find the resulting dry-bulb temperature, wet-bulb temperature, and relative humidity.

4-36 A boiler generates 3000 lb of steam per hour at a pressure of 150 psig, steam temperature 465°F, feedwater temperature 180°F; what is the total equivalent evaporation from and at 212°F/hr?

4-37 Tests of a six-cylinder 4-in. bore and 3¹/₈-in. stroke aircraft engine at full throttle show a brake horsepower of 79.5 at 3400 rpm. The compression ratio is 8:1, the specific fuel consumption is 0.56 lb/bhp-hr. The higher heating value of the fuel is 19,800 Btu/lb. Determine (1) brake mean effective pressure; (2) brake torque; (3) brake thermal efficiency; (4) brake engine efficiency based on cold air standard if the ideal efficiency e is $e = 1 - (1/r_k^{k-1})$, where r_k = compression ratio.

4-38 Find the rated brake mean effective pressure of a 225-hp nine-cylinder diesel aircraft engine. It is a four-stroke cycle engine, rated at 1950 rpm. Cylinder bore is 4¹³/₁₆ in.; stroke is 6 in.

4-39 Air is compressed in a diesel engine from an initial pressure of 13 psia and a temperature of 120°F to one-twelfth its initial volume. Calculate the final temperature and pressure, assuming the compression to be adiabatic.

4-40 A centrifugal compressor is to operate at 3600 rpm and handle 20,000 cfm of free air. Compression is according to the law $PV^{1.35}$ = constant. Intake is atmospheric (14.4 psia at 70°F) and compression is to 25 psig.

a. What is the final delivery in cubic feet per minute?

b. Would final temperature be lower or higher if compression were (1) isothermal? (2) adiabatic?

4-41 A 100-hp engine is being tested by loading it with a water-cooled Prony brake. When the engine delivers the full-rated 100 hp to the shaft and the Prony brake being cooled with tap water absorbs and transfers to the cooling water 95 percent of the 100 hp, determine:

a. How many Btu's of heat are transferred to the cooling water per second?

b. If the tap water is entering the Prony-brake shell at 65°F and leaves at 131°F, at what rate is tap water passing through the Prony-brake shell in cubic feet per second?

4-42 An open-cycle gas turbine receives air at 15 psia and 60°F. The compressor discharges at 75 psia. The temperature entering the turbine is 1250°F. Calculate:

a. Heat added in the combuster, Btu/lb

b. Efficiency of the cycle, percent

4-43   The internal energy of a gas is given by

$$u = 0.08t + 0.002t^2 \qquad (t = °F)$$

If 1 lb of this gas expands without heat flow but with a temperature drop from 300 to 100°F, how much work is done?

4-44   A mixture of gases at 100°F consists by weight of 80 percent nitrogen, 12 percent carbon dioxide, 7 percent oxygen, and 1 percent water vapor. Total pressure is 14,696 psia.

a. What is the dew point?

b. How much heat would be removed in reducing the temperature of 1 lb of mixture to the dew point at constant total pressure?

4-45   Five cubic feet of a gas with $C_p = 0.55$, $R = 96.2$, expands polytropically from 140 psia and 120°F to a pressure of 50 psia and a volume of 10 cu ft. Find:

a. t_2

b. Change in internal energy

c. Work done

d. Change in entropy

e. Heat added or given off

State in each answer, where it applies, the direction in which the change took place.

4-46   An Otto cycle engine has a compression ratio of 6:1. On test it uses 1 gal of gasoline in 15 min while developing 150 lb-ft torque at 1500 rpm. The gasoline has a specific gravity of 0.70 and a higher heating value of 19,100 Btu/lb.

a. What is the thermal efficiency of the engine?

b. What is the ratio of the thermal efficiency to the air-standard ideal thermal efficiency?

c. What is the specific fuel consumption in pounds per horsepower-hour?

4-47   A gas made up of 5 lb methane and 10 lb ethane is placed in a tank at 60°F. If the volume of the tank is 19 cu ft,

a. What is the resulting pressure in the tank?

b. What is the mole percent of the resulting mixture?

4-48   A throttling steam calorimeter has a pressure of 150 psia on the high-pressure side of its orifice and atmospheric pressure on the discharge side. The thermometer on the discharge side reads 280°F. Calculate the quality of the steam in the line.

4-49   Air is compressed in a diesel engine from an initial pressure of 13 psia and a temperature of 120°F to one-twelfth its initial volume. Calculate the final temperature and pressure, assuming the compression to be adiabatic.

4-50   Determine the heat-transfer coefficient of a wall consisting of 13 in. of brick and 1 in. of plaster. The specific conductivity for brick work and for plaster is 5.00. The surface resistances of brick and plaster are 4.02 and 1.40, respectively.

4-51   How many pounds of water can be evaporated from and at 212°F by the heat involved in the complete combustion of 1 lb of coal containing carbon, 65.2 percent; hydrogen, 4.92 percent; and oxygen, 8.6 percent?

4-52 The atmospheric air has 87°F dry-bulb temperature and 65°F wet-bulb temperature. Calculate:

a. Specific humidity in grains per pound of dry air
b. Dew-point temperature
c. Relative humidity

Barometric pressure is 14.7 psia.

4-53 Given a 500-kW generating set; efficiency of the engine and generator, 85 percent. Steam pressure, 150 psia; feedwater temperature, 180°F. The engine uses 20 lb of steam per indicated horsepower per hour. Evaporation from and at 212°F is 10 lb of water per pound of dry coal. Coal contains 13,000 Btu/lb. What is the heat efficiency of the plant?

4-54 A steel tube 1.315 in. OD and 1.049 in. ID transmits 38,000 Btu/(hr)(ft^2) of outside-wall area to the surroundings. Calculate the outside-wall temperature if the inner wall is at 212°F. ($k = 19.1$ Btu/(hr)(°F)(ft) for tube metal.)

4-55 A jet-propelled plane has a fuel flow of 0.9 lb/sec with a heating value of 20,000 Btu/lb. Air inlet flow is 40 lb/sec. The plane is flying at a speed of 460 mph, and an absolute jet velocity of 2100 ft/sec occurs. Compute:

a. Propulsive force
b. Useful horsepower
c. Overall thermal efficiency

4-50 Water at 200 psia and 300°F is introduced into an insulated chamber maintained at 50 psia. What percentage of the water so injected flashes into steam?

4-57 A tank contains air at 800 psia and 80°F. An amount of air measuring 700 ft^3 at 14.7 psia and 60°F is removed. The pressure in the tank is now found to be 300 psia when the temperature is 80°F. What is the volume of the tank?

4-58 Steam is supplied to a nozzle at 100 psia and 400°F. The nozzle exhaust pressure is 20 psia.

a. What is the ideal exit velocity?
b. What would the ideal exit velocity have been if air were used under the same conditions?

4-59 A furnace fired with a hydrocarbon fuel oil has a dry-stack analysis of CO_2, 12 percent; O_2, 7 percent; N_2, 81 percent. Calculate:

a. Composition of original fuel oil expressed as weight percent
b. Percentage of excess air
c. Cubic feet of air at standard conditions per pound of fuel

4-60 A hollow steel cylinder having an internal volume of 4 ft^3 contains air at the atmospheric pressure of 15 psi. How many cubic feet of the outside air must be pumped into the cylinder to raise the pressure to 45 psi? Assume that the temperature remains constant.

4-61 In an air cycle, process 1 to 2 is constant volume, 2 to 3 is adiabatic isentropic, and 3 to 1 is constant pressure. V_1 is 1 cu ft, V_3 is 2 ft^3, p_1 and p_3 are each 1 atm, and t_1 is 60°F (Fig. 4-P-61). Find each of the following:

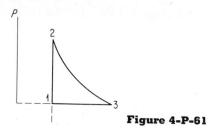

Figure 4-P-61

a. Heat supplied, Btu
b. Heat rejected, Btu
c. Efficiency of cycle, percent
d. Corresponding Carnot efficiency, percent

4-62 At 1, Fig. 4-P-62, water is supplied at the rate of 500 lb/min at 3 psig and 60°F. At 2, steam is supplied at 3 psig and 95 percent dry. At 3, the mixture leaves at 212°F as saturated liquid. Assuming no heat loss from the chamber, answer each of the following:

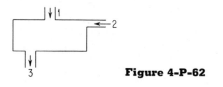

Figure 4-P-62

a. How many pounds of steam per minute are required at 2?
b. Find the entropy change across the system.

4-63 An ammonia refrigerator plant is to operate between a saturated liquid at 120 psia at the condenser outlet and a saturated vapor at 15 psia at the evaporator outlet. If 30 tons' capacity is desired, compute on the basis of an ideal cycle only:

a. Coefficient of performance
b. Work of compression Btu/lb
c. Refrigerating effect, Btu/lb
d. Pounds per minute of ammonia required for circulation
e. Ideal horsepower per ton of refrigeration
f. Pounds of condenser cooling water per minute, assuming a 15°F temperature rise

4-64 An ideally insulated tank of helium, pressure 3000 psig, temperature 60°F, is connected to an empty weather balloon through a valve. The system is mounted on a weighing platform. The whole has a tare of 200 lb, including valve and weather balloon. The weather balloon is inflated to 4 ft in diameter, pressure 16 psia, average temperature 60°F in air at 14.7 psia, 60°F. If the tank pressure at the end of inflation is 200 psig, find the volume of the tank.

MULTIPLE-CHOICE PROBLEMS

For each question select the correct answer from the five given possibilities.

4M- 1 In an office building containing 100 employees, the design allowed for 10 cfm per person of outside air for ventilation purposes. If the supply fan delivers 6000 cfm, what is the temperature of the air entering the cooling coil if the outside air is 105°F and the recirculated air is 80°F?
(a) 75.3°F   (d) 93.7°F
(b) 80.6°F   (e) 95.0°F
(c) 84.2°F

4M- 2 In order to determine the relative humidity of a sample of air, the easiest way to proceed would be as follows:
(a) Measure the wet-bulb and dry-bulb temperatures and use a psychrometric chart
(b) Measure the wet-bulb and dry-bulb temperatures and divide the smaller value by the larger
(c) Measure the wet-bulb and dry-bulb temperatures and use a Mollier chart
(d) Measure the barometric pressure and dry-bulb temperature and divide the smaller value by the larger
(e) Measure the barometric pressure and use a psychometric chart

4M- 3 A steam turbine uses steam having an initial pressure of 140 psia and an enthalpy of 1192 Btu/lb. The turbine exhausts steam at 23 psia and 1158 Btu/lb enthalpy. The energy used by the turbine is:
(a) 2350 Btu/lb   (d) 2342 Btu/lb
(b) 34 Btu/lb   (e) none of these
(c) 42 Btu/lb

4M- 4 If an ideal gas is compressed from a lower pressure to a higher pressure at constant temperature, which of the following is true?
(a) The work required will be zero.
(b) The volume remains constant.
(c) The volume will vary inversely as the absolute pressure.
(d) The volume will vary directly as the gage pressure.
(e) Heat is being absorbed by the ideal gas during the compression.

4M- 5 An adiabatic throttling process is one in which:
(a) the entropy is constant
(b) the enthalpy is constant
(c) the specific volume is constant
(d) the available energy is constant
(e) none of these

4M- 6 An adiabatic process is one in which:
  (a) the pressure is constant
  (b) internal energy is constant
  (c) no work is done
  (d) no heat is transferred
  (e) friction is not considered

4M- 7 How much heat is required to raise 10 grams of water from 0 to 1°C?
  (a) 10 Btu   (d) 5 joules
  (b) 1 Btu   (e) 10 calories
  (c) 1 calorie

4M- 8 If a gas is cooled at constant volume, which of the following will be true?
  (a) The pressure will remain constant.
  (b) The absolute pressure will increase.
  (c) The gage pressure will decrease by the same ratio as the temperature decrease in °F.
  (d) The absolute pressure will decrease by the same ratio as the absolute temperature decrease.
  (e) Cooling the gas has no effect on the pressure.

4M- 9 In the ordinary commercial steam turbine, the change in enthalpy of the steam as it passes through the reaction section:
  (a) occurs almost entirely in the nozzles of the turbine
  (b) occurs almost entirely in the turbine blading
  (c) is due to the condensation of the steam
  (d) is negligible
  (e) occurs across the throttling valve of the turbine

4M-10 The maximum thermal efficiency that can be obtained in an ideal reversible heat engine operating between 1540°F and 340°F is closest to:
  (a) 100 percent   (d) 40 percent
  (b) 60 percent   (e) 22 percent
  (c) 78 percent

5
MECHANICS OF
MATERIALS

The subject of mechanics of materials is an expansion of the more general subject of mechanics; it includes study of the effects of the elastic properties of structural materials. All materials deform when subjected to a stress. Fortunately, most of the important engineering materials exhibit unit deformations, or strains, which are very nearly proportional to the applied stresses within the working limits of the materials. This fact, expressed in the statement of Hooke's law that stress and strain are proportional, is the basis of the study of mechanics of materials.

The various phases of this subject which have been treated in past fundamentals examinations have included general considerations of stress and strain, effects of temperature on stress and strain, stresses resulting from torque loading, loaded beams, composite beams (principally reinforced concrete), shear and moment diagrams, riveted and welded joints, columns, and some consideration of combined stresses.

5-1. MODULUS OF ELASTICITY

The ratio of stress (in pounds per square inch) to strain (in inches per inch), called "Young's modulus" or the "modulus of elasticity," is constant for a given material up to the proportional limit. The modulus of elasticity, the proportional limit, the yield point, etc., are all determined experimentally for a given material. Curves illustrating these properties for some of the common engineering materials are shown in Fig. 5-1.

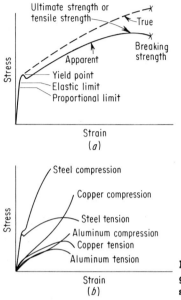

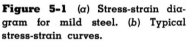

Figure 5-1 (a) Stress-strain diagram for mild steel. (b) Typical stress-strain curves.

5-2. CREEP STRENGTH

When a metal deforms slowly under the long-time application of a stress which is well below its yield strength, the deformation is known as "creep." Creep is particularly important in high-temperature applications, as in turbine blades where the temperatures are high and the stresses are also high. A typical curve showing deformation of a material vs. time under load is given in Fig. 5-2.

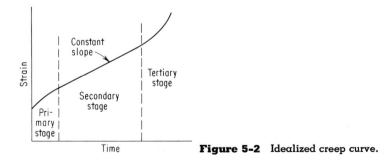

Figure 5-2 Idealized creep curve.

5-3. FATIGUE AND ENDURANCE LIMIT

Fatigue is the term used to describe the failure or rupture of a metal part under repeated application of a load which is well below the permissible load as calculated from ordinary static stress considerations. This is of particular importance in the design of parts which are to be subjected to cyclic loading. Figure 5-3 shows some curves of stress vs. number of cycles to cause failure for a few engineering materials. For most steels the endurance limit, or fatigue stress, is that stress which will withstand 10 million cycles without breaking. For nonferrous metals it may be necessary to flex the metal through many more than 10 million cycles to determine the limiting stress at which rupture will occur, and some metals may fail to show any endurance limit at all, the value of stress at rupture constantly declining as the number of stress cycles is increased. As a rough rule of thumb, the endurance limit for steels is generally considered to be about half the ultimate strength, though this will vary for different steels and may actually give stress values too high for the steel being considered. For nonferrous metals, many of which have no endurance limit, the endurance strength may be as low as 20 percent and is rarely more than 40 percent of the tensile strength.

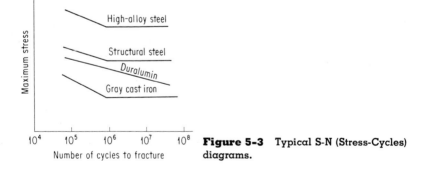

Figure 5-3 Typical S-N (Stress-Cycles) diagrams.

5-4. STRESS CONCENTRATION

Notch sensitivity, or the stress concentration factor, is also extremely important in the design of members subject to cyclic loading. Any discontinuity or change of section, e.g., a hole, groove, notch, or bend, is a stress raiser. The localized stresses in some cases may reach three or more times the value of the calculated average stress in the member. Rough machining also produces localized stress concentrations, and the endurance limit of a hardened steel part with a rough file finish may be only 60 percent of the endurance limit of a similar part of the same material with a polished finish.

5-5. STRESS AND STRAIN

The general relationship of $E = S/\delta$ is all that is necessary to solve many of the simpler problems in mechanics of materials. To illustrate the use of this equation we can examine a few examples from past examinations. One such problem asked:

- A steel rod 10 ft long has a cross-sectional area of 0.40 in.^{2} and is carrying a load of 6000 lb. Assume the modulus of elasticity for this steel to be 30,000,000 psi and determine:

  a. Total elongation produced in the rod by this load

  b. Load required to cause the steel rod to just reach its elastic limit of 82,000 psi.

For a 6000-lb load the stress would be

$$\frac{6{,}000 \text{ lb}}{0.40 \text{ in.}^2} = 15{,}000 \text{ psi}$$

$$\delta = \frac{S}{E} = \frac{15{,}000}{30{,}000{,}000} = 5 \times 10^{-4} \text{ in./in.}$$

a. The total elongation would then equal

$$(10 \times 12) \text{ in.} \times (5 \times 10^{-4}) \text{ in./in.} = 0.060 \text{ in.}$$

b. The load to cause a stress of 82,000 psi would be

$$82{,}000 \text{ psi} \times 0.40 \text{ in.}^2 = 32{,}800 \text{ lb}$$

Another, somewhat typical example is illustrated by the problem:

- A steel tape for measuring distance is 0.30 in. wide and 0.015 in. thick. It is exactly 100 ft long when supported throughout its length and pulled with a force of 15 lb. What will be its length if the chainmen pull with a force of 50 lb? E for steel $= 30 \times 10^6$ psi.

The cross-sectional area of the tape is $0.30 \times 0.015 = 0.0045$ in.^{2}. If the chainmen pull with a force of 50 lb, this is $50-15$, or 35 lb more than the force required to stretch the tape to a length of 100 ft. Since the total applied load is still well within the proportional limit, the additional stretch is proportional to the additional load. The added load of 35 lb produces 35 lb/0.0045 in.^{2} = 7780 psi additional stress.

$$\delta = \frac{S}{E} = \frac{7780 \text{ psi}}{30 \times 10^6 \text{ psi}} = 2.59 \times 10^{-4} \text{ in./in.}$$

so the total length of the chain when subjected to a 50-lb pull will be 100 ft + (100 × 12) in. × (2.59 × 10⁻⁴) in./in. = 100 ft + 0.311 in., or 100.0259 ft.

Another problem from a past example is:

- Two parallel wires 6 in. apart, one of copper and one of steel, are used to support a load P. The area of the copper wire is twice the area of the steel wire. Determine the distance x from the steel wire for applying the load so that the wires will remain equal in length (Fig. 5-4).

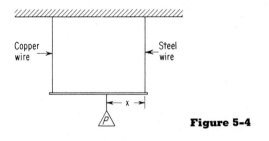

Figure 5-4

$$E_s = 30 \times 10^6 \text{ psi}$$
$$E_{cu} = 10 \times 10^6 \text{ psi}$$

Since $\delta = S/E$ and δ_s is to equal δ_{cu}, $S_s/E_s = S_{cu}/E_{cu}$, which means that the stress in the steel wire must be three times the stress in the copper wire for the lengths of the two to remain equal. Since load equals stress times area of cross section, the load on the steel wire will equal $S_s \times A_s$. Applying the ratios given, $S_s = 3S_{cu}$ and $A_s = \frac{1}{2}A_{cu}$, we have that the steel wire must support $\frac{3}{2}$ as much load as the copper wire for the lengths of the two to remain the same. From $\Sigma F_y = 0$, we find that the copper wire will have to support $\frac{2}{5}$ of P and the steel wire $\frac{3}{5}$ of P. Taking moments about S gives $6 \times \frac{2}{5}P = Px$, or $x = 2.4$ in.

A slightly different type of example is illustrated in another past examination question:

- The lower chord in a panel of a bridge truss consists of three eyebars, each 8 by 1 in. in cross section. The two outside bars are each 20 ft 0 in. long, center to center of pinholes, and the middle bar is 20 ft ⅛ in. long. The total tension in the three bars is 360,000 lb. Disregarding the bending of the pins, and assuming that they are parallel to each other and that they fit loosely in the pinholes before load is applied, calculate the unit stress in the bars. $E = 30 \times 10^6$ psi.

First, let us draw a figure (Fig. 5-5). Since the two outside bars are identical, the elongation in both will be the same. The inner bar is $1/8$ in. longer, so the total elongation of it will be $1/8$ in. less. The stress in the two outer bars will be $E\delta_1$, and the stress in the inner bar will be $E\delta_2$. The total load will equal the sum of the products of the stresses times the areas stressed, or

$$360,000 = (8 \times 1) \times E\delta_1 + (8 \times 1) \times E\delta_1 + (8 \times 1) \times E\delta_2$$

which gives $360,000/8E = 2\delta_1 + \delta_2$, where δ_1 and δ_2 are in in./in. or ft/ft. Because of the different lengths of the bars and since the total elongation equals length times δ_1, we have

$$20 + 20\delta_1 = 20.0104 + 20.0104\delta_2 \quad (20.0104 \text{ ft} = 20 \text{ ft } 1/8 \text{ in.})$$

which gives $\delta_1 = \delta_2 + 0.00052$. We then have

$$\frac{360,000}{8 \times 30 \times 10^6} = 0.0015 = 2\delta_1 + \delta_2$$

and $\delta_1 = 0.000673$ ft/ft or in./in.; $\delta_2 = 0.000153$ ft/ft or in./in. The stress in the two outer bars is

$$S = E\delta_1 = 0.000673 \times 30 \times 10^6 = 20,200 \text{ psi}$$

and the stress in the inner bar is $0.000153 \times 30 \times 10^6 = 4600$ psi.

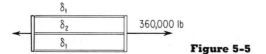

Figure 5-5

To check: load held by outer bars,

$$20,200 \text{ psi} \times 16 \text{ in.}^2 = 323,000 \text{ lb}$$

load held by inner bar, 4600 psi $\times 8$ in.$^2 = 37,000$ lb, and the sum equals the total applied load of $360,000$ lb.

Another typical example is the following:

■ A standard test specimen 0.506 in. in diameter by 2 in. gauge length shows an increase in length of 0.00067 in. when tension is increased from 500 to 2500 lb. Calculate the modulus of elasticity.

The change in stress equals

$$\frac{2500 - 500}{(0.506)^2 \times \pi/4} = 10,000 \text{ psi}$$

$$\frac{0.000667}{2} = 0.0003335 \text{ in./in. change in strain}$$

$$E = S/\delta = \frac{10,000}{0.000333} = 30 \times 10^6 \text{ psi}$$

5-6. THERMAL STRESS

When the effects of temperature (thermal expansion) are included, the picture is changed a little. In most cases it will simplify the calculations if strain due to stress and the strain due to temperature are calculated separately and are then added algebraically to obtain the net or final result. This can best be illustrated by means of examples. One past examination problem was:

■ A 3-ft-long copper bar having a circular cross section 1 in. in diameter is arranged as shown in Fig. 5-6 with a 0.001-in. gap between its end and the rigid wall at room temperature. If the temperature increases 60°F, find the stress in the rod. (Assume the coefficient of thermal expansion for copper is 9.3×10^{-6} in./(in.)(°F) and its modulus of elasticity is 17×10^6 psi.) List the assumptions you had to make in order to arrive at your answer.

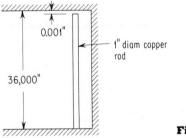

0.001"

1" diam copper rod

36,000"

Figure 5-6

The unrestrained increase in length due to the increase in temperature would have been

$$[9.3 \times 10^{-6} \text{ in./(in.)(°F)}] \times 60°F \times 36 \text{ in.} = 0.0201 \text{ in.}$$

The bar can expand 0.001 in. and then will be restrained. This is the same (stresswise) as if it had been allowed to increase 0.0201 in. and was then compressed 0.0191 in. The resulting stress would be $S = E \times \delta$.

$$S = 17 \times 10^6 \text{ psi} \times 0.0191 \text{ in.}/36 \text{ in.} = 9020 \text{ psi}$$

The assumptions that would have to be made are:

1. The walls must be perfectly rigid, i.e., undergo neither expansion nor deformation.
2. The bar (column) must remain straight.
3. The yield point of the copper was not exceeded.

Problems concerned with the effects of shrinkfits are not uncommon. An example of this type is as follows:

■ A steel liner is assembled in an aluminum pump housing by heating the housing and cooling the liner in dry ice. At 70°F before assembly, the liner outside diameter is 3.508 in. and the housing inside diameter is 3.500 in. After assembly and inspection, several units were rejected because of poor liners. It is desired to salvage the housing by heating the complete unit to a temperature which would cause a difference (clearance) in diameter of 0.002 in. between liner and housing and permit free removal of the liner. Determine the temperature at which this may be possible.

Thermal coefficient of linear expansion: steel, 0.0000065 in./(in.)(°F); aluminum, 0.000016 in./(in.)(°F).

At 70°F the interference is 0.008 in. A clearance of 0.002 in. is desired, so the expansion of the aluminum housing must be 0.010 in. more than the expansion of the steel liner, or the unit expansion of the aluminum must be $0.010/3.5 = 0.00286$ in./in. greater than that of the steel. The differential expansion is $0.000016 - 0.0000065 = 9.5 \times 10^{-6}$ in./(in.)(°F). The required temperature increase would then be

$$\frac{0.00286 \text{ in./in.}}{9.5 \times 10^{-6} \text{ in./(in.)(°F)}} = 301°F$$

so the assembly would have to be heated to a temperature of 371°F.

5-7. TORSION

Pure torsional loading (couple only, no bending) produces a shearing stress in a shaft with the magnitude of the stress, in any cross section, being proportional to the distance from the center of the shaft (Fig. 5-7). This follows from the fact that the deformation at any point equals $\rho \, d\theta$, and

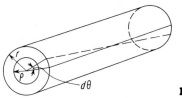

Figure 5-7

since stress is proportional to strain the stress increases from zero at the center to a maximum at the outside. Let S_s be the maximum shearing stress at the outer fiber. Then S_s/r will be the stress 1 in. from the center and $(\rho/r) \times S_s$ will be the stress at a distance ρ from the center. Force equals stress times area, so the force resisting the applied torque due to the area dA equals $(\rho/r)S_s\,dA$; the resisting moment about the axis of the bar due to this force is equal to $(\rho^2/r)S_s\,dA$, and $(S_s/r)\int\rho^2\,dA$ is the total moment about the axis due to all the internal shearing forces. The quantity $\int\rho^2\,dA$ is, we recall from Chap. 2, Mechanics, the moment of inertia of the cross-sectional area about the axis through its center or the polar moment of inertia of the area, which quantity is ordinarily represented by the symbol J. This gives us the relationship that torque $= (S_s/r) \times J$, and the maximum stress in the shaft due to the applied torque is $S_s = rT/J$. The stress at any other distance ρ from the center of the shaft equals

$$\frac{\rho}{r} \times \frac{rT}{J} = \frac{\rho T}{J}$$

Within the proportional limit, $\delta_s = S_s/E_s$, which gives the relationship $S_s = E_s\delta_s = (E_s r\theta)/l$, since, from Fig. 5-8, $\delta_s = r\theta/l$. This also gives us $T = (E_s J\theta)/l$.

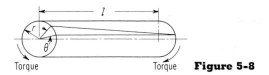

Torque   Torque   **Figure 5-8**

A problem taken from a past examination is:

■ A solid round shaft $3\frac{1}{2}$ in. in diameter transmits 100 hp at 200 rpm. Determine:
  a. Torsional stress in the outer fibers
  b. Increase of stress if the shaft is made hollow with $1\frac{1}{2}$ in. ID
  c. Stress in the inner fibers of the hollow shaft

The power transmitted by a shaft equals $2\pi NT/33,000$ (Fig. 5-9), which means $T = (33,000 \times 100)/(2\pi \times 200) = 2630$ ft-lb, or 31,500 in.-lb. The

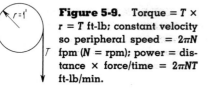

Figure 5-9. Torque $= T \times r = T$ ft-lb; constant velocity so peripheral speed $= 2\pi N$ fpm (N = rpm); power = distance × force/time $= 2\pi NT$ ft-lb/min.

polar moment of inertia of a circle is $(\pi r^4)/2$, so $S_s = (r$ in. $\times T$ in.-lb)$/(J$ in.$^4)$ $= 2T/(\pi r^3)$.

$$S_s = \frac{2 \times 31,500}{\pi \times (1.75)^3} = 3740 \text{ psi}$$

For a hollow shaft of the same outer diameter but with an inner diameter of $1\frac{1}{2}$ in.,

$$J = \frac{\pi(r_{OD})^4}{2} - \frac{\pi(r_{ID})^4}{2} = \frac{\pi}{2}(1.75^4 - 0.75^4) = 14.25$$

$$S_s = \frac{1.75 \times 31,500}{14.25} = 3,870 \text{ psi} \qquad \text{stress in outer fiber}$$

Increase of stress $= 3870 - 3740 = 130$ psi. The stress in the inner fibers of the hollow shaft will equal

$$\frac{r_{inner}}{r_{outer}} \times S_{outer} = \frac{0.75}{1.75} \times 3870 = 1660 \text{ psi}$$

One other past examination problem will be used as an example. Referring to Fig. 5-10, the problem asks:

■ *a.* Determine the torque M_B such that the maximum unit shear stress is the same in both parts. Disregard stress concentrations.
b. Determine the angle of twist in *BC*.

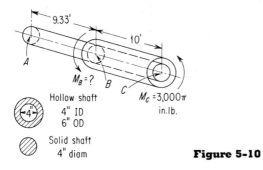

Figure 5-10

The torque $M_c = 3000\pi$ in.-lb. The stress in the outer fibers of the shaft BC may be found from the equation:

$$S_s = \frac{rT}{J} = \frac{3 \times 3,000\pi}{(\pi/2)(3^4 - 2^4)} = 277 \text{ psi}$$

To produce the same stress in the solid portion of the shaft would require a torque of

$$T = \frac{JS_s}{r} = \frac{(\pi/2)(2^4)277}{2} = 1110\pi \text{ in.-lb}$$

so torque in the amount of $3000\pi - 1110\pi$, or 1890π in.-lb, must be taken off at B, giving $M_B = 1890\pi$ in.-lb.

The angle of twist in length BC may be determined from the relationship $T = (E_s J \theta)/l$, or

$$\theta = \frac{T \times l}{JE_s} = \frac{3000\pi \times (10 \times 12)}{(\pi/2)(3^4 - 2^4) \times 12 \times 10^6} \frac{\text{in.-lb} \times \text{in.}}{\text{in.}^4 \times \text{lb/in.}^2} = 0.000923 \text{ radians}$$

0.000923 radians $\times 57.3°/$radian $= 0.0529°$.

5-8. BEAMS

The general relationship for determining the stress in a beam due to bending is $S = (Mc)/I$, where $M = $ moment, in.-lb, $I = $ moment of inertia of the cross section, and $c = $ distance from the neutral axis to the outermost fiber. The stress at any other point than the outer fiber is proportional to its distance from the center and would be $S = (y/c)(Mc/I) = (My)/I$.

The values of c and I can be determined from the geometry of the beam. The moment M depends upon the loading of the beam. As can be seen, the maximum stress will occur at the location of the greatest moment (for a beam of constant cross section).

The transverse shear stress at any point is equal to the magnitude of the shearing force divided by the cross-sectional area at that point. It is seldom possible to determine by a glance just where the maximum shear and the maximum moment will occur in a loaded beam, so it is usually desirable to construct both the shear diagram and the moment diagram for the beam under consideration; then the points of maximum shear and maximum moment will be readily apparent.

5-9. SHEAR DIAGRAM

The shear at any section of a beam is the algebraic sum of all the external forces on *either* side of the section; it is considered positive if the segment of the beam on the left of the cross section tends to move up with respect to the segment on the right. The algebraic sum of all the forces on *both* sides of the section will, of course, equal zero, since the system is in static equilibrium. It is usually simplest to add all the shearing forces to the left of the section being considered; this sum is equal to the shearing force acting at that point. If this process is repeated for different points over the length of the beam, a complete shear diagram may be constructed.

5-10. MOMENT DIAGRAM

The moment diagram may be constructed in a similar manner; one must remember that the bending moment at any section of a beam is the algebraic sum of the moments of all the external forces on either side of the section. When calculating the bending moment, use the segment (or side) for which the arithmetic will be the simplest. A positive moment is one which tends to cause the beam to be concave on the upper side (top fibers in compression). An easy method of arriving at the correct sign in calculating bending moment is to give plus signs to the moments of upward forces and minus signs to the moments of downward forces. This method will give the correct sign to the algebraic sum of the moments, whether the forces to the right of the section or to the left of the section are used.

Examples of shear and moment diagrams for a few of the commoner types of loading are shown in Fig. 5-11.

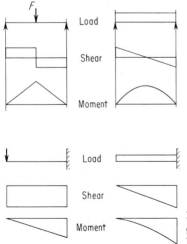

Figure 5-11 Shear and moment diagrams for some of the commoner types of loading.

For combined loadings the shear and moment diagrams may be constructed by adding (algebraically) the individual shear and moment diagrams due to the individual loadings. This will give the total values for shear and moment for all parts of the beams. Each load will produce the same effect as if it had acted alone; it is unaffected by the other loads. Another method is to calculate the moments at different points on the beam and draw a smooth curve. Care must be taken, however, to make certain that the points of maximum total shear and maximum total moment are included. These points of maximum magnitude will be more easily determined by sketching the component shear and moment diagrams and determining the total shear and moment for all points of maximum magnitude (both positive and negative) on the component diagrams. This is illustrated in Fig. 5-12.

There are a few relationships which are of value in constructing and in checking shear and moment diagrams. One of these is that the derivative of the moment M with respect to distance is equal to the shear V, or $V = (dM)/(dx)$. This means that the slope of the moment curve at any point is equal to the magnitude of the shear at that point. The relationship may also be written

$$\int_{M_1}^{M_2} dM = \int_1^2 V\,dx \qquad \text{or} \qquad M_2 - M_1 = \int_1^2 V\,dx$$

which means that the difference between the values of the moments at points 1 and 2 is equal to the area under the shear diagram between points 1 and 2. Similarly we have $W = (dV)/(dx)$ where W is the load at any point and $V_2 - V_1 = \int_1^2 W\,dx$. The use of these relationships is illustrated in Fig. 5-12.

A few examples are in order here:

■ A wooden beam is made up of two timbers, one 4- by 6-in. and one 4- by 8-in., as shown in Fig. 5-13.
   a. Locate its horizontal neutral axis.
   b. Compute its moment of inertia about this axis.
   c. If this section is used as a uniformly loaded beam in a simple span 32 ft long, what maximum total load could it sustain, assuming a maximum bending stress of 1200 psi?

The location of the horizontal neutral axis can be obtained by taking moments about the lower edge.

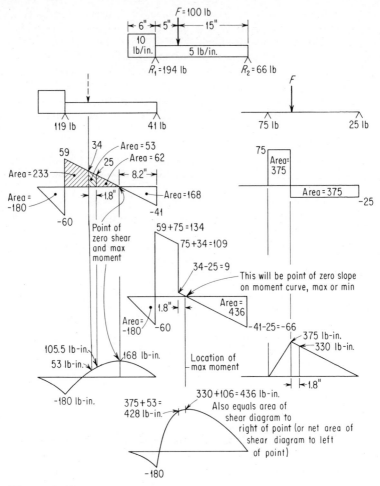

Figure 5-12

$$\bar{y} = \frac{(4 \times 8) \times 2 + (6 \times 4)(4 + 3) + (4 \times 6)(4 + 6 + 2)}{(4 \times 8) + (6 \times 4) + (4 \times 6)}$$

$$\bar{y} = \frac{\text{moments of individual areas}}{\text{total area}}$$

$\bar{y} = 6.5$ in. from the bottom of the beam, 7.5 in. from the top.

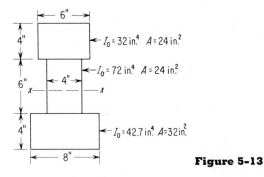

Figure 5-13

Using the parallel-axis theorem and $I_{\text{rect}} = (bh^3)/12$,

$$I_{x-x} = 32 + (6 \times 4)5.5^2 + 72 + (6 \times 4)0.5^2 + 42.7 + 32 \times 4.5^2$$
$$= 1527 \text{ in.}^4$$

$S = Mc/I$; $M_{\max} = (wl^2)/8$; $c = 7.5$.

$$\text{Total load} = w \times l = \frac{S \times I}{c} \times \frac{8}{l} = \frac{1200 \times 1527 \times 8}{7.5(32 \times 12)} = 5090 \text{ lb}$$

for a compressive stress in the topmost fibers of 1200 psi.

If the tensile stress (or bending stress) in the lower fibers should control (1200 psi tensile stress),

$$w \times l = \frac{1200 \times 1527 \times 8}{6.5(32 \times 12)} = 5880 \text{ lb}$$

The allowable compressive stress in wood parallel to the grain is slightly less than the allowable tensile stress, however, and the smaller load would control.

Another past examination problem is as follows:

■ For the loaded beam shown in Fig. 5-14,
  a. Calculate reactions R_1 and R_2.
  b. Draw shear and moment diagrams with the magnitude of maximum points indicated.
  c. Determine section modulus of beam if maximum allowable stress is 20,000 psi.

Referring to the figure, the reaction R_2 may be found by taking moments about R_1.

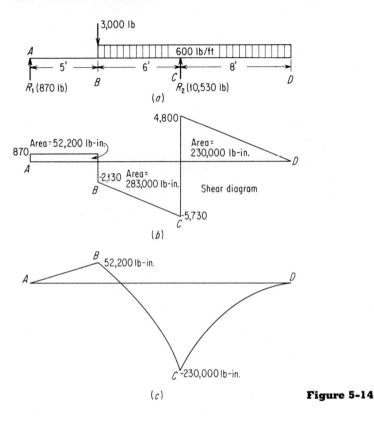

Figure 5-14

This gives $R_2 = \dfrac{5 \times 3000 + 12(600 \times 14)}{11} = 10{,}530$ lb

Taking moments about R_2 gives

$$R_1 = \frac{3000 \times 6 - 1 \times (600 \times 14)}{11} = 870 \text{ lb}$$

Adding the forces in the y direction gives

$$10{,}530 + 870 - 3000 - (600 \times 14) = 0$$

which checks.

The shear and moment diagrams are drawn directly below the given figure. From the moment diagram we see that the maximum moment is $-230{,}000$ lb-in.

$$S = \frac{Mc}{I} = \frac{M}{Z}$$

where $Z = I/c$, section modulus

$$Z = \frac{M}{S} = \frac{230,000 \text{ lb-in.}}{20,000 \text{ lb/in.}^2} = 11.5 \text{ in.}^3$$

The four important points on the beam have been labeled A, B, C, and D. Note in the shear diagram that the slope from A to B is zero since the load on the load diagram from A to B is zero. At B is a concentrated load, and the shear changes perpendicularly at that point. From B to C the load is constant and the slope angle on the shear diagram is also constant. At point C the load changes abruptly because of the reaction R_2, and the shear also changes by the same amount. From C to D the load is constant, and the slope of the shear diagram is also constant. (Remember that loads on the *left* in an *upward* direction give a positive shear and loads *down* on the *right* give a positive shear.)

Looking at the moment diagram we see that the slope at any point equals the magnitude of the shear diagram at that point. Thus the slope is constant from A to B and makes an abrupt change at B to a steadily increasing negative slope. It makes another abrupt change at C from a large negative slope to a large positive slope which decreases to no slope (tangent to abscissa) at D.

Another example will help to illustrate these principles further. A past examination problem was as follows:

- A cantilever beam 12 ft long is fixed at the left end and carries a uniformly distributed weight of 200 lb/lin ft and a concentrated load of 2000 lb at a point 8 ft from the right end. Draw the shear and moment diagrams and show the values of shear and moment at concentrated load and ends. The diagrams must show the correct shapes.

A figure must be drawn as the first step toward solution. The figure and calculations are shown in Fig. 5-15.

5-11. BEAM SLOPE AND DEFLECTION

We have reviewed shear, bending moment, and stress in beams; this brings us to the problem of determining slopes and deflections. The basic relationship for determining deflection or slope is $EI(d^2y/dx^2) = M$, to a close approximation. The radius of curvature $R = EI/m$.

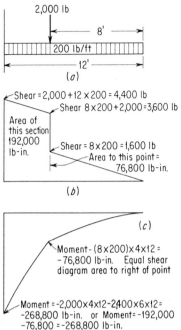

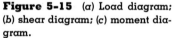

Figure 5-15 (a) Load diagram; (b) shear diagram; (c) moment diagram.

For a beam simply supported at the ends with a concentrated load at the center, this gives us the equation (see Fig. 5-16)

$$y_{max} = \frac{Fl^3}{48EI}$$

for the maximum deflection which occurs at the point of load (midpoint of beam). This relationship can be used to solve a past examination problem which asked:

■ An I beam 14 in. deep on 20-ft span is simply supported at the ends. It has a moment of inertia of 440 in.[4].

   a. What load may be hung midway between the supports without producing a deflection of more than $1/4$ in.?

   b. What is the intensity of the stress produced?

   c. What total uniformly distributed load would produce the same deflection?

   d. What would then be the maximum bending stress? Use $E = 30,000,000$ psi.

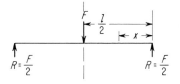

Figure 5-16 Taking right half of beam and considering center point stationary, right end free to move up.

$$M_x = \frac{F}{2}x \qquad EI\frac{dy}{dx} = \int M\,dx = \int \frac{F}{2}x\,dx = \frac{Fx^2}{4} + C_1$$

$$\frac{dy}{dx} = 0 \text{ at } x = \frac{l}{2} \qquad C_1 = \frac{-Fl^2}{16}$$

$$EIy = \int \frac{F}{4}x^2\,dx - \int \frac{Fl^2}{16}\,dx = \left(\frac{Fx^3}{12}\right) - \left(\frac{Fl^2}{16}\right)x + C_2$$

$$y = 0 \text{ at } x = \frac{l}{2} \qquad C_2 = \frac{-Fl^3}{48}$$

$$y = \left(\frac{F}{EI}\right)\left(\frac{x^3}{12} - \frac{l^2x}{16} + \frac{l^3}{48}\right) \qquad y_{max} = \frac{Fl^3}{48EI} \text{ at } x = 0$$

First, draw a figure (Fig. 5-17). Disregarding the weight of the beam, we obtain for the load

$$F = \frac{48EIy}{l^3} = \frac{48 \times 30 \times 10^6 \times 440 \times 0.25}{(20 \times 12)^3} = 11{,}460 \text{ lb}$$

$$S = \frac{Mc}{I} \qquad M = \frac{11{,}460}{2} \times (10 \times 12) = 688{,}000 \text{ lb-in.}$$

$$c = \frac{14}{2} = 7 \text{ in.} \qquad S = \frac{688{,}000 \times 7}{440} = 10{,}950 \text{ psi}$$

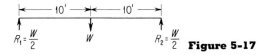

Figure 5-17

For a uniformly distributed load, $y_{max} = (5wl^4)/(384EI)$, which gives

$$w = \frac{384EIy}{5l^4} = \frac{384 \times 30 \times 10^6 \times 440 \times 0.25}{5(20 \times 12)^4} = 76.4 \text{ lb/in.}$$

Unit algebra gives

$$\frac{\text{lb/in.}^2 \times \text{in.}^4 \times \text{in.}}{\text{in.}^4} = \text{lb/in.}$$

The total uniformly distributed load would then equal

$$76.4 \text{ lb/in.} \times (20 \times 12) \text{ in.} = 18{,}300 \text{ lb}$$

which could have been determined without the intermediate step by solving directly for $W = w \times l = \dfrac{384EIy}{5l^3}$. The maximum moment would occur at the midpoint of the beam and would equal

$$\frac{W}{2} \times \frac{l}{2} - \frac{W}{2} \times \frac{l}{4} = \frac{W \times l}{8}$$

$$S = \frac{Mc}{I} = \frac{18{,}300 \times (20 \times 12) \times 7}{8 \times 440} = 8740 \text{ psi}$$

Another example which will help to illustrate a slightly different aspect of the subject of beam deflection is the following past examination problem:

- A simply supported beam 10 ft long carries three equal concentrated loads located respectively 3, 5, and 7 ft from the left end. Calculate the ratio of the maximum deflection to the deflection under the load located 3 ft from the left end.

Again the first step is to construct a figure (Fig. 5-18). The loading is symmetrical about the center, so the maximum deflection will occur at the center. The maximum deflection will equal the sum of the deflections at the center of the beam due to each of the three individual loads. Each of these three deflections will be the same as if each load acted by itself; an easier solution is first to determine the deflection at the center due to the symmetrically placed loads P_1 and P_3, then to determine the deflection due to the load P_2, and add these two deflections to obtain the maximum deflection.

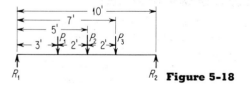

Figure 5-18

The deflection at the center due to loads P_1 and P_3 can be calculated from the equation

$$\Delta_{\max} = \frac{Pa}{24EI}(3l^2 - 4a^2)$$

where $a = 36$ in.

$$l = 120 \text{ in.}$$

giving $\Delta_{\max} = 57,000P/EI$.

The deflection at the center of the beam due to P_2

$$\Delta_{\max} = \frac{Pl^3}{48EI} = 36,000\frac{P}{EI}$$

The total deflection at the center of the beam would then be equal to $93,000P/EI$, which is the sum of the partial deflections due to loads P_1, P_2, and P_3.

The deflection under the load located 3 ft from the left end would also be made up of two components. The deflection due to the two equal symmetrically applied loads P_1 and P_3 would equal

$$\frac{Pa}{6EI}(3la - 4a^2) = 46,600\frac{P}{EI}$$

The deflection at the same point due to the force P_2 would equal

$$\frac{Pa}{48EI}(3l^2 - 4a^2) = 28,500\frac{P}{EI}$$

The total deflection under the left load would then equal $75,100P/EI$, and the required ratio would be 1.24.

Another type of beam problem occasionally asked is illustrated by the example below:

■ The beam illustrated in Fig. 5-20 has a 10 WF 54 section. Determine the reaction R and the maximum bending stress in the beam.

5-12. STATICALLY INDETERMINATE BEAMS

This is an example of a statically indeterminate beam. It is called statically indeterminate because the number of unknown reactions exceeds those which may be determined by the static relationships alone. In this case there are three unknown reactions—the vertical reactions at the two ends and the couple at the left end. There being only vertical forces, we have only two equations based on the static relationships $\Sigma F_y = 0$ and $\Sigma M = 0$. To obtain the required answer, we resort to a handbook for a method of determining the reaction R. For a single concentrated load, beam fixed at one end, supported at the other,

$$R = \frac{Fb^2}{2l^3}(a + 2l)$$

Applying this equation twice, once for each concentrated load, gives us the value for R.

$$R_1 = \frac{15,000 \times (6)^2}{2 \times (9)^3}(3 + 2 \times 9) = 7780 \text{ lb}$$

$$R_2 = \frac{15,000 \times (3)^2}{2 \times 9^3}(6 + 2 \times 9) = 2220 \text{ lb}$$

Note that the units of length cancel out, and it makes no difference whether we use lengths in feet or in inches, as long as we are consistent. $R = R_1 + R_2 = 10,000$ lb, where R_1 is due to the right-hand 15,000-lb load and R_2 is due to the left of the two loads.

With this value for R we can determine the upward reaction at the wall, $R_w = 15,000 + 15,000 - 10,000 = 20,000$ lb. Having the loading of the beam, we can construct the shear and moment diagrams. These are shown in Fig. 5-19 with the loading diagram. The direction of the beam has been reversed, with the free end placed at the left to simplify the analysis.

The shear diagram may be constructed in the manner which has been discussed previously. The moment diagram, however, is not quite so simple as those constructed previously. To construct the moment diagram, start at the left end of the beam where the moment is zero. We see from the shear diagram that the moment diagram will have a constant positive slope to the right to the point of application of the first load. Furthermore, the magnitude of the intercept of the moment curve at this point equals the algebraic sum of the areas under the shear diagram to the left of this point, which is 10,000 lb $\times$ 36 in. $= 36 \times 10^4$ lb-in. The moment curve then takes a constant negative slope to the point of application of the next load, at which point the moment equals

$$36 \times 10^4 - (5000 \times 36) = 18 \times 10^4 \text{ lb-in.}$$

At this point the negative slope increases and remains constant to the end of the beam, where the moment equals

$$18 \times 10^4 - (20,000 \times 36) = -540,000 \text{ lb-in.}$$

which is the maximum moment along the length of the beam.

The section modulus of a 10 WF 54 beam is (from a table) 60.4 in.[3].
Since stress $= (Mc)/I = M/Z$, where $Z = I/c$,

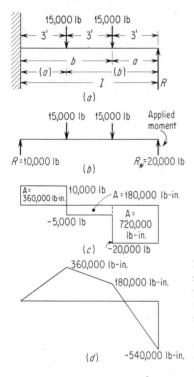

Figure 5-19 (c) Shear diagram; (d) moment diagram. The moment at the wall may also be calculated from the relationship $M = -[Fab(l + a)]/(2l^2)$, which gives, for the two applied loads $-(240,000 + 300,000) = -540,000$ lb-in., which checks the values obtained by the other method.

$$S = \frac{540,000}{60.4} = 8950 \text{ psi}$$

The required answers are $S_{\max} = 8950$ psi and $R = 10,000$ lb.

5-13. SHEAR STRESSES IN BEAMS

A loaded beam will strain in some fashion. When it does, it flexes and the longitudinal fibers are placed in tension or compression every place except at the neutral axis. The effect of flexure on a simple beam is shown in Fig. 5-20. If the beam were made of two parallel pieces in contact, but not fixed to one another, which were originally the same length, the ends of the top member would overlap the ends of the bottom member as shown in Fig. 5-21. The bottom-most fibers of the top member would strain in tension and elongate. The uppermost fibers of the bottom member would strain in compression and reduce in length. If the two members were joined together in

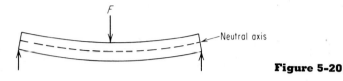

Figure 5-20

such a manner that the differential strain were prevented, there would be a definite longitudinal shearing stress at the junction, the neutral axis of the beam shown in Fig. 5-20.

The longitudinal shear stress is related to the transverse (vertical in this case) shear stress by the relationship

$$S_s = \frac{VQ}{It} \qquad \text{lb/in.}^2$$

or

$$\text{shear force} = \frac{VQ}{I} \qquad \text{lb/in.}$$

where Q is termed the shear flow.

Figure 5-21

V = vertical shear load at the point being considered

$$Q = \int y \, dA = \bar{y}A$$

where $\bar{y}$ = distance from the neutral axis of the beam to the neutral axis of area A.

I = transverse moment of inertia
t = length (width) of section being considered.

Take, for example, a rectangular cross-section beam as shown in Fig. 5-22. The longitudinal shear stress will be the greatest at the neutral axis. Using the above relationship for longitudinal shear,

$$Q = \bar{y}A = \frac{h}{4} \times \frac{bh}{2} = \frac{bh^2}{8}$$

$$I = \frac{bh^3}{12}$$

$$t = b$$

$$S_s = \frac{VQ}{It} = V \frac{bh^2/8}{(bh^3/12) \times b} = \frac{3V}{2bh}$$

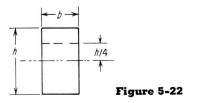

Figure 5-22

The transverse shear stress equals V/bh, so the longitudinal shear stress at the neutral axis is 50 percent greater than the transverse shear stress for a rectangular beam.

For a circular beam, see Fig. 5-23.

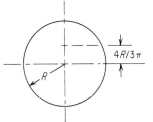

Figure 5-23

$$Q = \bar{y}A = \frac{4r}{3\pi} \times \frac{\pi r^2}{2} = \frac{2r^3}{3}$$

$$I = \frac{\pi r^4}{4}$$

$$t = 2r$$

$$S_s = \frac{VQ}{It} = V\frac{2r^3/3}{(\pi r^4/4) \times 2r} = \frac{4V}{3\pi r^2}$$

So the longitudinal shear at the neutral axis of a beam with a circular cross section is $33\frac{1}{3}$ percent greater than the transverse shear.

As an example of the type of question asked in a previous fundamentals examination, take the beam shown in Fig. 5-24. The beam is made of two pieces of 2 by 6 plank (actual dimensions) fastened together with nails. If one nail is capable of withstanding a shearing force of 150 lb, what should be the spacing of the nails?

First determine the location of the neutral axis of the beam by taking moments about an axis touching the bottom edge of the beam.

$$\bar{y} = \frac{(6 \times 2) \times 3 + (6 \times 2) \times 7}{2 \times (6 \times 2)} = 5.00 \text{ in.} \qquad (\bar{y} \text{ for beam})$$

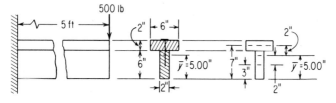

Figure 5-24

The transverse moment of inertia equals

$$I = \frac{2 \times 6^3}{12} + 12 \times 2^2 + \frac{(6 \times 2^3)}{12} + 12 \times 2^2 = 136.0 \text{ in.}^4$$

$$Q = \bar{y}A = 2 \times 12 = 24 \text{ in.}^3 \qquad (\bar{y} = \text{distance to neutral axis of area } A)$$

The shear force $V = 500$ lb.

$$\text{Shear force} = \frac{VQ}{I} = \frac{500 \times 24}{136} = 88.24 \text{ lb/in.}$$

One nail will withstand 150 lb, so a nail would be required every $150/88.24 = 1.70$ in.

If the crosspiece were glued in place, what would be the shear stress in the glue?

$$S_s = \frac{VQ}{It} = \frac{500 \times 24}{136 \times 2} = 44.1 \text{ psi}$$

If the T beam were made of a 2×8 plank with pieces of 2×2 nailed to the sides, as shown in Fig. 5-25, what should be the spacing of the nails? Assume each nail would withstand a shearing force of 150 lb.

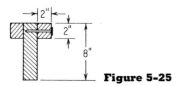

Figure 5-25

The dimensions of the beam are the same as for the previous case, so the location of the neutral axis will be the same and the value of the moment of inertia will be the same.

This case is different from the previous case in that the shear force will be divided and act at two junctions instead of just one.

Calculate the shear flow acting on one side piece.

$$Q = \bar{y}A = 2 \times (2 \times 2) = 8 \text{ in.}^3$$

$$\text{Shear force} = \frac{VQ}{I} = \frac{500 \times 8}{136} = 29.4 \text{ lb/in.}$$

The required nail spacing for each of the 2-in. by 2-in. side pieces would equal 150/29.4 = 5.10 in.

5-14. COMPOSITE BEAMS

The beams so far considered have all been of one homogeneous material. There are many beams, however, that are made of two materials, principally reinforced-concrete beams and steel-reinforced wood beams. Such beams are called "composite beams"; because of the difference in the moduli of elasticity of the two materials, they require a slightly different type of analysis than that used for homogeneous beams. One type of composite beam is shown in Fig. 5-26. It is made up of a 6- by 8-in. timber 10 ft in length, with a 6-in.-wide strip of steel $1/8$ in. thick attached to the bottom edge. To withstand the applied load, the beam will deflect, the upper part of the wooden portion will compress, and the lower part of the wood and the steel will stretch. The stress in any part of the beam will equal the unit deformation in that particular portion times the modulus of elasticity of the material of which that portion is made. Since the steel on the bottom of the beam deforms the same amount as the wood fibers on the bottom of the beam, the steel will be stressed in the ratio E_s/E_w times the stress in the bottom-most wood fibers. Since $E_s = 30 \times 10^6$ and $E_w = 10^6$, the stress in the steel will be thirty times the stress in the wood. This means that the tensile load in the steel will be considerably greater than that in the wood, and the neutral axis will be below the geometric center of the beam.

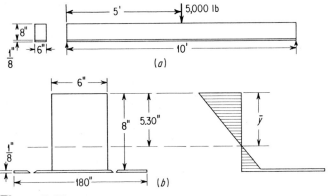

Figure 5-26

Another way of looking at this problem is that since

$$\text{Force} = \text{stress} \times \text{area}$$

and Stress $= E \times \delta$, the steel strip could, theoretically, be replaced by a piece of wood with an area equal to $E_s/E_w \times$ area of steel, or thirty times the area of the steel, but with the same thickness. The equivalent wooden beam is shown in the figure, and the internal-force diagram showing the distribution of $S \times A$ over the cross section of the beam is also shown. To determine the stresses in the wood and in the steel, we need to know the distance to the neutral axis y. This can be determined by taking moments about the upper edge of the beam; one must remember that a slice of the equivalent beam section would balance on a fulcrum placed at the neutral axis.

$$(6 \times 8) \times 4 + (1/8 \times 180) \times 8^{1}/_{16} = (6 \times 8 + 1/8 \times 180) \times y$$
$$y = 5.30 \text{ in.}$$

$S = Mc/I$, where I is the moment of inertia of the equivalent beam about the neutral axis. Utilizing the parallel-axis theorem,

$$I_x = \frac{6 \times 8^3}{12} + 48 \times (1.30)^2 + \frac{180 \times (1/8)^3}{12} + (180 \times 1/8)(2.70 + 1/16)^2$$
$$I_x = 256 + 81.8 + 0.0293 + 171.7 = 509 \text{ in.}^4$$
$$M_{max} = \frac{5000}{2} \times (5 \times 12) = 150,000 \text{ lb-in.} \qquad c = 5.36$$

Maximum compressive stress in wood $S = \dfrac{150,000 \times 5.36}{509} = 1580 \text{ psi}$

The tensile stress in the bottom-most fibers of the equivalent beam would be

$$\frac{150,000 \times 2.77}{509} = \frac{2.77}{5.36} \times 1580 = 816 \text{ psi}$$

and the stress in the steel would then equal $30 \times 816 = 24,500$ psi, since $\delta_{wood} = \delta_{steel}$ and $E_{steel} = 30 \times E_{wood}$.

The commonest type of composite beam is the reinforced-concrete beam, which can also be handled by the method described. One difference, however, is that the concrete is assumed to withstand no tensile stress and the portion of the concrete on the tension side of the neutral axis is disregarded in the stress calculations. A reinforced-concrete beam is shown in Fig. 5-27. It is required to determine the allowable concentrated load which the beam will hold if $n = 12$ and the maximum allowable stresses are 1000 psi for the concrete and 20,000 psi for the steel. $n = E_{steel}/E_{concrete}$.

The equivalent beam cross section is drawn; as the concrete will take only compressive loading, the effective concrete area will extend only to the neutral axis. The location of the neutral axis is found as before.

$$10kd \times \frac{kd}{2} = (20 - kd) \times 24$$

$$(kd)^2 + 4.8kd - 96 = 0$$

$$kd = \frac{-4.8 + \sqrt{23 + 384}}{2} = 7.7$$

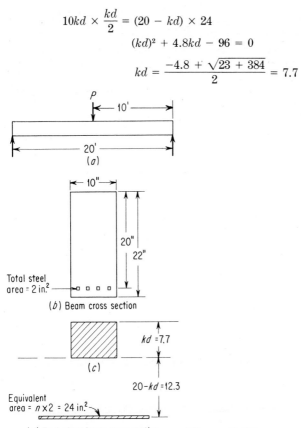

(a)

(b) Beam cross section

(c)

(d) Equivalent beam cross section

Figure 5-27

The moment of inertia

$$I = \frac{10 \times (7.7)^3}{12} + (10 \times 7.7)\left(\frac{7.7}{2}\right)^2 + 24(20 - 7.7)^2 = 5150 \text{ in.}^4$$

Note that the moment of inertia of the steel portion about its own neutral axis is disregarded since it is so small compared with the other components. $S = Mc/I$.

$$M_{\text{concrete}} = \frac{1000 \times 5,150}{7.7} = 669,000 \text{ lb-in.} \qquad \text{for 1000 psi stress}$$

$$M_{\text{steel}} = \frac{20,000 \times 5,150}{n \times 12.3} = 698,000 \text{ lb-in.} \qquad \text{for 20,000 psi stress}$$

Compressive stress in the concrete controls. Note the use of the n in the denominator for determining the allowable moment for the steel.

$$M_{max} = \frac{P}{2} \times \frac{L}{2} \qquad P = \frac{4 \times 669,000 \text{ lb-in.}}{20 \times 12 \text{ in.}} - 11,150$$

5-15. COMPOUND STRESSES

We shall restrict ourselves here to a brief discussion of one method of determining the principal stresses and the maximum shearing stress for a condition of combined loading. The method we shall use is the one based on Mohr's circle.

As a basis for illustration let us use a problem from a past examination:

- The diagram of Fig. 5-28 (Mohr's circle) is used to find the maximum shearing stress in a member that is in combined tension, torsion, and circumferential tension, viz., a pipe under pressure with torsion and axial tension applied.

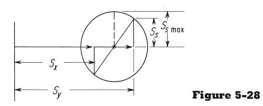

Figure 5-28

a. Derive the equation for the maximum shearing stress from this diagram in terms of $S_{s\,max}$ = maximum shearing stress; S_s = stress due to torsion; S_x = stress due to pure tension; S_y = stress due to circumferential tension.

b. Draw a stress diagram for a unit area of the external pipe surface.

This problem has been given in a simplified manner. Let us look at it in the more general sense to obtain maximum benefit from this illustration.

5-16. MOHR'S CIRCLE

First determine the stresses acting at a section on the outer surface of the pipe. S_x = axial load divided by the cross-sectional area, $S_y = Pd/2t$ (assuming a thin-walled pipe), and $S_s = Tc/J$. Having determined these stresses, we can construct Mohr's circle for the given case. The general procedure is shown by means of Fig. 5-29. First lay out the coordinate axes. From the

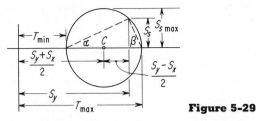

Figure 5-29

origin, point 0, lay off a distance $(S_x + S_y)/2$ along the abscissa. This gives the location of C, the center of the circle. Note here that $S_x + S_y$ is the algebraic sum of the two perpendicular stresses; if one of these stresses had been compressive we should have had the arithmetic difference instead of the arithmetic sum. At a distance from the origin equal to S_y draw a vertical line and at a distance equal to S_s above the abscissa mark a point on this line. This point lies on the circumference of the circle. Now draw the circle.

The maximum shearing stress is equal to the radius of the circle. From the figure, the trigonometric relationship is

$$S_{s\max} = \sqrt{S_s{}^2 + \left(\frac{S_y - S_x}{2}\right)^2}$$

The principal stresses can also be quickly obtained and are shown on the figure as $T_{\min}$ and $T_{\max}$. The directions of these principal stresses are also readily obtained from the figure. The lines drawn from the ends of the horizontal diameter to the construction point on the circumference give these angles, and the principal stresses act perpendicularly to these lines; i.e., $T_{\min}$ acts at an angle of β and $T_{\max}$ acts at an angle of α with the direction of S_y. The maximum shearing stress acts on a plane at an angle of 45° to these two principal planes.

The required stress diagram for a unit area of the external pipe surface shows the directions of the different stresses and would ordinarily be drawn before Mohr's circle was constructed. The diagram is shown in Fig. 5-30.

This discussion of Mohr's circle and compound stresses has been very brief but it provides an elementary working knowledge of the method which is adequate to handle successfully the bulk of this type of problem so far given in fundamentals examinations.

5-17. RIVETED AND WELDED JOINTS

Riveted joints can be lumped into two general classes—those in which the line of action of the applied load passes through the centroid of the rivet group and those which are subject to eccentric loading.

The simpler case, in which the line of action of the applied load passes

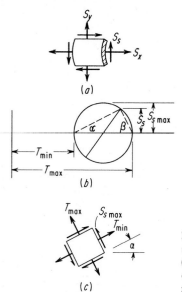

(a)

(b)

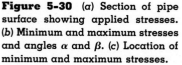

(c)

Figure 5-30 (a) Section of pipe surface showing applied stresses. (b) Minimum and maximum stresses and angles α and β. (c) Location of minimum and maximum stresses.

through the centroid of the rivet group, may be illustrated by a past examination problem, illustrated in Fig. 5-31. The allowable load could be limited by tension in the plates, bearing stresses in the plates, or shearing stress in the bolts.

The controlling area in tension is the same for the left and right members and equals $\frac{3}{4}(5\frac{1}{2} - 2 \times \frac{3}{4}) = 3$ in.^{2}. The maximum allowable load from tension considerations then equals

$$3 \times 20,000 = 60,000 \text{ lb}$$

The bearing area in single shear equals

$$4 \times 2(\frac{3}{4} \times \frac{3}{8}) = 2\frac{1}{4} \text{ in.}^2$$

This is the same as the bearing area in double shear, so the maximum allowable load as limited by bearing equals

$$2\frac{1}{4} \times 32,000 = 72,000 \text{ lb}$$

The shear area equals eight times the cross-sectional area of one bolt, or $8 \times \pi/4 \times (\frac{3}{4})^2 = 3.53$ in.^{2}. The maximum allowable load as limited by shearing stress is then $3.53 \times 15,000 = 53,000$ lb.

The allowable load is then limited by shear and equals 53,000 lb.

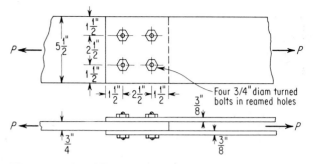

Figure 5-31 Allowable unit stresses, in pounds per square inch.

Tension, structural steel, net section 20,000
Shearing, turned bolts in reamed holes 15,000
Bearing, turned bolts in reamed holes:
    Double shear 40,000
    Single shear 32,000

In this case each of the four rivets has been assumed to withstand one-quarter of the applied load; the analysis is applicable only when the line of action of the applied force passes through the centroid of the rivet group. When the load is applied eccentrically, there will be a torque loading, as well as a portion of the applied load to be withstood by each rivet, and since different rivets will ordinarily be at different distances from the centroid of the rivet group, the stresses in the rivets will vary. This can be illustrated by means of another example selected from a past examination:

■ What is the maximum allowable load P if the allowable unit shearing stress in the rivets is 13,000 psi? Assume that bearing stress does not govern. All rivets are $^3/_4$ in. in diameter. The problem is illustrated in Fig. 5-32.

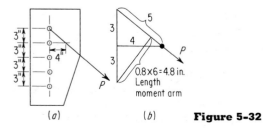

Figure 5-32

First we must determine the centroid of the rivet group. This is just the centroid of the rivet areas and may be determined in the same way that we have determined centroids of areas before. In this case the rivet group is perfectly symmetrical about the center rivet, so the centroid of the group is at the center of the center rivet.

Next the length of the moment arm D must be calculated. We recognize the familiar 3-4-5 right triangle, which means that $\alpha = 37°$ and $D = 6 \cos \alpha = 6 \times 0.8 = 4.8$ in.

As the eccentric force P is applied to the plate, the plate will turn and the rivets will deform slightly, since stress is proportional to strain and there can be no stress without a corresponding strain. Assuming that all rivets fill their holes completely, we know that the deformation of rivet (1) due to the twisting of the plate will equal $6/3$ times the deformation of rivet (2) (Fig. 5-33). This means that the stress in rivet (1) due to the torque loading will be twice the stress in rivet (2). Because of the symmetry of the joint, the stresses in rivets (1) and (5) due to resisting the applied torque will be equal, as will the stresses in rivets (2) and (4). The stress in rivet (3) due to the torque will equal zero. Since all rivets are the same size, the resisting forces will be proportional to the strains (or the stresses). This gives us

$$4.8P = 6 \times 2 \times TF_{(1)} + 3 \times 2 \times TF_{(2)}$$

where $TF_{(1)}$ and $TF_{(2)}$ are the forces exerted by rivets (1) and (2) resisting the applied torque. We know that $TF_{(1)}$ is twice $TF_{(2)}$, since the deformation of rivet (1) is just twice that of rivet (2). This gives us that $4.8P = 15 TF_{(1)}$, or $TF_{(1)} = 0.32P$.

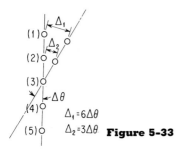

$\Delta_1 = 6\Delta\theta$
$\Delta_2 = 3\Delta\theta$ **Figure 5-33**

To this must be added, vectorially, $P/5$, since each rivet must withstand one-fifth of the applied force directly. The two resisting force vectors are shown in Fig. 5-34. The sum of the resisting horizontal force components is $0.32P + 0.12P = 0.44P$, and the vertical force component is $0.16P$.

The total shearing force acting on rivets (1) and (5) is then $P\sqrt{0.44^2 + 0.16^2}$ = $0.468P$.

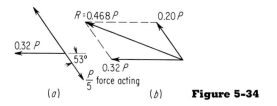

(a)   (b)   **Figure 5-34**

Since the total permissible shearing force in any one rivet is $13,000 \times \pi/4 \times (^3/_4)^2 = 5,740$ lb, the maximum allowable value of $P = 5,740/0.468 = 12,300$ lb.

The majority of the controlling features discussed above also apply to welded joints, and it might be wise to review the important relationships here. First the minimum shearing stress will occur when the line of action of the force passes through the centroid of the rivet or weld area. If the load is applied eccentrically, it will provide a moment about the centroid, and the centroidal point will act as a pivot point about which the riveted or welded joint will tend to rotate. Stress and strain are proportional; the greatest stress due to the applied moment will occur in the rivet or portion of weld farthest from the centroid, since the greatest strain will occur at that point.

To illustrate the application of these principles to the problem of a welded joint, let us look at a problem taken from a past fundamentals examination:

- A 3- by 3- by $^1/_2$-in. angle is to be welded to each side of a gusset plate, as shown in Fig. 5-35. Design the size and length of welds necessary to

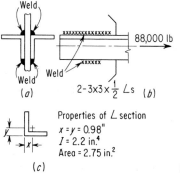

Figure 5-35

hold safely a total tension force of 88,000 lb. The allowable shearing stress in the weld is 11,300 psi.

The shear area of a welded joint is taken as the smallest section of the joint, and the fillet weld is made the same thickness as the plate being welded. This is shown in Fig. 5-36. The effective thickness of the welds would then be $1/2 \sin 45° = 0.354$ in. The line of action of the force passes through the centroids of the angles; it must also pass through the centroid of the weld area to provide a joint free from any torque effects. The welds will then be stressed in pure shear; since the moments of the two weld areas about the centroid must be equal in order to have no moment resulting from unequal strain we have

$$0.98W_u = 2.02W_L \quad \text{or} \quad W_u = 2.06W_L$$

From stress considerations,

$$88,000 \text{ lb} = 11,300 \text{ psi} \times \text{effective weld area}$$
$$\text{Effective weld area} = 7.8 \text{ in.}^2 = (W_u + W_L) \times (0.500 \times 0.707)$$
$$W_u + W_L = 22 \text{ in.}$$
$$W_L = 7.24 \text{ in.} \qquad W_u = 14.76 \text{ in.}$$

Make the welds $7^1/2$ in. and 15 in. long.

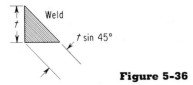

Weld

$t \sin 45°$

Figure 5-36

If the line of action of the force should not pass through the centroid of the weld area, the maximum stress will equal the vector sum of two component forces, as was the case for the eccentrically loaded riveted joint.

SAMPLE PROBLEMS

5- 1 See Fig. 5-P-1. A cast-iron punch-press frame is reinforced with a $1/4$- by 6-in. steel plate on the inner face, as shown. Allowable stresses, in pounds per square inch, are:

$$
\begin{array}{lr}
\text{Cast-iron compression} = & 15,000 \\
\text{Cast-iron tension} = & 5,000 \\
\text{Steel tension} = & 20,000 \\
E_{\text{cast-iron}} = & 18,000,000 \\
E_{\text{steel}} = & 30,000,000 \\
\end{array}
$$

Find the maximum safe bending moment for the frame.

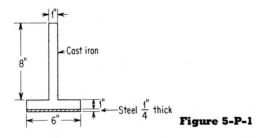

Cast iron

8"

1"

Steel $\frac{1}{4}$" thick

6"

Figure 5-P-1

5- 2 The equation for the elastic curve of a simply supported beam of length L carrying a uniformly distributed load is $EI(d^2y/dx^2) = (wlx/2) - (wx^2/2)$. Find the equation for the slope of the beam at any point.

5- 3 Three wires, each having a cross-sectional area of 0.20 in.² and the same unstressed length of 200 in. at 60°F, hang side by side in the same plane. The outer wires are copper. The middle wire, equidistant from each of the others, is steel.

    a. If a weight of 1000 lb is gradually picked up by the three wires, what part of the load is carried by each?

    b. What must the temperature become for the entire load to be carried by the steel wire?

$$E_{\text{steel}} \qquad\qquad = 30 \times 10^6 \text{ psi}$$
$$E_{\text{copper}} \qquad\qquad = 17 \times 10^6 \text{ psi}$$
Coefficient of expansion of steel $= 6.5 \times 10^{-6}$ psi*
Coefficient of expansion of copper $= 9.3 \times 10^{-6}$ psi*
*psi given in problem. Units should be in./(in.)(°F).

5- 4 See Fig. 5-P-4. What is the maximum tensile stress at section A-A?

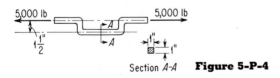

5,000 lb   5,000 lb

$1\frac{1}{2}$"

Section A-A **Figure 5-P-4**

5- 5 Draw a shear and a moment diagram for a cantilever beam with a concentrated load at the end and a uniform load along its length. Explain the relation between the shear and moment diagram. What is the relation between the maximum and the average shearing stresses over a particular cross section?

5- 6 *a.* Draw a moment diagram and a shear diagram for the 22 ft 0 in. beam loaded as shown in Fig. 5-P-6.

    b. What is the magnitude of the maximum moment? What is its location (in feet from the left end)?

5- 7 A mine hoist weighing 5 tons empty is descending at a speed of 20 mph when

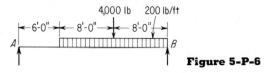

Figure 5-P-6

the hoist mechanism jams and stops suddenly. If the cable is steel having a cross section of 1 in.² and the hoist is 5000 ft below the hoisting drum, what is the stress produced in the cable? (E for hoist cable $= 12 \times 10^6$ psi)

5- 8 Locate the neutral axis by calculating the distance y on the reinforced-concrete beam section as shown in Fig. 5-P-8.

$$
\begin{aligned}
E_{\text{steel}} &= 30{,}000{,}000 \text{ psi} \\
E_{\text{concrete}} &= 2{,}000{,}000 \text{ psi} \\
A_s = \text{area of steel} &= 3.00 \text{ in.}^2
\end{aligned}
$$

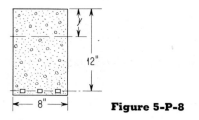

Figure 5-P-8

5- 9 A select oak timber finished sized to 6 by 8 in. is 12 ft long and is employed as a properly embedded cantilever beam with a load P (Fig. 5-P-9). If $E = 1.8 \times 10^6$, compute:

a. Deflection y_2 of the free end of the beam
b. Maximum tensile stress at y-y.

Allowable stress − oak tensile = 1000 psi; shear = 125 psi.

Figure 5-P-9

5-10 A steel column is an H section 50 ft long and supports a concentric load of 350,000 lb. The column section has the following dimensions: $A = 49.09$ in.²; $I_{1-1} = 790.2$ in.⁴; $d_{1-1} = 15^{1}/_{8}$ in.; $d_{2-2} = 15^{1}/_{8}$ in.; $I_{2-2} = 2{,}020.8$ in.⁴. Employing the American Institute of Steel construction method, determine whether the column design is safe, when

$$
\frac{P}{A} = \frac{18{,}000}{1 + (1/18{,}000)(l/r)^2}
$$

5-11 See Fig. 5-P-11. Draw shear and moment diagrams, showing values at all breaks in diagrams.

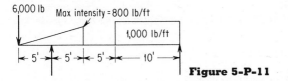

Figure 5-P-11

5-12 The load deflection curve for a spring is shown in Fig. 5-P-12. When the load reaches 150 lb the spring is fully closed and no further deflection results. The load deflection curve can be represented by the equation

$$d = 4 - \frac{(L - 200)^2}{10,000}$$

where d = deflection, in.
L = load, lb

a. How much energy is stored in the spring when the load is 150 lb? When it is 300 lb?

b. At what load will the spring constant be 0.032 in./lb?

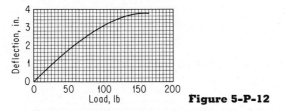

Figure 5-P-12

5-13 a. Draw a shear diagram and a moment diagram for the 10-ft 0-in. beam loaded as shown in Fig. 5-P-13.

b. Using a homogeneous steel beam with an allowable unit stress of 20,000 psi, compute the required section modulus.

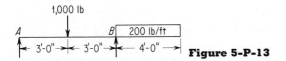

Figure 5-P-13

5-14 A beam 40 ft long is supported 10 ft from the left end and 5 ft from the right end. The beam is loaded as follows: a concentrated load of 6000 lb at the left end and a uniform load of 600 lb/ft from the left end to the right end. What are the shear diagram, moment diagram, maximum shear, maximum moment, and location of maximum moment and shear with respect to the left end of the beam?

5-15 A 6-in. by 12-in. by 14-ft timber beam is supported at the left end and also at 4 ft from the right end. It has a load of 2000 lb at the right end, a load of 2000 lb 3 ft from the left end, and a load of 800 lb/ft between supports. Find the maximum moment in magnitude and position, and the maximum shear.

5-16 A beam 40 ft long rests on simple supports at the ends and has the following loads:

Dead load—300 lb/lin ft
Uniform live load—700 lb/ft
Concentrated live load—10,000 lb

    a. Find the maximum and the minimum bending moment that can be developed at a point 16 ft from the left support.

    b. Find the maximum and the minimum shear that can be developed 16 ft from the left support.

5-17 A steel wire accurately graduated in feet was used to measure the depth of a deep well. The cross-sectional area of the wire was 0.03 in.² and a 250-lb weight was attached to the wire to lower it into the well. According to the graduated scale, the measurement read 2000 ft. What should the corrected measurement be? A steel bar 1 in.² in cross-sectional area and 1 yd long weighs 10 lb.

5-18 Design a steel shaft of the proper size to transmit 40 hp at a speed of 240 rpm if the allowable working-unit stress in the shaft is $s = 9000$ psi. Show all computations.

5-19 Concrete is to be mixed with a proportion by volume of 1:2:3 (1 cement, 2 sand, 3 aggregate, and 6 gal of water per sack of cement). The cement has 51.3 percent voids and weighs 94 lb/ft³, the sand has 35.4 percent voids and weighs 111 lb/ft³, the gravel has 32.0 percent voids and weighs 108 lb/ft³. Find the amount of each ingredient by weight necessary to make 1 yd³ of concrete. (The water necessary may be given in gallons or pounds.)

5-20 A beam 20 ft long carries a uniform load of 2000 lb/lin ft, including its own weight, on two supports 14 ft apart. The right end of the beam cantilevers 2 ft; the left end cantilevers 4 ft. Draw the shear diagram and compute the positions and amounts of the maximum bending moments.

5-21 What uniform load W (lb/ft) will cause a simple beam 10 ft long to deflect 0.3 in. if it is also supported at mid-span by a spring whose modulus $K = 30,000$ lb/in.? Assume $I = 100$ in.⁴, $E = 3 \times 10^7$ psi, $d = 10$ in. At no load the spring just touches the beam (Fig. 5-P-21).

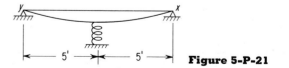

Figure 5-P-21

5-22 A 2-in. shaft is transmitting 30 hp at 200 rpm. A pulley which delivers the entire amount of power is keyed to the shaft by a ³/₈- by ¹/₂- by 3-in. flat key. The shaft is in pure torsion. Determine the shear and crushing stresses in the key.

5-23 Find the shear just to the right of the left-hand reaction, and the bending moment under the 5000-lb load (Fig. 5-P-23).

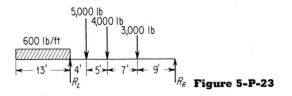

Figure 5-P-23

5-24 For the loaded beam shown in Fig. 5-P-24,
  a. Plot shear diagram.
  b. Plot moment diagram.
  c. Find maximum tensile unit stress.

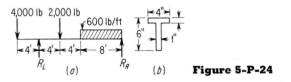

Figure 5-P-24

5-25 Where does the maximum bending moment occur in a beam with a 24-ft span that carries a load of 24,000 lb uniformly spread over its entire length and a further load of 12 tons uniformly spread over a section that starts 6 ft from the left support and extends 8 ft to the right? What is the maximum moment?

5-26 A steel gear is to be assembled on a steel motor shaft by heating the gear and cooling the shaft with dry ice. At 70°F the shaft measures 7.000 in. in diameter and the hole in the gear measures 6.996 in. in diameter. To what temperature must the shaft be cooled so that the gear may be assembled on the shaft with 0.005 in. clearance when the gear is warmed to 140°F? The coefficient of linear expansion for steel is 0.0000065 in./(in.)(°F).

5-27 *a.* Draw the shear and moment diagram for the load condition shown in Fig. 5-P-27.
  b. On these diagrams give the location and amount of the critical values.

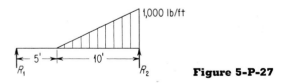

Figure 5-P-27

5-28 A machine part is loaded as a cantilever beam. Two possibilities of design are being considered which would utilize different materials and different cross sections. Show by your calculations which design will be the more rigid. Length of each = 6 in. (Fig. 5-P-28).

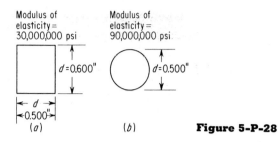

Modulus of
elasticity =
30,000,000 psi

Modulus of
elasticity =
90,000,000 psi

$d = 0.600''$

$d = 0.500''$

d
$0.500''$

(a)

(b)

Figure 5-P-28

5-29 A rigid horizontal bar 10 ft long, of uniform cross section and weighing 300 lb, is supported by two vertical wires, one of steel and one of copper. Each wire is 5 ft long and $^1/_{100}$ in.^{2} in cross section. The copper wire is attached 1 ft from one end of the bar, and the steel wire at such a distance X from the same end that both wires stretch by the same amount. Find the tension in each wire and the distance X.

Young's modulus for steel = 30×10^6 psi; for copper = 15×10^6 psi.

5-30 A block of steel would increase its volume by 330 ppm when heated from 20 to 30°C at constant pressure. What increase of pressure would be necessary to keep the block from expanding when its temperature is thus raised? The compressibility of steel is 4.4×10^{-8} in.^{2}/lb.

5-31 A weight of 3500 lb is supported by one piece of material A and two pieces of material B, as shown in Fig. 5-P-31. Each of the three pieces has a cross section of 2 in.^{2} and a height of 6 in. The weight is symmetrically placed on the supporting blocks. Young's modulus for material A is 3×10^7 psi and for material B is 2×10^7 psi.

a. What is the stress in the piece of material A?

b. What is the stress in each piece of material B?

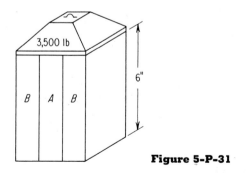

3,500 lb

6"

B | A | B

Figure 5-P-31

5-32 A solid circular shaft has a uniform diameter of 2 in. and is 10 ft long. At its midpoint 65 hp is delivered to the shaft by means of a belt passing over a pulley. This power is used to drive two machines, one at the left end of the shaft consuming 25 hp, and one at the right end consuming the remaining 40 hp. Determine the maximum shearing stress in the shaft and also the relative angle of twist between the two extreme ends of the shaft if the steel shaft turns at 200 rpm.

5-33   An eccentrically loaded riveted joint is shown in Fig. 5-P-33. The applied load is 18,000 lb acting with an eccentricity of 8 in. from the geometric center of the group of six $^3/_4$-in. rivets. Determine the maximum shearing stress in the rivets.

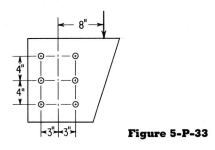

Figure 5-P-33

5-34   A rectangular wooden beam is loaded as shown in Fig. 5-P-34.
  a. If the allowable shearing stress parallel to the grain is 68 psi, is the maximum shearing stress in the beam within the allowable limit?
  b. If the allowable bending stress is 720 psi, is the bending stress in the beam within the allowable limit?

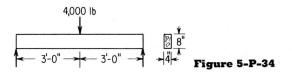

Figure 5-P-34

5-35   A 4-in.-diameter shaft transmits 1000 hp at 2000 rpm. If the same power is transmitted at 10,000 rpm,
  a. What diameter of shaft is required to maintain the same maximum shearing stress?
  b. Which shaft has the greatest angle of twist per unit length?

MULTIPLE-CHOICE PROBLEMS

For each question select the correct answer from the five given possibilities.

5M- 1 If a simple coil spring is compressed by an axial load, which of the following is true? (The spring is not compressed sufficiently to bring the sides of the coils into contact.)

(a) The maximum stress occurs at the center of the spring wire.

(b) The spring wire is in torsion.

(c) If the diameter of the spring is increased while the wire size and number of turns are kept constant, the force necessary to compress the spring a given amount will be increased.

(d) If the number of turns in the spring is increased, the spring will be harder to compress with load.

(e) None of these statements are correct.

5M- 2 If a homogeneous simple wooden beam fails under load with a crack parallel to the grain at the end supports, the failure was most likely caused by excessive:

(a) compressive stresses

(b) bearing stresses

(c) shearing stresses

(d) tension stresses

(e) none of these

5M- 3 The term $I = bd^3/12$ refers to:

(a) radius of gyration

(b) section modulus

(c) instantaneous center

(d) moment of inertia

(e) product of inertia

5M- 4 In Euler's formula $P/A = \pi^2 E/(L/R)^2$ the L/r is commonly known as the:

(a) slenderness ratio

(b) proportional limit

(c) stiffness ratio

(d) rigidity factor

(e) Poisson's ratio

5M- 5 In a uniformly loaded simple beam, the maximum vertical shearing force occurs:

(a) at either end support

(b) at the section of maximum moment

(c) at the center of the span

(d) on the bottom fiber of the beam

(e) on the top fiber of the beam

5M- 6 If a simple beam is loaded with a uniform and a concentrated load, which of the following statements is true?

(a) The maximum bending moment is indeterminate.

(b) The area of the shear diagram to the left of a section represents the bending moment at that section to some scale.

(c) The maximum shear, as represented in the shear diagram, is located at the same section as the maximum moment.

(d) The maximum bending moment always occurs at the support which carries the most load.

(e) None of these statements are correct.

5M- 7 The moment curve for a simple beam with a concentrated load at midspan takes the shape of a:

(a) triangle

(b) semicircle

(c) semi-ellipse

(d) parabola

(e) rectangle

5M- 8 In a loaded horizontal, homogeneous, rectangular beam the horizontal shearing stress at any cross section ($\perp$) perpendicular to the neutral axis:

(a) is a constant across that section

(b) has a maximum value of $3/2$ the average vertical shearing stress on the cross section at the neutral axis

(c) is equal to zero

(d) can be neglected if the beam is short

(e) has a maximum value at the outer fibers of the beam

5M- 9 A steel test specimen is $5/8$ in. ϕ at root of thread. If it is to be stressed to 50,000 psi tension, what load must be applied?

(a) 12,790 lb   (d) 25,600 lb

(b) 15,340 lb   (e) 31,250 lb

(c) 16,320 lb

5M-10 What load must be applied to a 1-in.-round steel bar 8 ft long ($E = 30,000,000$ psi) to stretch the bar .05 in.?

(a) 7200 lb   (d) 25,600 lb

(b) 9850 lb   (e) 15,000 lb

(c) 8600 lb

6

ELECTRICITY AND ELECTRONICS

The coverage of electricity and electronics in the fundamentals examinations of recent years has been expanded considerably beyond that in the examinations at the time the first edition of this book was published. Some of the problems contained in the more recent examinations have covered fairly specialized topics such as single-stage, resistance-coupled amplifiers; two-stage, triode amplifiers; cathode followers; vacuum tube and pentode amplifiers; ac signal generators; and transistor circuitry and amplifiers. These subjects fit well into an electrical engineer's background, but are somewhat foreign to other branches of engineering. There is not sufficient space in a book such as this to provide adequate coverage of the above listed subjects as well as a large number of other electrical engineering subjects which could be considered as being equally as important as those listed. Neither could those many different electrical engineering topics be considered as review for most non-electrical engineers. Another important consideration is the fact that while one or two of these diverse subjects might appear on any one examination, the odds in favor of any one particular subject's being treated in any given examination are very small. So time spent on studying (in many cases for the first time) these different subjects would not be spent efficiently. The material covered in this chapter, therefore, is limited to those topics which are of more general importance—those subjects which have been frequently treated in past examinations and which have the greatest probability of being treated in future examinations. However, if you are an electrical engineer, it would be advisable to review amplifiers, signal generators, and transistor circuitry on your own. You just might pick up a few easy points in your specialty.

The subject which has received the greatest amount of emphasis in past examinations is alternating-current circuit analysis, including power factor and phase angle. Following in importance have been resistance and Ohm's law, dc circuitry, motor-generator theory, three-phase circuitry, voltmeters, batteries, and basic electrical relationships.

6-1. ELECTRON THEORY

Many of the various electrical phenomena can be explained by means of the electron theory. Current is the flow of electrons. Voltage differential between two points is the potential difference between those two points and is equal to the work done per unit charge against electrical forces when a charge is transported from one point to the other. Also remember that like charges repel and unlike attract. The force of repulsion or attraction is proportional to the product of the two charges divided by the square of the distance between them.

6-2. OHM'S LAW AND RESISTANCE

Whenever there is a flow of electrons through a conductor, there is a resistance to that flow. The resistance is described by Ohm's law, $V = IR$. The electrical resistance between two points is proportional to the distance and inversely proportional to the cross-sectional area of the conductor. As an illustration, let us review a past examination problem.

■ A copper wire $1/2$ in. in diameter and 1000 ft long has a resistance of 0.05 ohms at 20°C. What is the resistance of 500 ft of $1/2$-in. square wire of similar material?

$$\frac{R_{\text{square}}}{R_{\text{round}}} = \frac{A_r}{A_s} \times \frac{L_s}{L_r}$$

$$R_S = 0.05 \times \frac{0.196}{0.250} \times \frac{500}{1000} = 0.0196 \text{ ohms}$$

It should also be remembered that the equivalent resistance for a group of resistances in series is equal to the sum of the individual resistances: $R = R_1 + R_2 + R_3 + \cdots$. The equivalent resistance for a group of resistances in parallel can be determined from the relationship $1/R = 1/R_1 + 1/R_2 + 1/R_3 + \cdots$.

For illustration, take the past examination problem:

■ Direct current flows through the circuit sketched in Fig. 6-1a. If the impressed voltage on terminals A and B is 25 volts, what is the line current?

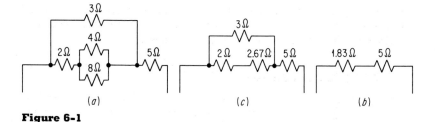

Figure 6-1

We have a series-parallel circuit here, and the easiest way to determine the current is to determine the equivalent circuit resistance and apply Ohm's law. Do this in steps. The first step is to calculate the equivalent resistance for the parallel circuit of 4 ohms and 8 ohms. $1/R = (1/4) + (1/8)$; $R = 2.67$, which gives the equivalent circuit shown in Fig. 6-1b. Adding 2 plus 2.67 gives 4.67 ohms, series resistance in the branch. This gives another parallel circuit, for which the equivalent resistance is

$$\frac{1}{R_2} = \frac{1}{3} + \frac{1}{4.67}$$

$R_2 = 1.83$ ohms, and we have the equivalent circuit shown in Fig. 6-1c. The line current is $I = V/R_T = 25/6.83 = 3.66$ amp.

A very important application of resistance and Ohm's law is the Wheatstone bridge. An example of this is afforded by a past examination problem:

■ In the circuit shown in Fig. 6-2, what is the required resistance of X so that $V_1 = 0$? With the value of X found above and 120 V applied, what is the voltage V_2?

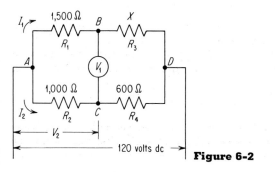

Figure 6-2

First add the letters A, B, C, and D to the figure to aid in identification of the points. For a general discussion also label the resistances R_1, R_2, R_3, and R_4, and the currents I_1 and I_2.

For voltage V_1 to be zero, the voltage drops I_1R_1 and I_2R_2 must be equal. This means that I_1R_3 and I_2R_4 must also be equal, giving $I_1R_1 = I_2R_2$ and $I_1R_3 = I_2R_4$. Dividing one equation by the other gives $(R_1/R_3) = (R_2/R_4)$.

$$R_3 = R_1\frac{R_4}{R_2} = 1500\frac{600}{1000} = 900 \text{ ohms}$$

To determine the voltage differential V_2, calculate either I_1 or I_2. Since the voltage difference between B and C is zero ($V_1 = 0$), there will be a drop of 120 volts across the branch ABD and across the branch ACD. Since $V = IR$,

$$I_1 = \frac{120}{1500 + 900} = 0.05 \text{ amp}$$

and the voltage drop from A to B equals $0.05 \times 1500 = 75$ volts.

To check, $I_2 = 120/(1000 + 600) = 0.075$ amp, giving

$$V_2 = 0.075 \times 1000 = 75 \text{ volts}$$

as before.

6-3. BATTERY INTERNAL RESISTANCE

Closely associated with the general subject of circuit resistance is the internal resistance of a battery. Any battery will have some internal resistance which will serve to reduce the effective potential as the current increases. A battery can be considered as consisting of a source of emf with a resistance, the internal resistance of the battery, in series. This equivalent circuit is shown in Fig. 6-3. The usefulness of this concept of an equivalent circuit may be illustrated by a past examination problem.

Figure 6-3

■ Two batteries, identical in every way, are connected in series across a 5-ohm resistor, and the current is found to be 0.2 amp. The same cells are then connected in parallel with the same 5-ohm resistor, and the current is found to be 0.16 amp across the 5-ohm resistor. What are the open-circuit voltage and the internal resistance of the batteries?

First draw the equivalent circuits of the two hookups (Fig. 6-4). For the series circuit we have

$$2V - 2I_1R_i - 5I_1 = 0$$
$$2V = 0.4R_i + 1.00$$

For the parallel circuit

$$V - \tfrac{1}{2}I_2R_i - 5I_2 = 0$$
$$V = 0.08R_i + 0.80$$

Combining these two equations gives

$$0.2R_i + 0.50 = 0.08R_i + 0.80$$
$$R_i = 2.50 \text{ ohms}$$

which is the internal resistance of the batteries; $V = 1.00$ volt, open-circuit voltage.

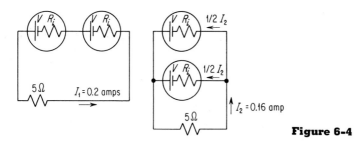

Figure 6-4

6-4. METER INTERNAL RESISTANCE

A closely associated subject is that of electrical metering and internal resistance of meters. A past problem asked:

- An electrical milliameter, having an internal resistance of 500 ohms, reads full scale when 1 mA (milliampere—0.001 amp) of current is flowing.

  a. If the meter is to be used as an ammeter reading 0.1 amp full scale, what size (ohms) resistor should be used in parallel with it?

  b. If used as a voltmeter reading 100 volts full scale, what size (ohms) resistor should be used in series with the milliameter?

An ammeter is essentially a voltmeter which measures the voltage drop across a resistance or shunt. The milliameter in the problem reads full scale when measuring 0.5 volt differential:

$$V = 0.001 \times 500 = 0.5 \text{ volt}$$

To be used as an ammeter reading 0.1 amp full scale, it would be connected as shown in Fig. 6-5a. Since a current of 0.001 amp will flow through the meter when it reads full scale, 0.099 amp must flow through the shunt resistance. The voltage drop across the two parallel resistances must equal 0.5 volt. Then, by Ohm's law, $R_1 = 0.5/0.099 = 5.05$ ohms.

Another way of looking at this is that the total resistance of the parallel circuit must equal 0.5 volt/0.1 amp = 5.0 ohms. Using the parallel-resistance rule, this gives $\frac{1}{5} = 1/R_1 + 1/500$, or $R_1 = 1/0.198 = 5.05$ ohms.

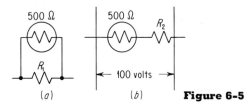

(a) (b) **Figure 6-5**

To use the milliammeter as a voltmeter reading 100 volts full scale, it should be connected as shown in Fig. 6-5b. The total resistance in the meter circuit is $(500 + R_2)$ ohms. The current through the meter must be 0.001 amp for full-scale reading at 100 volts. $100 = 0.001R$; $R = 100,000$ ohms. The total resistance $R = 500 + R_2$; $R_2 = 99,500$ ohms required.

6-5. KIRCHHOFF'S LAWS

Two principles which hold in the analysis of electric circuits are known as Kirchhoff's Laws. These may be stated as follows:

1. The algebraic sum of all currents flowing into (and out of) any junction of an electric circuit equals zero.

2. The algebraic sum of all voltages and voltage drops around a closed circuit equals zero.

The first law states that the sum of all currents merging at any one point must equal zero. That is, the sum of the currents entering a point equals the sum of the currents leaving that point.

The second law states that the sum of all voltages added into a circuit loop from batteries, generators, or other sources must equal the voltage drops in the loop due to the resistance elements in that loop.

These two laws or principles can best be illustrated by an example. In the circuit in Fig. 6-6 it is desired to determine the current through the 30-ohm resistance.

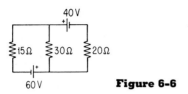

60 V   **Figure 6-6**

Redraw the circuit as in Fig. 6-7 and assume the directions of the branch currents as shown. In this case it has been assumed that the positive direction of current flow is out of the positive terminal of each battery. If one or more of the assumed directions is incorrect, the value obtained for that current when the circuit equations are solved will be negative. A negative value indicates that the current actually flows in a direction opposite to that which was assumed.

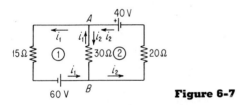

60 V   **Figure 6-7**

Currents i_1 and i_2 meet at junctions A and B. From the first law the sum of all currents meeting at a junction must equal zero. That is, i_1 enters junction B, so current in the amount of i_1 must leave. Similarly i_2 leaves junction B, so current in the amount of i_2 must enter junction B. Thus the current in the 30-ohm resistance, going in the positive direction around loop 1, i.e., in the direction of the assumed current flow, will equal $i_1 - i_2$, since that is the net current that flows through that branch. Similarly, as loop 2 is traversed in the direction of assumed current flow, the current through the 30-ohm resistance will equal $i_2 - i_1$. This is the same current in amperes, but in the opposite direction.

Summing the voltages and voltage drops around loop 1 gives:

$$60 - (i_1 - i_2) \times 30 - i \times 15 = 0$$

which reduces to

$$4 = 3i_1 - 2i_2$$

Note that the voltage drop through the 30-ohm resistance equals 30 times the net current flowing through that resistance, $i_1 - i_2$. This will also be true when the voltage drops through loop 2 are calculated.

Summing the voltages and voltage drops around loop 2 gives:

$$40 - (i_2 - i_1) \times 30 - i_2 \times 20 = 0$$

which reduces to

$$4 = 5i_2 - 3i_1$$

Solving the two equations simultaneously gives:

$$i_1 = 3.11 \text{ amp}$$
$$i_2 = 2.67 \text{ amp}$$

The current through the 30-ohm resistor would equal

$$i_1 - i_2 = 3.11 - 2.67 = 0.44 \text{ amp}$$

A somewhat more complex problem is given in Fig. 6-8. The directions of the currents may be assumed as shown. The current around loop 1 is taken as i_1, around loop 2 as i_2, etc. Both of the currents i_1 and i_2 flow through branch BD, but in opposite directions. So the net current flowing through branch BD equals $i_1 - i_2$ in the direction shown. Similarly the net current flowing through branch AD equals $i_3 - i_1$ in the direction shown. And the net current flowing through branch DC equals $i_3 - i_2$ in the direction shown.

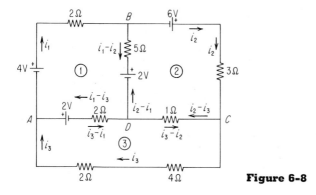

Figure 6-8

The equations for the summation of the voltages around each of the three loops can now be written. It should be remembered that current flows out of the positive pole of a battery and into the negative pole. Similarly, if current flows through a resistance, a voltage drop equal to iR will result, but if the assumed current direction is opposite to the direction in which the loop is being analyzed, the product of iR will add to the assumed voltage.

For loop 1:

$$4 - 2i_1 - 5(i_1 - i_2) - 2 + 2(i_3 - i_1) + 2 = 0$$
$$4 = 9i_1 - 5i_2 - 2i_3$$

For loop 2:

$$6 - 3i_2 + 1(i_3 - i_2) + 2 + 5(i_1 - i_2) = 0$$
$$8 = -5i_1 + 9i_2 - i_3$$

For loop 3:

$$-2 - 2(i_3 - i_1) - 1(i_3 - i_2) - 6i_3 = 0$$
$$2 = 2i_1 + i_2 - 9i_3$$

Solving these three equations simultaneously gives

$$i_1 = 1.4847 \text{ amp}$$
$$i_2 = 1.7449 \text{ amp}$$
$$i_3 = 0.3009 \text{ amp}$$

The current flowing through branch BD (through the 5-ohm resistor) equals

$$i_1 - i_2 = -0.264; \text{ or } 0.264 \text{ amp}$$

Similarly, the current flowing through branch AD would equal 1.180 amp and the current through branch DC would equal 1.444 amp.

6-6. INDUCTANCE AND CAPACITANCE

In addition to resistance, ac circuits usually contain inductance and/or capacitance.

An illustrative problem is the following:

■ A single-phase load takes 855 watts and 9.75 amp from a source when a 60-cycle voltage of 120 volts is applied. It is known that the current lags the voltage. Determine:

  a. Resistance of the load

  b. Impedance of the load

  c. Reactance of the load

  d. Power factor of the load

  e. Power-factor angle

  f. Inductance or capacitance, whichever predominates

6-7. POWER FACTOR

The power taken by an electric load is measured in watts where watts = volts × amperes × power factor. The power factor equals the in-phase component of the current divided by the total current, or watts divided by voltamperes. In this case the power factor pf = $855/(120 \times 9.75)$ = 73 percent. The impedance of the load $Z = E/I$ = $120/9.75$ = 12.3 ohms. The impedance also equals $\sqrt{(X_L - X_C)^2 + R^2}$, where X_L is the inductive reactance and X_C is the capacitive reactance. The power factor is equal to R/Z. The resistance of the load is then

$$0.73 \times 12.3 = 8.98 \text{ ohms}$$

and the impedance is 12.3 ohms. The reactance $X = X_L - X_C$ is equal to $X = \sqrt{Z^2 - R^2}$ = $\sqrt{151.3 - 80.8}$ = 8.40 ohms. The power factor, already determined, equals 0.73, or 73 percent. The power-factor angle, more commonly called "phase angle," φ, equals the arc cosine of the power factor. In this case

$$\varphi = \cos^{-1} 0.73 = 43.1° \text{ lagging}$$

6-8. PHASE ANGLES—LEADING OR LAGGING

It is given that the current lags the voltage. From this we know that the inductance exceeds the capacitance. There are different methods of remembering this. One is by remembering the phrase, "*ELI* the *ICE* man." The *ELI* indicates that E, or voltage, comes before, or leads, I the current for an inductive load L. *ICE* indicates that current I comes before, or leads, the voltage E for a capacitive load C. The current lags; therefore the voltage leads and the inductance predominates. $X_L = 2\pi f L$, or $L = X_L/(2\pi f)$, where L is in henrys (H).

$$L = \frac{8.40}{2\pi \times 60} = 0.0223 \text{ H}$$

or 22.3 mH. Capacitive impedance $X_C = 1/2\pi f C$, where C is in farads.

6-9. VECTOR DIAGRAMS

Figure 6-9a shows the vector diagram of the current and the phase angle. Note that the in-phase component of the current will equal watts/volts = $855/120$ = 7.12 amp.

$$I \times \cos \phi = 9.75 \times 0.73 = 7.12 \text{ amp}$$

The time diagram is shown in Fig. 6-9b. Note that the current starts, or

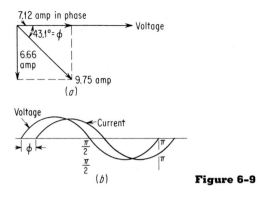

7.12 amp in phase

43.1° = ϕ

6.66 amp

9.75 amp

(a)

Voltage

Current

ϕ

$\frac{\pi}{2}$

$\frac{\pi}{2}$

π

π

(b)

Figure 6-9

passes through zero, at a later time, $\omega t = 0$. This indicates that the current lags, since it arrives at similar points (maximum, zero, minimum) after the voltage, or later than the voltage.

6-10. RESONANT CIRCUITS

Resonance occurs when the inductive impedance of a circuit equals the capacitive impedance. In the case of a series circuit the resulting impedance will equal the resistance for resonance (series resonance) and the current will be a maximum

$$Z = \sqrt{R^2 + (X_L - X_C)^2} = R \qquad \text{for resonance}$$

In the case of resonance the current will be in phase with the voltage and the power factor will equal one. For series resonance the inductive reactance equals the capacitive reactance. For the case just considered, if the current had been limited only by the resistance, it would have been $120/8.98 = 13.36$ amp, and the power would have equaled $W = I^2R = 13.36^2 \times 8.98 = 1605$ watts, instead of 855 watts as before.

As an example, a past problem asked:

■ Find the resonant frequency of a series circuit consisting of a 0.005-μF capacitor and a 10-mH inductance.

$X_L = 2\pi fL$, and $X_C = 1/(2\pi fC)$. For resonance, $X_L = X_C$,

$$f = \frac{1}{2\pi\sqrt{LC}} = \frac{1}{2\pi\sqrt{0.005 \times 10^{-6} \times 10 \times 10^{-3}}} = 22{,}500 \text{ cps}$$

Parallel resonance (that is, resonance in a parallel circuit) also occurs when the inductive impedance equals the reactive impedance and the phase angle equals zero. However, in general, in the case of a parallel circuit, this condition will not occur when the inductive and capacitive reactances are equal. Take, for example, the circuit shown in Fig. 6-10. The inductive and capacitive reactances are equal, but the current is not in phase with the voltage. In order for the parallel circuit shown to be resonant, holding the

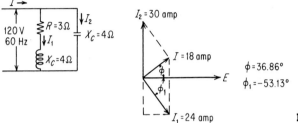

Figure 6-10

inductive branch constant, the capacitive reactance must be increased until the leading branch current is reduced to $19.2a$, the lagging component of the lagging branch current.

$$X_c = 120/19.2 = 6.25$$

The resulting resonant circuit with the corresponding phasor diagram is shown in Fig. 6-11.

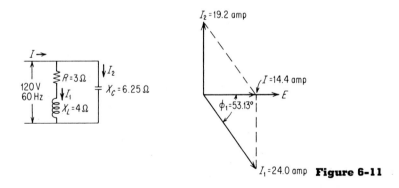

Figure 6-11

Alternatively, if the capacitive reactance were held constant, the inductive reactance would have to be reduced to 2.01 ohms. The new circuit,

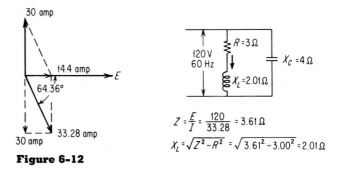

Figure 6-12

phasor diagram, and calculations are shown in Fig. 6-12. Note how construction of the phasor diagram simplifies the analysis of the circuit.

6-11. CAPACITORS IN SERIES AND PARALLEL

An equivalent capacitor to replace three capacitors in parallel is equal to the sum of the three capacitors. Looking at Fig. 6-22 we see that the voltage across the three capacitors is the same and equals V. The current is made up of the three branch currents: $I = I_1 + I_2 + I_3$. From $E = IX_C$, we get $I = V/X_C$; since $X_C = 1/(2\pi fC)$, $I = V \times 2\pi fC$. Similarly $I_1 = V \times 2\pi fC_1$, $I_2 = V \times 2\pi fC_2$, and $I_3 = V \times 2\pi fC_3$, which combine to give $C = C_1 + C_2 + C_3$.

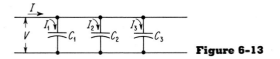

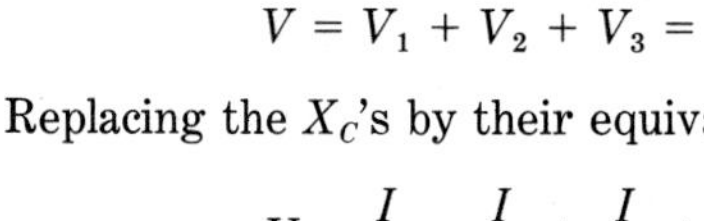

Figure 6-13

The case of the series circuit is shown in Fig. 6-14. Here

$$V = V_1 + V_2 + V_3 = IX_{C1} + IX_{C2} + IX_{C3} = IX_C$$

Replacing the X_C's by their equivalents and canceling like terms gives

$$V = \frac{I}{C} = \frac{I}{C_1} + \frac{I}{C_2} + \frac{I}{C_3} \qquad \frac{1}{C} = \frac{1}{C_1} + \frac{1}{C_2} + \frac{1}{C_3}$$

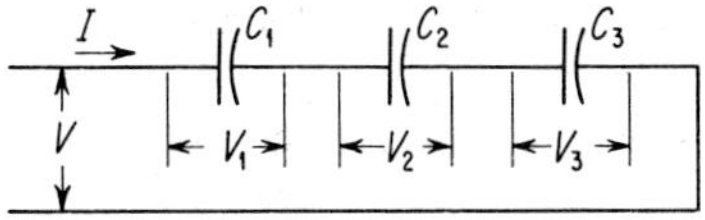

Figure 6-14

6-12. INDUCTANCES IN SERIES AND PARALLEL

A circuit of inductances in series is shown in Fig. 6-15. Here $V = V_1 + V_2 + V_3$, or $IX_L = IX_{L_1} + IX_{L_2} + IX_{L_3}$; since $X_L = 2\pi fL$, this gives

$$L = L_1 + L_2 + L_3$$

for inductances in series, which we could reason from the fact that three (or more) coils connected end to end are equivalent to one coil divided into three (or more) sections.

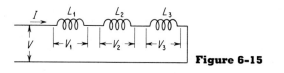

Figure 6-15

Inductances in parallel (Fig. 6-16) give us $I = I_1 + I_2 + I_3$ and $V/X_L = V/X_{L_1} + V/X_{L_2} + V/X_{L_3}$, which reduces to

$$\frac{1}{L} = \frac{1}{L_1} + \frac{1}{L_2} + \frac{1}{L_3}$$

for inductances in parallel.

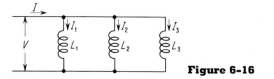

Figure 6-16

The relationships for resistances in series and parallel can be derived in the same way.

6-13. COMPLEX NUMBER NOTATION

Complex number notation is commonly used to describe the electrical vector quantities. Electrical vectors are now frequently called "phasors" to differentiate them from force vectors. Since a phasor has either a leading component or a lagging component (Fig. 6-17), it is convenient to use the complex number notation to describe it. The out-of-phase component corresponds to the imaginary part, and the in-phase component corresponds to the real part of the complex number. Thus in Fig. 6-17 $I_1 = 5 + j2$, $I_2 = 3 - j3$, $I_3 = 4$. $I = I_1 + I_2 + I_3 = 12 - j$. An example is afforded by a past examination problem:

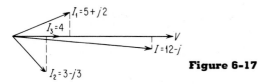

Figure 6-17

■ The admittance of an a-c circuit is given as $Y = 4 - j3$.
   a. What is the simplified complex expression for the impedance?
   b. What is the magnitude of the pure resistance in the circuit?

The admittance of a circuit is defined as the reciprocal of the impedance, or $Y = 1/Z$. In this problem, then,

$$Z = \frac{1}{Y} = \frac{1}{4 - j3}$$

Simplification of this fraction, which is an example of division by a complex number, is effected by multiplying both numerator and denominator by the conjugate of the denominator. The conjugate of a complex number is formed by changing the sign of the complex term. For this case the conjugate of the denominator would be $4 + j3$. Simplification is achieved as follows:

$$Z = \frac{1}{4 - j3} \times \frac{4 + j3}{4 + j3} = \frac{4 + j3}{16 + 9} = 0.16 + j0.12$$

This is the answer to part (*a*). The answer to part (*b*) is $R = 0.16$ ohm. Note that the $+j0.12$ indicates a lagging phase angle (current lags voltage or voltage leads current) and a predominantly inductive circuit (*ELI*, voltage leads current in an inductive circuit). The phase angle $\phi = \tan^{-1} 0.12/0.16 = 36.9°$ (Fig. 6-18). The magnitude of the impedance equals

$$\sqrt{0.12^2 + 0.16^2} = 0.200 \text{ ohm}$$

Figure 6-18 shows the reactive component above the axis, which is in accordance with our convention. The voltage will equal current times impedance. This gives an identically shaped diagram for the voltage, shown also in Fig. 6-18, where the voltage leads the current, or the current lags the voltage.

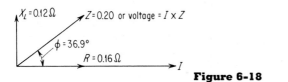

Figure 6-18

The complex number notation serves to simplify many of the ac circuit calculations. For a series circuit, the total impedance $Z = Z_1 + Z_2 + Z_3$, which follows directly from the fact that $Z = \sqrt{R^2 + (X_C - X_L)^2}$. Parallel-circuit analysis is much more troublesome, however; it can be handled more simply if one remembers that

$$\frac{1}{Z} = \frac{1}{Z_1} + \frac{1}{Z_2} + \frac{1}{Z_3} + \cdots$$

where all the Z's are in the complex number notation, than if one uses a step-by-step analysis.

Phasors can also be written in polar form where the length of the phasor and the angle it makes with the abscissa are specified. Referring to Fig. 6-17, the current $I_1 = 5 + j2$. This can be written as $I_1 = 5.385\underline{/21.80°}$ where $5.385 = \sqrt{5^2 + 2^2}$ and is the length of the phasor. The phase angle $\theta = \tan^{-1}(2/5) = 21.80°$. Phasor multiplication and division are much simpler to perform with the quantities in the polar form than with quantities in the rectangular-coordinate form.

A phasor in polar form is written as $P = M\underline{/\theta}$ where M equals the magnitude of the phasor and θ is the angle through which the vector is rotated from zero, the right-hand horizontal axis. A positive angle denotes counterclockwise rotation; a negative angle denotes clockwise rotation. Again referring to Fig. 6-17, the current $I_2 = 3 - j3$ can also be written as $I_2 = 4.24\underline{/-45°}$.

The product of two phasors P_1 and P_2 equals

$$P_1 \times P_2 = (M_1 \times M_2)\underline{/\theta_1 + \theta_2}$$

$$\text{Division of two phasors } \frac{P_1}{P_2} = \left(\frac{M_1}{M_2}\right)\underline{/\theta_1 - \theta_2}$$

This can be illustrated by the previous sample problem:

$$Z = \frac{1}{Y} = \frac{1}{4 - j3} \qquad (4 - j3 \text{ can be written } 5\underline{/-36.87°}$$

$$Z = \frac{1}{5\underline{/-36.87°}} = \frac{1\underline{/0°}}{5\underline{/-36.87°}} = 0.20\underline{/36.87°}$$

This can be transformed to rectangular coordinates:

$$Z' = 0.20 \cos 36.87° + j0.20 \sin 36.87° = 0.16 + j0.12$$

As an example of parallel-circuit analysis, let us review the past examination problem:

■ Two parallel branches of an electrical circuit, one consisting of 5 ohms resistance and 0.023 H inductance, and the other of 8.66 ohms resistance and 530 μF capacitance, take 30 amp from a 60-cycle main.

a. What is the current in each branch?

b. What are the instantaneous values of the currents in each branch when the current in the main has an instantaneous value of 36.7 amp?

First sketch the circuit (Fig. 6-19) and calculate the individual impedance.

$$X_L = 2\pi fL = 2\pi \times 60 \times 0.023 = 8.67 \text{ ohms}$$

$$X_C = \frac{1}{2\pi fC} = \frac{1}{2\pi \times 60 \times 530 \times 10^{-6}} = 5.00 \text{ ohms}$$

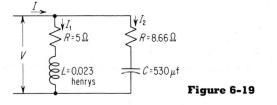

Figure 6-19

These give $Z_1 = 5 + j8.67$ and $Z_2 = 8.66 - j5.00$.

$$\frac{1}{Z} = \frac{1}{5 + j8.67} + \frac{1}{8.66 - j5.00} = \frac{5 - j8.67}{100} + \frac{8.66 + j5.00}{100}$$

$$= 0.1366 - j0.0366$$

$$Z = \frac{1}{0.1366 - j0.0366} = \frac{0.1366 + j0.0366}{0.02} = 6.83 + j1.83$$

or

$$Z = \sqrt{6.83^2 + 1.83^2} = 7.07 \text{ ohms}$$

The applied voltage $V = IZ = 30 \times 7.07 = 212$ volts. The same voltage acts across each branch, so

$$I_1 = \frac{V}{Z_1} = \frac{212}{5 + j8.67} = 10.6 - j18.36$$

or

$$I_1 = \sqrt{10.6^2 + 18.36^2} = 21.2 \text{ amp}$$

$$I_2 = \frac{V}{Z_2} = \frac{210}{8.66 - j5.00} = 18.36 + j10.6$$

or

$$I_2 = \sqrt{18.36^2 + 10.6^2} = 21.2 \text{ amp}$$

These could also have been calculated using the polar form for the phasors.

$$I_1 = \frac{212}{5 + j8.67} = \frac{212\underline{/0°}}{10\underline{/+60°}} = 21.2\underline{/-60°} = 10.6 - j18.27$$

$$I_2 = \frac{210}{8.66 - j5.00} = \frac{210\underline{/0°}}{10\underline{/-30°}} = 21.0\underline{/30°} = 18.19 + j10.5$$

which are the values required for part (a).

The current in the main is given as 30 amp. This is the effective, or root mean square, current. The maximum current would equal

$$I_{\text{eff}} \times \sqrt{2} = 30 \times 1.414 = 42.4 \text{ amp}$$

The line current would equal the vector sum of the two branch currents $I = I_1 + I_2 = 28.96 - j7.76$. The phase angle would be $\tan^{-1} 0.268 = 15°$ lagging. Similarly for the two branch currents, $\phi_1 = 60°$ lagging and $\phi_2 = 30°$ leading. These three currents can then be plotted, since they are all of the form $I = I_{\text{max}} \sin (\omega t + \phi)$. The main current will have an instantaneous value of 36.7 amp when $\sin (\omega t - 15) = 36.7/42.4 = 0.866$, or $\omega t - 15 = 60°$ and $\omega t = 75°$.

The instantaneous value of branch current

$$I_1 = 21.2\sqrt{2} \sin (\omega t - 60°)$$

giving $I_1 = 30 \sin 15° = 7.76$ amp.

The instantaneous value of branch current

$$I_2 = 21.2\sqrt{2} \sin (\omega t + 30°)$$

giving $I_2 = 30 \sin 105° = 29$ amp.

Taking the voltage as the reference phasor, plot the effective currents as in Fig. 6-20.

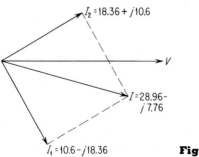

Figure 6-20

The construction of phasor diagrams can be of considerable aid in working ac problems and in understanding the effect of changing different components. The author strongly recommends that you draw the phasor diagram whenever possible. An example showing how this can simplify an analysis is afforded by the not-uncommon requirement of correcting a power factor. Take, for example, a 230-volt, 60-Hz, single-phase, 50-hp motor which has a power factor of 70 percent. It is desired to correct this to 90 percent. What size capacitor is required?

The line current initially equals

$$\frac{50 \times 746}{230 \times 0.70} = 231.7 \text{ amp}$$

and the phase angle equals

$$\cos^{-1} 0.700 = 45.6°$$

The corrected phase angle will equal $\cos^{-1} 0.900 = 25.8°$. The final current must equal

$$\frac{50 \times 746}{230 \times 0.90} = 180.2 \text{ amp}$$

The phasor diagram is shown in Fig. 6-21. From this diagram we can see that I_c, the capacitor current, must equal

$$(231.7 \sin 45.6° - 180.2 \sin 25.8°)$$

or $(162.2 \tan 45.6° - 162.2 \tan 25.8°)$. Either of these gives

$$I_c = 87.2 \text{ amp}$$
$$X_c = \frac{230}{87.2} = 2.64 \text{ ohms}$$
$$C = 1/(2\pi f X_c) = 0.001005 \text{ F, or } 1005 \ \mu\text{F}$$

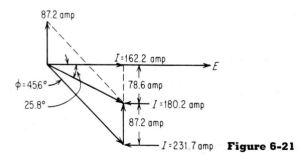

Figure 6-21

It might also be of interest to note that since the power loss due to line resistance equals I^2R, the installation of the capacitor would reduce the line power loss by almost 40 percent.

A somewhat more involved calculation utilizing the same principles can be illustrated with the analysis of the circuit shown in Fig. 6-22. Determine the load current, the voltage across the load, and the current in each branch.

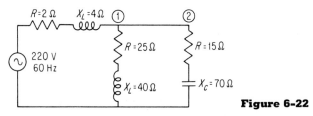

Figure 6-22

First determine the total impedance.
Line impedance:

$$Z_L = 2 + j4 = 4.47\underline{/63.43°}$$

Load 1 impedance:

$$Z_1 = 25 + j40 = 47.17\underline{/57.99°}$$

Load 2 impedance:

$$Z_2 = 15 - j70 = 71.59\underline{/-77.91°}$$

For the two parallel loads,

$$\frac{1}{Z} = \frac{1}{Z_1} + \frac{1}{Z_2} = \frac{1}{25 + j40} + \frac{1}{15 - j70} = (0.01124 - j0.01798) +$$
$$(0.00293 + j0.01366)$$

$$Z_{\text{load}} = \frac{1}{0.01417 - j0.00432} = 64.57 + j19.69 = 67.51\underline{/16.96°}$$

or

$$\frac{1}{Z} = \frac{1\underline{/0°}}{47.17\underline{/57.99°}} + \frac{1\underline{/0°}}{71.59\ \underline{-/77.91°}} =$$

$$0.02120\underline{/-57.99°} + 0.01397\underline{/77.91°}$$

$$\frac{1}{Z} = (0.01124 - j0.01798) + (0.00293 + j0.01366) = 0.01481\underline{/-16.95°}$$

$$Z_{\text{load}} = \frac{1\underline{/0°}}{0.01481\underline{/-16.95°}} = 67.52\underline{/16.95°} = 64.58 + j19.68$$

Total impedance

$$Z_T = (64.58 + j19.68) + (2 + j4) = 66.58 + j23.68$$
$$Z_T = 70.67\underline{/19.58°}$$

$$I = \frac{E}{Z} = \frac{220}{66.50 + j23.68} = 2.93 - j1.04 \quad \text{or} \quad 3.11\underline{/-19.54°}$$

or $$I = \frac{220\underline{/0°}}{70.67\underline{/19.58°}} = 3.11\underline{/-19.58°}$$

The line current equals 3.11 amp and lags the voltage by 19.6°.
The voltage across the load equals the applied voltage less the drop
across the line impedance.
Drop across line impedance $= IZ_{\text{line}}$

$$E = (2.93 - j1.04)(2 + j4) = 10.02 + j9.64 = 13.90\underline{/43.89°}$$

$$E = (3.11\underline{/-19.58°}) \times (4.47\underline{/63.43°}) = 13.90\underline{/43.85°}$$

Voltage across the load $E_L =$

$$(220 + j0) - (10.02 + j9.64) = 209.98 - j9.64$$

The voltage across the load can also be calculated by multiplying the
current by the impedance of the load.

$$E_{\text{load}} = IZ_{\text{load}} = (2.93 - j1.04)(64.57 + j19.69) = 209.67 - j9.46$$

or

$$E_{\text{load}} = (3.11\underline{/-19.58°}) \times (67.51\underline{/16.96°}) = 209.96 - 2.62° =$$
$$209.74 - j9.60$$

The branch currents can be calculated similarly.

$$I_1 = \frac{E_{\text{load}}}{Z_1} = \frac{209.98 - j9.64}{25 + j40} = 2.19 - j3.88$$

or

$$I_1 = \frac{209.96\underline{/-2.62°}}{47.17\underline{/57.99°}} = 4.45\underline{/-60.61°} = 2.18 - j3.88$$

$$I_2 = \frac{E_{\text{load}}}{Z_2} = \frac{209.98 - j9.64}{15 - j70} = 0.746 + j2.84$$

or

$$I_2 = \frac{209.96\underline{/-2.62°}}{71.59\underline{/-77.91°}} = 2.93\underline{/75.29°} = 0.744 + j2.83$$

$$I_1 + I_2 = (2.18 - j3.88) + (0.746 + j2.84) = 2.93 - j1.04$$

which is the load current.

$$\sqrt{2.93^2 + 1.04^2} = 3.11 \qquad \phi = \tan^{-1}(1.04/2.93) = 19.54°$$

One type of problem which appears rather frequently is one wherein the examinee is asked what capacitance must be added to a circuit to raise the power factor to a particular value. Such a problem could be as follows (see also Fig. 6-21).

■ A single-phase, 10-hp, 230-volt 85 percent efficient induction motor operates at full load with a 75 percent power factor. The power factor drops to 65 percent when the motor is partially loaded to 4 hp, where the efficiency drops to 68 percent. The frequency of the supply current is 60 Hz. How large a capacitor would have to be added to the circuit to raise the power factor to 95 percent for either condition?

The circuit can be sketched as shown in Fig. 6-23. The equivalent circuit of the motor will be a resistance and an inductance in series. The capacitor will be connected in parallel with the motor load.

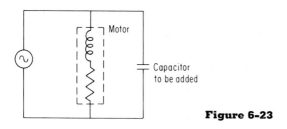

Figure 6-23

At full load the in-phase component of the current would equal

$$\frac{10 \times 746}{230 \times 0.85} = 38.16 \text{ amp}$$

The phase angle $\phi = \cos^{-1} 0.75 = 41.41°$ lagging.
The out-of-phase component of the current is

$$38.16 \times \tan 41.41° = 33.65 \text{ amp}$$

It is desired to raise the power factor to 95 percent which would mean a phase angle of $\phi = \cos^{-1} 0.95 = 18.19°$ and an out-of-phase component equal to $38.16 \times \tan 18.19° = 12.54$ amp, so the current through the capacitor would have to equal

$$33.65 - 12.54 = 21.11 \text{ amp}$$

and the impedance

$$X_c = \frac{230}{21.11} = 10.90 \text{ ohms}$$

This can be shown graphically (see Fig. 6-24). For the unloaded condition,

$$\text{In-phase current} = \frac{4 \times 746}{230 \times 0.68} = 19.08 \text{ amp}$$

The phase angle $\phi = \cos^{-1} 0.65 = 49.46°$.

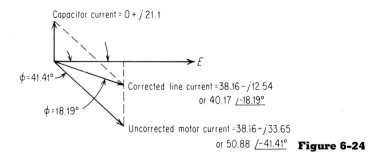

Capacitor current = $0 + j\,21.1$

$\phi = 41.41°$

$\phi = 18.19°$

Corrected line current = $38.16 - j12.54$
or $40.17 \; \underline{/-18.19°}$

Uncorrected motor current = $38.16 - j33.65$
or $50.88 \; \underline{/-41.41°}$ **Figure 6-24**

The out-of-phase component would equal

$$19.08 \times \tan 49.46° = 22.31 \text{ amp}$$

The out-of-phase component must be corrected to

$$19.08 \times \tan 18.19° = 6.27 \text{ amp}$$

which requires a current equal to $22.31 - 6.27 = 16.04$ amp through the capacitor. The required impedance

$$X_c = \frac{230}{16,04} = 14.34 \text{ ohms}$$

This condition is shown graphically in Fig. 6-25.

The capacitive component lowers the total current through the system and thus reduces the line losses (I^2R losses in the transmission line). It also reduces the voltage drop from the source to the load.

The product of reactive current and voltage is termed *reactive power* or *wattless power* and is measured in reactive volt-amperes (var, voltamperes reactive). The reactive power for the fully loaded motor equals

$$EI \sin \phi = 230 \times 50.88 \times \sin 41.41° = 7740 \text{ var, or } 7.74 \text{ kvar}$$

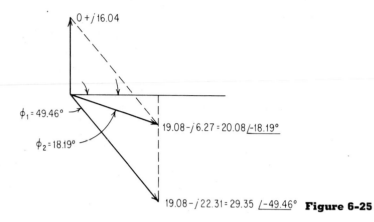

$0 + j\,16.04$

$\phi_1 = 49.46°$

$\phi_2 = 18.19°$

$19.08 - j\,6.27 = 20.08 \underline{/-18.19°}$

$19.08 - j\,22.31 = 29.35 \ \underline{/-49.46°}$ **Figure 6-25**

This is termed "positive reactive power." Capacitive reactance is termed "negative reactive power." The capacitive or negative reactive power required for the fully loaded motor to correct the system power factor to 95 percent would equal

$$230 \times 21.11 = 4855 \text{ var, or } 4.86 \text{ kvar}$$

Similarly the reactive power for the unloaded motor would equal

$$230 \times 29.35 \times \sin 49.46° = 230 \times 22.30 = 5130 \text{ var, or } 5.13 \text{ kvar}$$

and the corrective factor required equals

$$230 \times 16.04 = 3689 \text{ var, or } 3.69 \text{ kvar}$$

The impedance of a capacitor is

$$X_c = \frac{1}{2\pi f c}$$

For the case of the fully loaded motor, $X_c = 10.90$ ohms and

$$C = \frac{1}{2\pi f X_c} = \frac{1}{2\pi \times 60 \times 20.70} = 2.43 \times 10^{-4}\text{F, or } 243 \ \mu\text{F}$$

For the case of the unloaded motor,

$$C = \frac{1}{2\pi \times 60 \times 16.04} = 1.65 \times 10^{-4} \text{ F, or } 165 \ \mu\text{F}.$$

The situation which occurs with the loaded motor controls; the size of the capacitor required to correct the power factor to 95 percent for the case equals 243 μF.

6-14. TRANSFORMERS

A transformer is a device for increasing or decreasing an ac voltage. The losses in a transformer are usually divided into two groups: iron losses, or core losses, and the I^2R losses due to the resistance of the windings. The iron losses are made up of the hysteresis loss and the eddy-current loss. The hysteresis loss is proportional to the frequency of the flux reversals and a power of the flux density. It can be expressed by the equation

$$P_h = k_1 f B^{1.6} \qquad \text{watts}$$

where the 1.6 power = a representative value. (The power of the flux density, called "the Steinmetz exponent," may vary from approximately 1.5 to over 2.0, with 1.6 being the value usually taken.)

f = frequency, Hz

k_1 = a constant depending upon the characteristics of the core

B = flux density, lines or kilolines/in.^{2}

The eddy-current loss in the core may be represented by the equation:

$$P_e = k_2 f^2 B^2 \qquad \text{watts}$$

where the symbols have the same meaning as before except that k_2 is a different constant.

The copper loss equals $I_1{}^2R_1 + I_2{}^2R_2$, where I_1, R_1, I_2, and R_2 are the current and resistance in the primary and secondary windings. The resistances are usually small and can be approximated by a dc test. The ac resistance will, however, be several percent higher, because of eddy-current losses in the copper.

The efficiency of a transformer is equal to

$$\frac{\text{output}}{\text{input}} \qquad \text{or} \qquad \frac{\text{input} - \text{losses}}{\text{input}}$$

The voltage regulation of a transformer in percent equals (no-load secondary voltage minus full-load secondary voltage) divided by (full-load secondary voltage).

$$\text{Regulation} = \frac{V_2' - V_2}{V_2} \qquad \text{percent}$$

Single-phase transformers are made with two windings and also with just one winding. When made with a single winding, they are termed "autotransformers." Schematics of a single-winding transformer are shown in Fig. 6-26 connected for voltage step-down and step-up. In each case $E_2 = E_1(N_2/N_1)$.

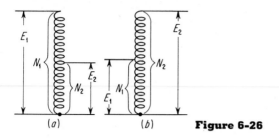

(a)   (b)   **Figure 6-26**

The common part of the winding carries both currents I_1 and I_2, but these currents buck each other, so this part of the winding is more lightly loaded than the other part. The current diagrams for the two cases are shown in Fig. 6-27 for an ideal autotransformer.

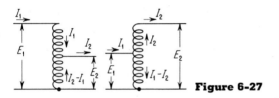

Figure 6-27

To illustrate, look at a past examination problem.

- A 2400/240-volt single-phase additive transformer is connected as shown in Fig. 6-28.
  a. What is the voltage at the load; i.e., is the transformer increasing or decreasing the voltage?
  b. At the instant the load current is 40 amp, what is the current in the supply circuit, assuming no transformer losses?

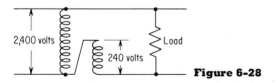

Figure 6-28

First redraw the diagram in the form of an autotransformer, as shown in Fig. 6-29. It is then immediately apparent that it is acting as a step-up transformer and is increasing the voltage. When the load current is 40 amp,

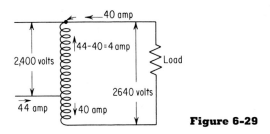

Figure 6-29

the supply current will be (2640/2400) × 40 = 44 amp. It is interesting that in this case the current in the primary would be only 44 − 40 = 4 amp.

6-15. GENERATORS

The generated voltage of the armature of a generator is:

$$E = \frac{N}{\text{paths}} \times \phi \times \text{poles} \times \frac{\text{rpm}}{60} \times 10^{-8}$$

where N = total number of inductors
ϕ = magnetic flux per pole

As an example, a past problem asked the following:

■ It is desired to design a dc generator which will generate an average electromotive force of 230 volts. The armature used has 180 series inductors between brushes. How many poles must be used if the flux per pole is 1.5×10^6 maxwells and the desired speed is 520 rpm?

A line of induction is called a maxwell, so $\phi = 1.5 \times 10^6$.

$$E = 180 \times 1.5 \times 10^6 \times \text{poles} \times {}^{520}/_{60} \times 10^{-8} = 230 \text{ volts}$$

giving $P = 9.82$, or 10 poles. The speed could be reduced slightly or, more probably, the flux lowered a small amount to give the voltage desired.

6-16. DC MOTORS

The action of a motor is quite similar to the action of a generator, and the construction of a dc motor is very nearly the same as the construction of a generator. The purpose of an electric motor is to convert electric power to mechanical power. It does this by utilizing the force exerted on a charge which lies in a magnetic field. The field is supplied by the field coils of the motor and the charge lies in the conductors wound on the armature. The force on a charge $F = qvB$, where v is the velocity of the charge. The charge q that is of interest here is due to the number of electrons affected by the magnetic field. The number of electrons in the conductor, which are also in the magnetic field, is proportional to the current in the conductor divided by the charge velocity. The equations describing this are:

$$i = nq_e vA$$
$$N = nlA$$
$$f = q_e vB$$
$$F = N \times f = nlAq_e Bv = ilB$$

where i = current

n = number of moving charges per unit volume (number of electrons)

q_e = charge on an electron

v = velocity with which the electrons are moving

f = force on one electron

N = number of electrons

l = length of the conductor in the magnetic field

A = cross-sectional area of the conductor

B = magnetic-field flux density

Since the force on a conductor times the distance to the axis gives torque, it follows that motor torque is proportional to the armature current times the field. For constant field, then the torque will be proportional to the armature current.

The static armature current in a dc machine will equal the armature voltage divided by the resistance of the armature winding. Since the armature resistance is ordinarily low, this can mean a dangerously high current. The armature voltage during motor operation equals the applied voltage minus the back emf or the counter emf. The back emf is the voltage generated in the armature by virtue of the fact that as it rotates, the conductors in the armature winding cut the lines of force in the field. This generates a voltage in the armature winding which opposes the applied voltage and gives a net armature voltage equal to the difference of these two. This limits the armature current. The back emf equals speed times flux times a constant. $V = kNB$.

Applied voltage = back emf + (armature current × armature resistance)

This explains why reduced voltages are necessary to start large motors. When the armature is not rotating there is no back emf and the current is limited by the applied voltage. The positioning of starting resistors for dc motors is shown in Fig. 6-30.

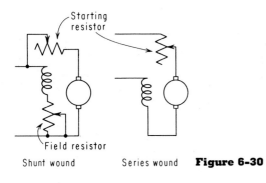

Shunt wound   Series wound   **Figure 6-30**

6-17. SHUNT MOTORS

The field current of the shunt-wound motor is essentially constant. If the load is reduced, the motor will tend to speed up a little until the back emf has increased enough to balance the loss in $I_a R_a$.

$$V = \text{back emf} + I_a R_a$$

As the load reduces, the required torque will reduce, thus reducing the armature current, and the speed must increase to increase the back emf enough to make up the difference. This also explains what happens when the field current of a shunt motor is reduced. A reduction in field current will reduce the flux density and the rotor must travel faster to generate an equivalent back emf. If the field current should be reduced to zero the speed would, theoretically, become infinite.

6-18. SERIES MOTORS

In the case of the series motor a reduction in load would cause a reduction in torque and thus in armature current. Since the armature current is also the field current, this would also cause a reduction in the flux density, and the increase in speed required to maintain a balance between $I_a R_a$ and back emf would rise as a power function. The curves for series and shunt motors are shown in Fig. 6-31. Compound-wound motors would have characteristics in between these two; a characteristic curve for this type is also shown in Fig. 6-31.

This explains why the load should never be removed from a series motor and why such a motor should be directly connected to its load or through gears and not through belts which might slip, come off, or break. It also explains why a shunt-wound motor should not be started or run with zero field current.

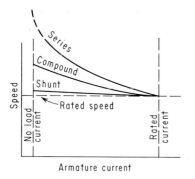

Figure 6-31

As an example of the application of some of these principles, a past examination problem was as follows:

- A dc series-wound motor has a field resistance of $R_f = 0.80$ ohm and an armature resistance of $R_a = 0.50$ ohm. At 1200 rpm the motor delivers 5 hp with a line voltage and line current of 100 volts and 46 amp, respectively.

  a. What is the efficiency of the motor?

  b. What is the counter emf generated by the armature at 1200 rpm?

  c. If the load is decreased so that the speed rises to 1500 rpm, what is the line current?

  d. What is the available torque for condition (*c*), assuming constant efficiency?

Given the following relations: (1) torque varies directly as the armature current squared; (2) speed varies directly as the armature voltage; (3) flux varies directly as the armature current.

First we have the output of the motor, 5 hp, and the input, 100 volts × 46 amp. The efficiency is

$$\frac{5.00 \times 746}{46 \times 100} = 81 \text{ percent}$$

since 746 watts equal 1 hp, which is the answer to part (*a*). This is a series-wound motor, so there is a drop through the field equal to 46 × 0.80 = 36.8 volts. The voltage across the armature is then 100 − 36.8 = 63.2 volts. The drop across the armature due to the armature resistance $I_a R_a$ = 23 volts. The counter electromotive force must then equal 63.2 − 23 = 40.2 volts, which is the answer to part (*b*).

If the speed increases to 1500 rpm, the counter emf must rise, and taking the second given relationship, armature voltage (counter emf) would rise to $(1500/1200) \times 40.2 = 50.2$ volts. As before, we have

$$iR_f + iR_a + \text{back emf} = 100 \text{ volts}$$

so $i = 49.8/1.3 = 38.3$ amp, line current (c).

Since for this case it is given that torque varies as the armature current squared, the torque would equal $(38.3/46)^2 = 0.695$ percent of the original torque. The initial torque from hp $= 2\pi NTq/33,000$ equals 21.9 ft-lb; so the torque for condition (c) is $0.695 \times 21.9 = 15.2$ ft-lb, answer to (d).

The relationship for horsepower is derived from the work done per revolution times rpm, N. Torque Tq is measured in foot-pounds or pounds force at a distance of 1 ft. The work done per revolution is then force acting times distance through which it acts 2π Tq ft-lb. This multiplied by rpm gives foot-pounds per minute, which, divided by 33,000 ft-lb/min hp, gives horsepower.

We might look at this problem a little further, since the second of the given relationships does not seem to be quite in accord with the theory. It states that speed varies directly as the armature voltage. This would imply that the armature voltage varies directly as the speed; but while this is correct for a shunt motor with its constant field, it is not correct for a series motor where the flux is not constant but varies with the current. Since back emf is proportional to $N \times B$ and the field flux B is proportional to the field current (which is also the armature current), we have that back emf $= KNi$, and we can determine the constant K from the data given for the first case when $N = 1,200$ rpm, $i = 46$ amp, and the back emf was determined to be 40.2 volts.

This gives $K = 40.2/(1200 \times 46) = 7.29 \times 10^{-4}$. When the speed increases to 1500 rpm, the back emf will equal

$$7.29 \times 10^{-4} \times 1500 \times I_2 = 1.093 \times I_2$$

The sum of the back emf and the IR drops must equal the line voltage of 100 volts. The IR drops equal $I_3(0.80 + 0.50)$, giving $1.30I_2 + 1.093I_2 = 100$, or $I_2 = 41.7$ amp (c), instead of the 38.3 amp calculated previously. This is the correct current based on the theory of series-wound motors. The 38.3 amp would be correct for a shunt-wound motor. The back emf would then equal $1.093 \times 41.7 = 45.6$ volts, instead of 50.2 volts, as calculated previously.

The answer for part (d) would then be $(41.7/46)^2 \times 21.9 = 18$ ft-lb.

Let us examine the three given relationships in a little more detail.

The first states that torque varies directly as the armature current squared. Torque is proportional to the product of armature current times field flux. Since field flux is proportional to field current (which equals armature current in a series motor) this relationship is correct for a series motor. For a shunt motor the field current and flux are constant and the torque will be proportional to the armature current.

The second relationship, as we have already noted, is true for a shunt motor but not for a series motor.

The third relationship states that flux varies directly as the armature current. This we know to be correct for a series motor, since the armature current equals the field current for this type of motor. For a shunt motor the flux is essentially constant and does not vary as the armature current. This last relationship would appear to be at variance with the second given relationship; we must assume that a mistake was made in the statement of the problem, or that an attempt was made to simplify it and the simplification introduced a contradiction to motor theory.

6-19. UNIVERSAL MOTORS

Series-wound motors will also operate on alternating current and are frequently used in the small sizes. Sparking is more of a problem, however, because of voltages induced by the alternating current, and ac series motors have a greater number of armature coils and commutator segments than series motors constructed for use with direct current. Series motors built for use with alternating current will also operate satisfactorily with direct current; they are often called "universal" motors. The relationships for ac series motors are the same as for dc series motors.

6-20. INDUCTION MOTORS

The commonest type of motor used on alternating current is the induction motor. Induction motors are made in two types, squirrel cage and wound rotor; they differ only in the construction of the rotor. The principle of operation of these two types is the same. The wound rotor is used to supply a higher starting torque and a limited amount of speed control. There is no electrical connection between the rotor circuits of an induction motor and the stator circuits of the supply line. The currents in the rotor are induced currents. The windings in the rotor are short-circuited and thus act like the secondary of a transformer, with the stator windings acting as the primary. When the rotor is stationary, the frequency of the induced current is the same as the line frequency. As the rotor gains speed, the frequency of the induced current will reduce, until the point at which if the rotor were to be turned at the synchronous speed, the frequency and the induced current would fall to zero, since no lines of flux would be cut. We recall that induced voltage is proportional to the rate at which the lines of flux are cut. If the rotor were to rotate at the synchronous speed, the conductors would be in phase with the alternating field and no lines of flux would be cut. Therefore an induction motor cannot operate at synchronous speed and there must be a certain amount of slip. This slip usually runs between 3 and 6 percent, where slip equals 1 minus the ratio of actual speed to synchronous speed, or

$$\text{Slip} = 1 - \frac{\text{actual speed}}{\text{synchronous speed}}$$

The synchronous speed in rpm equals $120f/p$, where f is the frequency and p is the number of poles. This formula is primarily for single-phase current but can be generalized by dividing the number of poles by the number of phases, so that p should actually be poles per phase.

We have here the same sort of starting problem that we had with dc motors, viz., excessively high starting current. When the rotor is not moving, its conductors are being cut by the greatest number of lines of flux per second; thus, when the voltage is a maximum the current is also a maximum. The torque should also be a maximum, but in a single-phase motor, for instance, two equal and opposite forces act at the positive and negative poles; thus there is no operating torque. Various methods are used to start single-phase induction motors.

The basic torque equation $\text{Tq} = KBI$ must be modified for alternating current to include the effects of phase angle. Since we are working with an inductive circuit, the current lags behind the voltage. The voltage is in phase with the flux, so there is a phase angle between the space distribution of the flux and the current that is equal to the phase angle between the voltage and the current. The torque is then proportional not only to the product of the flux and the current but to the cosine of the phase angle as well. In other words, torque is proportional to the product of voltage times the component of the current which is in phase with the voltage: $\text{Tq} = KBI \cos \phi$.

Under operating conditions, the slip is large enough to provide a frequency of the rotor voltage large enough to produce a reactance which cannot be neglected.

Starting torque of an induction motor is given by the equation

$$\text{Tq} = \frac{KBE_2R_2}{R_2{}^2 - X_2{}^2}$$

where $R_2{}^2 - X_2{}^2$ = the square of the rotor impedance, which is constant for zero rotor speed

K = a constant depending upon the physical characteristics of the motor

R_2 = resistance of the rotor

E_2 = rotor voltage

B = flux of the rotating field

This means that starting torque is proportional to $B \times E_2$. The flux is proportional to the stator voltage, and, of course, the stationary rotor voltage is also proportional to the stator voltage, so the starting torque is proportional to the square of the stator or applied voltage.

6-21. SYNCHRONOUS MOTORS

Synchronous motors operate on a different basis than induction motors. The rotor may be considered as a permanent magnet, and it follows the rotating field at synchronous speed. If the load is increased to the point where the torque required to maintain the speed exceeds the torque supplied by the motor, the rotor will slow down and stop. This is termed the breakdown, or pull-out, torque. If the load is maintained constant and the rotor field current is increased sufficiently, the current will lead the voltage and the synchronous motor will operate as a synchronous condenser.

6-22. THREE-PHASE CURRENT

Problems have been given in balanced three-phase circuit network analysis. An example of this is the problem:

- A three-phase, 60-Hz, 220-volts-between-wires, three-wire line supplies current to three single-phase heaters. Each heater has a resistance of 3 ohms and a reactance of 4 ohms. Determine:
  a. The power supplied to the three heaters connected in three-phase Δ
  b. The power which would be supplied by the same line if the heaters are connected in three-phase Y
  c. The power factor of three-heater load in (a) and (b)

Start by listing some of the important factors in balanced three-phase network analysis:

1. For a Y connection, line currents and phase currents are equal.

2. For a Y connection, the line voltages equal $\sqrt{3}$ times the phase voltages.

3. For a Δ connection, the line voltages and the phase voltages are equal.

4. For a Δ connection, the line currents equal $\sqrt{3}$ times the phase currents.

5. The instantaneous power for a three-phase system is constant and is equal to three times the average power per phase.

6. A balanced Δ connection may be replaced by a balanced Y connection if the circuit constants per phase obey the relation: Y impedance = $\frac{1}{3}\Delta$ impedance.

Solve part (b) first. Draw an equivalent four-wire circuit (Fig. 6-32).

Since this circuit is balanced, the neutral wire will carry no current and may be removed without affecting the circuit. We may simplify this circuit by

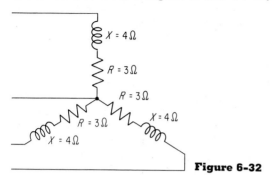

Figure 6-32

sketching one branch (Fig. 6-33). The voltage from line to neutral, phase voltage, will equal $220/\sqrt{3} = 127$ volts. The current will equal

$$\frac{V}{Z} = \frac{127}{3 + j4} = \frac{127}{3 + j4} \times \frac{3 - j4}{3 - j4} = \frac{381 - j508}{25} = 15.2 - j20.3$$

The line current would then equal $\sqrt{15.2^2 + 20.3^2} = 25.4$ amp. The phase angle would equal $\tan^{-1} (20.3/15.2) = 53.1°$. Since we do not know whether the impedance is reactive or inductive, we do not know whether the phase angle will be leading or lagging. The power per phase would equal $127 \times 25.4 \times \cos 53.1° = 1940$ watts, or since $Z = \sqrt{3^2 + 4^2} = 5$, line current equals $^{127}/_5 = 25.4$ amp; phase angle $= \tan^{-1} (X/R) = \tan^{-1} (^4/_3) = 53.1°$;

$$\text{Power} = I^2R = (25.4)^2 \times 3 = 1940 \text{ watts per phase}$$

The total power supplied would be three times the power per phase, or $3 \times 1940 = 5820$ watts.

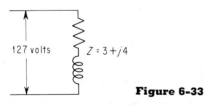

127 volts $Z = 3 + j4$

Figure 6-33

Total power can also be calculated from the relationship line-to-line voltage times line current times power factor times $\sqrt{3}$, which would be $220 \times 25.4 \times 0.60 \times 1.732 = 5820$ watts, which is the answer to part (b).

Part (a) may be determined by taking a Y network with one-third the impedance per phase and calculating as above, or it may be determined directly. The Δ network is shown in Fig. 6-34. First solve the Δ network as shown. The phase voltage equals the line voltage, or 220 volts.

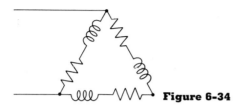

Figure 6-34

The phase impedance is again 5 ohms, so the phase current is $220/5 = 44$ amp. The phase angle is $\tan^{-1}(4/3) = 53.1°$, and the power per phase equals $220 \times 44 \times 0.60 = 5820$ watts. The total power supplied is then $3 \times 5820 = 17{,}460$ watts—the answer to part (a).

We can determine this also from the line current and line-to-line voltage. The line current for a Δ connection equals $\sqrt{3}$ times the phase current, or $44\sqrt{3} = 76.2$ amp. Total power would be $220 \times 76.2 \times 0.60 \times \sqrt{3} = 17{,}460$ watts.

The answer to part (c) is the cosine of the phase angle, which equals 0.60.

SAMPLE PROBLEMS

6- 1 A 60-cycle 13,200-volt transformer has a core loss of 525 watts. When it is operated at the same maximum flux density of 50 kilolines/in.^{2} on 25 cycles, the core loss is 140 watts.

  a. What was the test voltage on 25 cycles?

  b. What was the normal hysteresis loss at 60 cycles?

$$P_h = k_1 f B_m^{1.6} \qquad P_e = k_2 f^2 B_m^2$$

6- 2 A single-phase ac motor delivers 5 hp and is 80 percent efficient. It operates at 220 volts at a power factor of 0.707 lagging. What must be the reactance of a capacitor in parallel to bring the power factor to unity?

6- 3 The power required by a three-phase 220-volt motor is measured by two wattmeters. One reads 2000 watts and the other reads -400 watts. An ammeter in one of the lines reads 12 amp. Calculate the power factor.

6- 4 Calculate the current in each of the four resistors shown in Fig. 6-P-4.

6- 5 A dc shunt-connected motor is being supplied 220 volts and 54 amp. The armature has a resistance, at 75°C, of 0.08 ohm. The shunt field has a resistance of 55 ohms at 75°C. The stray power loss for this motor has been mea-

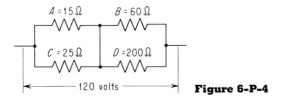

$A = 15\,\Omega$ $B = 60\,\Omega$

$C = 25\,\Omega$ $D = 200\,\Omega$

$\longleftarrow$ 120 volts $\longrightarrow$ **Figure 6-P-4**

sured and is 180 watts. For this input, assume the motor to be operating continuously at a constant temperature of 75°C and determine:

a. Shunt-field current
b. Efficiency of the motor at this load
c. Horsepower output for this load

6- 6 Two condensers of 6 and 12 μF, respectively, are charged by a 120-volt battery. The condensers are in series. After being charged the condensers are connected by joining the two positive terminals to each other and doing the same with the negative terminals. What will be the final voltage of each condenser? What will be the charge on each?

6- 7 A 110-volt 60-Hz supply is connnected to an unknown circuit. The circuit draws a current of 7.5 amp and uses 600 watts of power. The oscilloscope picture of current and voltage is shown in Fig. 6-P-7. Determine the following:

a. Resistance in the circuit
b. Impedance of the circuit
c. Power factor of the circuit
d. Phase angle of the circuit
e. Whether the current leads or lags the voltage
f. Reactance of the circuit
g. Kind of reactance in the circuit

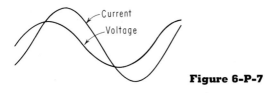

Current

Voltage

Figure 6-P-7

6- 8 A circuit consists of a 200-ohm resistor connected in series with the parallel combination of a coil and a 100-μF capacitor. The coil has a resistance of 10 ohms and an inductance of 1 henry. Calculate the voltage across the capacitor when this circuit is connected across a source of power delivering 125 volts at a frequency of $100/2\pi$ Hz.

6- 9 A group of small induction motors on a 230-volt 60-Hz feeder in an industrial plant requires a total power of 25 kW at 0.707 power factor lagging. Determine the size of the capacitor to be connected in parallel with this load to correct the power factor to 0.90 lagging.

6-10 A load of 100 kVA of induction motors on the three-phase 208-volt 60-Hz power supply to a shop is balanced and operates at a power factor of 80 percent. It is proposed to bring the power factor to 90 percent, lagging, by means of a Δ-connected bank of capacitors. Calculate the total kilovoltamperes of capacitors required in this bank.

6-11 A 1000-hp three-phase 2200-volt induction motor is loaded to rated capacity. The efficiency is 92 percent, and the power factor is 87 percent lagging.

 a. What is the current input to the motors?

 b. What total kilovolt of capacitors is required connected in parallel with the motor to bring the power factor to 100 percent?

6-12 A trouble lamp designed for plugging into the cigarette lighter receptacle on an automobile has a 50-ft cord of No. 18 copper wires. The lamp is rated 60 watts at 10 volts. Assuming that 12 volts is maintained at the receptacle and that the resistance of the lamp does not change with temperature, what is the wattage input to the lamp?

6-13 Given the circuit shown in Fig. 6-P-13, determine V_L, V_{R_1}, and V_{R_3} when:

 a. Switch A is open

 b. Switch A is closed

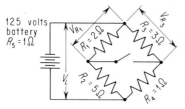

125 volts
battery
$R_5 = 1\,\Omega$

Figure 6-P-13

6-14 A coil has a reactance of 37.7 ohms and a resistance of 12 ohms. It is connected to a 110-volt 60-Hz line. Determine:

 a. Impedance of the coil

 b. Current through the coil

 c. Phase angle with respect to the supply voltage as a reference

 d. Power factor of the circuit

 e. Reading of a wattmeter connected to the circuit

6-15 A single-phase transmission line 12 miles long has a conductor with a resistance of 0.27 ohm/mile and a reactance of 0.62 ohm/mile. The voltage at the load end is 13,200 volts.

 a. Calculate the voltage required at the powerhouse when the load is 1500 kVA at unity power factor.

    b. Recalculate with a power factor of 0.8 lagging.

    c. Recalculate with a power factor of 0.8 leading.

6-16 Find the current flowing through a series circuit consisting of a 10-mH inductor, a 10-μF capacitor, and a 10-ohm resistor when 120 volts, 60 Hz is applied across the combination.

6-17 A 2-μF capacitor has an air dielectric between plates separated 0.5 cm. If the capacitor plates have a potential difference of 1000 volts, what will be the charge on the plates, the energy stored in it, and the voltage gradient? If a dielectric with a coefficient of 4 is placed between the plates with the same charge on them, what will be the voltage across the plates and the energy stored in the capacitor?

6-18 Calculate the three line currents and their phase angles. The loads are listed below. (See Fig. 6-P-18 for circuit.)

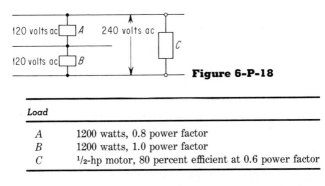

Figure 6-P-18

Load	
A	1200 watts, 0.8 power factor
B	1200 watts, 1.0 power factor
C	$^1/_2$-hp motor, 80 percent efficient at 0.6 power factor

6-19 A storage battery for farm lighting produces 64 volts across its terminals on open circuit. The battery has 0.035 ohm internal resistance and is connected through a pair of copper wires, which have a resistance of 1.835 ohms per 1000 ft, to a load 200 ft distant. The load draws 15 amp under the above conditions.

    a. Determine the voltage at the load 200 ft from the battery.

    b. Determine the voltage across the battery terminals when they are delivering this load.

    c. Determine the voltage regulation at the load.

6-20 Two batteries, which are identical in every way, are connected in series across a 5-ohm resistor, and the current is found to be 0.2 amp. The same cells are then connected in parallel with the same 5-ohm resistor, and the current is found to be 0.16 amp across the 5-ohm resistor. What are the open-circuit voltage and the internal resistance of the batteries?

6-21 A 60-Hz single-phase motor draws 8.5 amp at 120 volts and has an inductive power factor of 87 percent at this load.

    a. How much power in watts is the motor using?

    b. If a 150-μF capacitor is connected in parallel with the motor, what will the new power factor be?

6-22 A dc generator has two poles, with each pole face having an area of 12 in.². The flux density of the air gap is 40,000 lines/in.².

  a. Compute the average emf for one turn on the armature when the machine is running at 1200 rpm. Assume that there is no stray flux outside the pole face area.

  b. What is the relation between the average voltage, maximum voltage, and rms voltage?

6-23 The impedance of a series ac circuit is $Z = 4 + j3$. The circuit consists of a resistor, capacitor, and inductor.

  a. Which has the greater reactance, the capacitor or the inductor?

  b. If a 200-volt 60-Hz source is connected to the circuit, what is the magnitude of the current flowing?

6-24 The resistance of the armature of a 50-hp 550-volt shunt-wound dc motor is 0.35 ohm. The full-load armature current of this motor is 76 amp.

  a. What should the resistance of the starter be for the initial armature current to be 150 percent of the full-load armature current?

  b. If the field current under full load is 3 amp, what is the overall efficiency of the motor when it is delivering 50 hp?

  c. Compute the stray power losses at 50-hp load.

6-25 It is necessary to measure the voltage across a line known to be about 220 volts. Two voltameters are available: (a) one with a 150-volt scale, internal resistance of 15,000 ohms, and (b) another with a 100-volt scale and internal resistance of 12,000 ohms. If the line voltage were actually 225 volts, what would be the reading of each meter if the two are connected in series across the line?

6-26 A slide-wire bridge is set up to measure an unknown resistance R_x (Fig. 6-P-26). The voltage E impressed on the circuit is 2.5 volts, and the resistance R_s is 50.5 ohms. The slide wire has a resistance of 12.92 ohms/ft at 70°F. When the slide is adjusted so that the galvanometer reads zero, $L_1 = 15.0$ in., and $L_2 = 25$ in.

  a. What is the magnitude of the resistor R_x?

  b. What is the magnitude of the current flowing through R_x?

  c. What is the magnitude of the total current flowing through the battery?

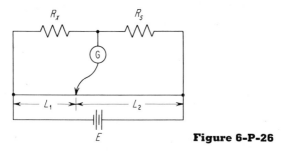

Figure 6-P-26

6-27 Specify the transformers required for a 500-hp 2300-volt three-phase motor of 90 percent power factor and 93 percent efficiency supplied from a 6600-volt three-phase line. If the above transformers are connected Δ and one of the three burns out, to what value must the load on the motor be decreased to prevent overloading the transformers?

6-28 The current in a certain circuit varies with time according to the relation $i = 4 + 2t^2$, where i is in amperes and t in seconds.

    a. How many coulombs of charge pass a point in the circuit in the time interval between $t = 5$ sec and $t = 10$ sec?

    b. What constant current would transport the same charge in the same time interval?

    c. What constant current would produce the same heating effect in the wire during the same interval?

6-29 Battery A has a no-load terminal voltage of 9 volts and an internal resistance of 2 ohms. Battery B has a no-load terminal voltage of 6.5 volts with an internal resistance of 1 ohm. When the positive terminals of the two batteries are connected together, the negative terminals are connected together, and a 3-ohm resistor is connected between positive and negative terminals, what current will flow through each battery?

6-30 An industrial plant has the following loads:

    25 kVA at 0.85 power factor inductive
    50 kW at 1.0 power factor
    250 hp in motors at 0.90 power factor inductive and 86 percent motor efficiency

    a. Draw a complete vector diagram, with each plant load labeled, and indicate by a vector the reactance needed to correct the power factor to zero.

    b. Calculate the magnitude of the reactance necessary to correct the plant power factor to one. (This may be done graphically, if you wish.)

6-31 Three single-phase electric furnaces are each rated at 2300 volts, 60 Hz, 2500 kVA, and 80 percent power factor lagging current. If these three furnaces were connected in Y to a three-phase 2300-volt 60-Hz supply, determine:

    a. Total kilowatts which they would take from the circuit

    b. Approximate temperature which will be reached in terms of normal temperature when connected three-phase Δ

6-32 For the circuit shown in Fig. 6-P-32 determine the equivalent resistance from A to B.

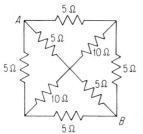

Figure 6-P-32

6-33 Three dry cells, each of 1.5 volts emf and 0.1 ohm internal resistance, are connected in series through two external resistances of 0.5 ohm each. A resistance of 1.0 ohm is connected from a point between the two external resistances to a point between the first and second dry cells. Find the current flowing in each branch of the circuit shown in Fig. 6-P-34.

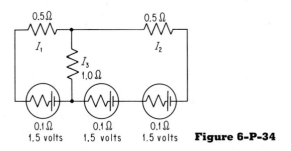

Figure 6-P-34

6-34 An induction motor draws 25 amp from a 125-volt source and has a power factor of 0.8. What will be the line current when a condenser of 300 μF is connected across the terminals of the motor? The frequency is 60 hertz.

6-35 A circuit containing 0.1 henry inductance and 20 ohms resistance in series is connected across 100-volt 25-hertz mains. Find:

  a. Circuit impedance
  b. Voltage across the inductance
  c. Power input
  d. Power factor

6-36 A series circuit consisting of a resistance of 50 ohms, a capacitance of 25 μF, and an inductance of 0.15 henry is connected across 120-volt 60-hertz mains. Find the line current.

6-37 A 30-hp 220-volt three-phase squirrel-cage induction motor with normal voltage applied to the stator has a starting torque of 1.25 times the full-load running torque.

  a. What would be the torque is 110 volts were applied for starting?
  b. What voltage should be applied to give full-load starting torque?

6-38 A 0.6 power factor inductive load takes 10 amp at 115 volts. What value of pure resistance in ohms may be placed in series with this load to make it operate normally from a 230-volt source?

6-39 A three-phase induction motor under test shows an input of 25 amp in each line. The voltage is 223 on each phase. The manufacturer's data show 85 percent power factor and 90 percent efficiency at this load. What is the motor output in horsepower?

6-40 *a.* In the circuit shown in Fig. 6-P-40, calculate the total current A_1 when 120 volts direct current has been applied for sufficient time for the current to have reached a steady value.

b. What will be the current A_1 if 120 volts, 60-hertz alternating voltage is applied?

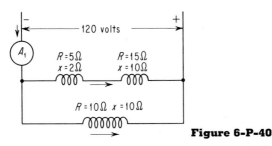

Figure 6-P-40

6-41 When a certain coil is connected to a 60-volt battery, 2 amp of current flows. When the same coil is connected to a 50-volt 60-hertz ac source, only 1 amp flows. What is the inductance of the coil in henrys?

6-42 Three resistors, $R_1 = 0.2$ ohm, $R_2 = 0.3$ ohm, and $R_3 = 0.6$ ohm, are connected in parallel across the terminals of a storage battery having an internal resistance of 0.02 ohm and an emf of 6 volts on open circuit. With this circuit closed, determine:

a. Total current supplied by the battery to the three resistors
b. Terminal voltage of the battery to this load
c. Total power in watts supplied to the three resistors
d. Current in amperes through the 0.2-ohm resistor

6-43 A 200-hp three-phase four-pole 60-hertz 440-volt squirrel-cage induction motor operates at full load, with an efficiency of 85 percent, a power factor of 91 percent, and a slip of 3 percent. For this full-load condition, determine the following:

a. Speed in rpm
b. Torque delivered in lb-ft
c. Line current fed to the motor

6-44 Three batteries have terminal voltages of 2.00, 1.95, and 1.84 volts, with internal resistances of 0.05, 0.075, and 0.064 ohm, respectively. The three batteries are connected in series with a 0.110-ohm resistance. (Assume the voltages and internal resistances remain constant.)
a. What is the voltage across the resistance?
b. What current is flowing through the resistance?

6-45 The wiring diagram for a dc voltmeter is shown in Fig. 6-P-45. When 0.588 volt is applied across AB, the meter registers a full-scale deflection. The scale is graduated from 0 to 15. What should the meter read when 20 volts is applied across OP?

6-46 A power company delivers 500 kW to factory A, 250 kVA at 0.85 power factor lagging to factory B, and 750 kVA at 0.97 power factor leading to factory C.

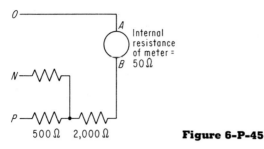

500 Ω   2,000 Ω   **Figure 6-P-45**

a. Draw a vector diagram showing the loads, power factor, and total kilovoltamperes.

b. What is the total power load in kilowatts for the three companies?

6-47 A dc shunt motor has the following constants:

Full-load speed, 1200 rpm
Rated voltage, 230 volts
Full-load current, 58 amp
Armature resistance, 0.15 ohm

Calculate the speed of the motor when it is delivering one-half the rated torque. Assume that the field remains constant.

6-48 Show how to connect a 5-kVA 2300/230-volt transformer to give a 10 percent boost on a 2300-volt power line. What load can be safely supplied by this booster transformer?

6-49 In the series circuit, Fig. 6-P-49, consisting of three elements, the current is I and the voltages are as follows: $E_1 = 120$ volts leading 80°; $E_2 = 240$ volts lagging 70°; $E_3 = 150$ volts leading 30°.

a. Compute the voltage E which is imposed on the circuit.

b. Draw the vector diagram, clearly indicating E.

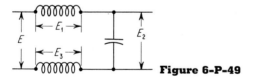

Figure 6-P-49

6-50 An electrical series circuit has a resistance of 4.5 ohms and an inductive impedance of 15 ohms.

a. If 240 volts direct current is impressed across the system, how many amperes will flow?

b. If 240 volts alternating current is impressed across the system, how many amperes will flow?

c. Compute the apparent power measured in kilovolt-amperes for case (b).

d. Compute the real power measured in kilowatts for case (b).

6-51   An alternating current has a sine waveform which in turn has a maximum
value of 200 volts (Fig. 6-P-51), and $e = E_m \sin \theta$.

a. Find for this waveform the instantaneous value of e when θ is $\pi/2$.

b. Find the effective voltage E_{eff}.

c. From 0 to π find the average voltage E_{avg} if the waveform is a true sine
wave.

d. What is the value of the form factor?

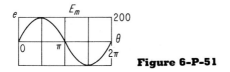

Figure 6-P-51

6-52   It is desired to supply a 250-watt 120-volt lamp 500 ft from the source of power
with full 120 volts over a circuit of No. 14 copper wire. What voltage is required at the source? Disregard reactance.

6-53   Calculate the amperes in each of the four resistors shown in Fig. 6-P-53.

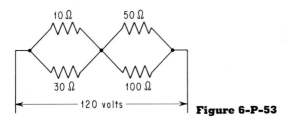

Figure 6-P-53

6-54   A load consists of three-phase motors and totals 750 kW at 80 percent power
factor lagging. A three-phase 300-kvar static capacitor is connected in parallel
with the load. What is the resulting power factor?

6-55   Referring to the sketch in Fig. 6-P-55, there is a difference in potential of 110
volts between A and D. Determine:

a. Drop in potential between B and C

b. Current in the reistance CD

c. Current flowing through the 200-ohm resistance between A and B.

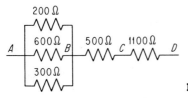

Figure 6-P-55

MULTIPLE-CHOICE PROBLEMS

For each question select the correct answer from the five given possibilities.

6M- 1 A direct-current electric heater has a resistance of 36.5 ± 0.5 ohm. The current flowing in the heater is 12.8 ± 0.2 amp. What is the maximum power (in watts) consumed under these conditions?

(*a*) 5980   (*d*) 467
(*b*) 481   (*e*) 5715
(*c*) 6253

6M- 2 The ratio of the actual power input (kW) divided by the apparent power input (kVA) in a single-phase alternating current is:

(*a*) the power factor
(*b*) the phase angle in degrees
(*c*) the impedance
(*d*) the reactive power
(*e*) the line current

6M- 3 The magnitude of the generated voltage (induced in the armature) in a shunt-wound generator is:

(*a*) directly proportional to the armature speed
(*b*) inversely proportional to the field current
(*c*) directly proportional to the power generated
(*d*) inversely proportional to the field flux
(*e*) none of these

6M- 4 A series alternating-current circuit consisting of a capacitor, inductor, and resistor is in resonance when:

(*a*) the in-phase current equals the out-of-phase current
(*b*) maximum current passes through the circuit
(*c*) minimum current passes through the circuit
(*d*) no current flows through the circuit
(*e*) the voltage across the resistor is equal to the combined voltage across the capacitor and inductor

6M- 5 An electric motor has a nameplate rating of the following: 220 volt, 23.0 amp, 5 hp, 3 phase, 60 hertz, 1750 rpm. Which of the following statements is true?

(*a*) The motor uses 220 volts, 23.0 amp at no load.
(*b*) The power input to the motor is 5 hp.
(*c*) The 1750 rpm is the no load speed.
(*d*) The motor will deliver 5 hp at full load.
(*e*) None of these is true.

6M- 6 The power consumed by a single-phase alternating-current circuit may be calculated if the

(*a*) resistance and impedance are known

(*b*) impedance and voltage are known
(*c*) inductive and capacitive reactance are known
(*d*) line current and line voltage are known
(*e*) none of these

6M- 7 An electric circuit consists of a resistance of 127 ohms in series with a
150-mH inductance and a capacitor. What must be the capacitance of the
capacitor in microfarads for the circuit to be resonant for a 100-hertz applied
current?
(*a*) 150
(*b*) 17
(*c*) 5
(*d*) 28
(*e*) 98

6M- 8 The phase angle of a single-phase circuit can be found by:
(*a*) measuring the voltage and the current flowing through the circuit
(*b*) measuring the watts consumed by the circuit
(*c*) measuring the capacitance and inductance in the circuit at a known frequency
(*d*) measuring the pure resistance, current, and voltage
(*e*) measuring the watts being consumed by the circuit, the voltage drop in
the circuit, and the current flowing through the circuit

6M- 9 In a parallel resonant (antiresonant) circuit, which of the following conditions must be true for all values of the branch elements?
(*a*) The susceptive elements are equal in magnitude and opposite in sign.
(*b*) The branch circuit impedances are conjugates.
(*c*) The voltage reactance drops are equal in magnitude.
(*d*) The reactive components are equal in magnitude, but opposite in sign.
(*e*) The branch circuit currents are conjugates.

6M-10 The rpm of an ac electric motor:
(*a*) varies directly as the number of poles
(*b*) varies inversely as the number of poles
(*c*) is independent of the number of poles
(*d*) is independent of the frequency
(*e*) is directly proportional to the square of the frequency

7
CHEMISTRY

A general knowledge of chemistry is presupposed; discussion here is limited to those portions of the subject which have been emphasized in past examinations and to a few of the more general fundamental relationships applicable to many engineering problems. The reader who finds that he or she has retained little (or less) of his introductory college chemistry course, or who had no such course, will find it of value to read a beginning text.

7-1. LAW OF LE CHATELIER

The law of Le Chatelier states, "When a system in equilibrium is subjected to a change (e.g., a change of temperature, concentration, or pressure), there is a shift in the point of equilibrium which tends to restore the original condition, or to relieve the strain." As an example, the combination of carbon and oxygen is exothermic and is represented by the equation $C + O_2 \rightarrow CO_2 + 97,000$ cal. If, after the reaction has reached a point of equilibrium, the system is heated, the point of equilibrium will shift to the left, using up heat and tending to lower the temperature of the system. The result will be a dissociation of some of the CO_2 which was previously formed and an increase in the concentrations of C and O_2.

An increase in the pressure acting on a system may also result in a shift in the point of equilibrium. For the reaction $N_2 + 3H_2 \rightarrow 2NH_3$, we have one volume of N_2 and three volumes of H_2 combining to form two volumes of NH_3, so an increase in pressure would cause the point of equilibrium to shift to the right. Such a shift would reduce the total equivalent volume of gas, as four volumes on the left combine to form two volumes on the right.

The shift in the point of equilibrium due to a change in concentration may be illustrated by means of the equilibrium constant, which is a constant for a given reaction at a given temperature and pressure. For a general chemical reaction $aA + bB \rightarrow cC + dD$, where we have a molecules of A reacting with b molecules of B, etc., the equilibrium constant $K = [(C)^c(D)^d]/[(A)^a(B)^b]$. For the previously used example, then, $K = (NH_3)^2/[(N_2)(H_2)^3]$, where $(NH_3)^2$ represents the concentration of NH_3 in the final mixture, in moles per liter, squared. Since K is a constant for a given temperature, we see that an increase in the concentration of either the N_2 or the H_2 must be followed by an increase in the concentration of the NH_3.

On the other hand, a chemical reaction will go to completion when one of the products is (1) insoluble, (2) volatile, or (3) only slightly ionized.

These conditions can be illustrated by the following chemical equations:

$$NaCl + AgNO_3 \rightarrow NaNO_3 + AgCl\downarrow$$
$$NH_4Cl + NaOH + heat \rightarrow NaCl + H_2O + NH_3\uparrow$$
$$HCl + NaOH \rightarrow NaCl + H_2O$$

In the last equation the water resulting is only slightly ionized so the reaction goes to the end.

7-2. DALTON'S LAW OF PARTIAL PRESSURES

Dalton's law of partial pressures states, "Each component of a mixture of gases exerts its pressure exactly as it would if the others were not present, and the sum of all their pressures makes up the total of the pressure exerted by the mixture of gases." The water-vapor pressure in gases collected over water is an example of a partial pressure. If oxygen is collected over water at a temperature of 64°F at a pressure of 780 mm Hg, the actual volume of oxygen at this temperature and pressure will be only 764.6/780 times the apparent volume, since the vapor pressure of water at 64°F is 15.4 mm Hg. This means that 15.4 mm of the measured pressure is due to water vapor, and the pressure of the oxygen is 780 − 15.4, or 764.6 mm. Then 15.4/780 × total volume is water vapor, and 764.6/780 × total volume is oxygen.

7-3. AVOGADRO'S LAW

Avogadro's hypothesis is an additional important relationship which should be kept in mind. It may be stated as follows: "Equal volumes of gases when subjected to the same conditions of temperature and pressure contain an equal number of molecules." This means that 1 liter of gas at 60°F and 14.7 psia will contain the same number of molecules regardless of the type of gas occupying that space—whether the gas is chlorine, hydrogen, or a mixture of gases, such as air. The number of molecules in 1 g-mole of gas, which occupies 22.4 liters at 0°C and 1 atm of pressure, is 6.024×10^{23} and is known as Avogadro's number.

The application of some of these principles can be illustrated by means of an example taken from a past engineer-in-training examination.

■ According to Avogadro's law, equal volumes of all gases at the same temperature and pressure contain the same number of molecules. Furthermore, 22.4 liters of a gas contain 6.024×10^{23} molecules at 0°C and 1 atm of pressure. How many molecules of nitrogen, N_2, are contained in 10 liters of nitrogen saturated with water vapor at 60°C and under a total absolute pressure of 2 atm? (Absolute zero may be taken as −273°C, and the vapor pressure for water is 3.1 psia at 60°C.)

7-4. VAPOR PRESSURE

This introduces the concept of vapor pressure, which is the pressure (absolute) exerted by a vapor when it is in equilibrium with its liquid. The vapor pressure of a given liquid is dependent only upon the temperature; it is not affected by the total pressure acting on the system nor by the amount of space above the surface of the liquid. The boiling temperature of a liquid is the temperature at which the vapor pressure is equal to the local atmospheric pressure.

The vapor pressure of the water at the given temperature is 3.1 psia, and the total pressure is $2 \times 14.7 = 29.4$ psia. From Dalton's law of partial pressures the pressure due to the nitrogen equals $29.4 - 3.1 = 26.3$ psia. Application of the universal gas law (discussed in Chap. 4, Thermodynamics) gives the equivalent volume of nitrogen at 0°C and 14.7 psia.

$$\frac{P_1 V_1}{T_1} = R = \frac{P_2 V_2}{T_2} \qquad V_2 = V_1 \frac{P_1}{P_2} \times \frac{T_2}{T_1}$$

Note that the units of temperature and pressure cancel out and that as long as the same units are used for P_2 as for P_1 (and for T_2 as for T_1) they can be any units desired so long as they are absolute units

$$V_2 = 10 \times \frac{26.3 \text{ psia}}{14.7 \text{ psia}} \times \frac{273 \text{ K}}{(273 + 60) \text{ K}} = 14.71$$

of pure nitrogen at 0°C and 1 atm. The total number of molecules of nitrogen contained in the original 10 liters would then equal

$$\frac{14.7}{22.4} \times 6.024 \times 10^{23} = 3.95 \times 10^{23} \text{ molecules}$$

An additional example illustrating a more detailed application of Dalton's law of partial pressure is the following:

- A tank having a capacity of 800 gal contains 500 gal of benzene at 35°F. Under these conditions the tank is sealed at atmospheric pressure. During storage the tank reaches 100°F. What is the gauge pressure in the tank? Assume that the air inside the tank is saturated with benzene vapor at the time the tank is sealed, and that the expansion of the tank due to the rise in temperature and pressure is negligible.

Coefficient of cubical expansion for liquid benzene = 0.00077 ft^3/(ft^3)(°F)

7.48 gal/ft^3
760 mm Hg = 1 atm

Temperature, °F	Vapor pressure, mm Hg
−9.5	10
71.2	100
109.4	200
141.9	400
163.7	600
176.2	760

Initially the pressure in the tank is equal to atmospheric pressure at 35°F. Assume this to be a standard atmosphere of 760 mm Hg. This is the total pressure in the tank at the time the tank is closed; it is made up of the pressure of the air in the tank and the pressure of the benzene vapor. To determine the amount of the pressure due to the benzene vapor at 35°F (and at the second temperature of 100°F), it is necessary to make a graph of vapor pressure vs. temperature from the data given. From the graph (Fig. 7-1) the vapor pressure of the benzene can be seen to equal 48 mm Hg at 35°F. The pressure due to the air at the time the tank is closed is then 760 − 48 = 712 mm Hg.

There are three things which affect the pressure: (1) The pressure of the air will increase because of the increase in the temperature. (2) The pressure of the air will increase because of the reduction in volume resulting from the expansion of the benzene. (3) The vapor pressure will increase as shown in Fig. 7-1. Dalton's law states that the change in the pressure of the air will be the same as if the air alone were occupying the space. Applying this reasoning, we can calculate what the final pressure of the air will be because of the change in temperature and volume, and then add the final vapor pressure of the benzene to the resulting air pressure; the sum will be the total pressure.

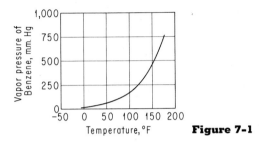

Figure 7-1

Volume of air at start = 800 gal − 500 gal = 300 gal

Final volume of air = 800 gal − 500(1 + 65 × 0.00077) gal = 275 gal
where 65 = change in temperature, °F

0.00077 = thermal coefficient of cubical expansion of the benzene

The change in absolute temperature is from 460 + 35 = 495°R to 460 + 100 = 560°R.

From the universal gas law, $PV = WRT$, the relationship $(P_1V_1)/T_1 = WR = (P_2V_2)/T_2$ follows directly, which gives

$$P_2 = P_1 \frac{T_2}{T_1} \frac{V_1}{V_2}$$

The final air pressure would equal:

$$P_2 = 712 \times \frac{560}{495} \times \frac{300}{275} = 880 \text{ mm Hg}$$

The total pressure would then equal 880 + 170 = 1050 mm Hg, with the vapor pressure of the benzene added, and the gauge pressure would equal 1050 − 760 = 290 mm Hg, or

$$\frac{290}{760} \times 14.7 = 5.61 \text{ psig}$$

7-5. SOLUTIONS

The concentration of a chemical solution may be given in a number of ways. The common methods are as follows:

$$\text{Weight percent} = \frac{\text{lb of solute}}{\text{lb of solvent}} \quad \text{or} \quad \frac{\text{g of solute}}{\text{g of solvent}}$$

(Note that this is not lb of solute per lb of *solution*.)

$$\text{Volume percent} = \frac{\text{ft}^3 \text{ of solute}}{\text{ft}^3 \text{ of solvent}} \quad \text{or} \quad \frac{\text{cm}^3 \text{ of solute}}{\text{cm}^3 \text{ of solvent}}$$

This is commonly used, for example, to express the amount of a gas dissolved in water or other liquid.

Concentration—may be either weight percent, mass percent (grams per gram), or volume percent.

Molar solution—the concentration of a solution may be given as the molarity (M) where the molarity of a solute equals the number of moles of solute per liter of *solution*. A 2 molar ($2M$) solution of sodium chloride (common salt) could be made by adding water to two moles of NaCl [$2 \times (22.991 + 35.457) = 116.896$ g] until the volume of the solution reached 1 liter. Note that a $2M$ solution would not be obtained by adding 116.896 g of NaCl to 1 liter of water, since the final volume of the solution would then be greater than 1 liter.

From the above, it can be seen that

Moles of solute = molarity $\times$ liters of solution

A typical type of question could be:

■ How many grams of potassium hydroxide (KOH) are contained in 750 ml of a 0.400 molar solution?

Moles of solute = $(0.400M) \times 0.750$ l = 0.300 moles
0.300 moles of KOH = $0.300(39.100 + 16.000 + 1.008) = 16.832$ g of KOH

Molal solution—Molality (m) of a solution equals the number of moles of solute dissolved in one kilogram of solvent:

$$\text{Molality} = \frac{\text{moles of solute}}{\text{kg of solvent}}$$

Citing the previous example, if two moles of NaCl (116.896 g) were dissolved in 2 kg of water, the result would be a 2 molal ($2m$) solution. Capital M indicates molarity; lowercase m indicates molality.

Normal solution—Normality (N) of a solution is the number of gram-equivalent weights of solute contained in one liter of solution.

$$\text{Normality} = \frac{\text{gram-equivalent weights}}{\text{liters of solution}}$$

A gram-equivalent weight is that amount of an element that will combine with 8.000 g of oxygen. Thus a gram-equivalent weight would equal the atomic weight of an element divided by the ionic charge number, or the molecular weight of a compound divided by the ionic charge number. For H_2SO_4, for example, one g-mole would equal

$$2 \times 1.008 + 32.066 + 4 \times 16.000 = 98.082 \text{ g}$$

But since the SO_4 ion has an ionic charge of -2, the gram-equivalent weight of H_2SO_4 would equal

$$\frac{98.082}{2} = 49.041 \text{ g}$$

Similarly, K_3PO_4 has a molecular weight of 212.275, but since each molecule splits up into three K^+ ions and one PO_4^{-3} ion, the gram-equivalent weight equals

$$\frac{212.275}{3} = 70.758 \text{ g}$$

Normality is similar to molarity, but the units used to express the quantity of solute are different.

7-6. CHEMICAL REACTIONS

A chemical equation is said to be balanced when the number of each of the different kinds of atoms on one side of the equation is exactly equal to the number of the same kinds of atoms on the other side of the equation.

$$C + O_2 \rightarrow CO_2$$
$$2H_2 + O_2 \rightarrow 2H_2O$$

are examples of balanced equations.

The weights of the different substances required for or produced by a chemical reaction will be in the same proportion as the molecular weights of the substances involved, as indicated by the balanced equation of the reaction. To illustrate this point, take an example from a past examination:

■ Hydrogen sulfide gas, H_2S, is used to precipitate lead sulfide, PbS, in a solution of hydrochloric acid.

$$\underline{\quad}H_2S + \underline{\quad}PbCl_2 \rightarrow \underline{\quad}PbS + \underline{\quad}HCl$$

Formula weights: H = 1.008, Pb = 207.2, S = 32.06, Cl = 35.46

a. Balance the chemical equation and state how many moles of H_2S are required to produce 1 lb of PbS.

b. How many pounds of H_2S are required to produce 1 lb of PbS?

c. How many cubic feet of H_2S gas at 70°F and atmospheric pressure are required to produce 1 lb of PbS?
The gas constant for H_2S is 45.3 ft-lb/(lb-°R).

To balance the equation, we see there are two atoms of hydrogen, two of chlorine, one of lead, and one of sulfur on the left, and only one of each on the right. The equation will be balanced if we put a 2 on the right-hand side, giving:

$$H_2S + PbCl_2 \rightarrow PbS + 2HCl$$

which shows that 1 mole of H_2S combines with 1 mole of $PbCl_2$ to give 1 mole of PbS and 2 moles of HCl.

The molecular weights can be determined from the formula weights given. $H_2S = 34$; $PbCl_2 = 278$; $PbS = 239$; $HCl = 36.5$.

To double check, $34 + 278 = 239 + 2(36.5)$.

The weight of H_2S to produce 1 lb of PbS can be determined by proportion: $x/34 = {}^1/_{239}$. $x = 0.142$ lb of H_2S.

The cubic feet of H_2S required can easily be determined by means of the universal gas law, $PV = RT$, which gives

$$V = \frac{45.3 \times (460 + 70)}{14.7 \times 144} = 11.34 \text{ ft}^3/\text{lb}$$

The volume of H_2S required is then equal to

$$0.142 \text{ lb} \times 11.34 \text{ ft}^3/\text{lb} = 1.61 \text{ ft}^3$$

Another method of calculating the volume of gas involves the fact that 1 lb-mole of gas occupies 359 ft³ at standard conditions. Since 1 lb-mole of H_2S weighs 34.06 lb, this would give a volume of $0.142/34.06 \times 359 = 1.50$ std ft³.

$$1.50 \times \frac{460 + 70}{460 + 32} = 1.615 \text{ ft}^3$$

at 70°F and 1 atm.

Another example is afforded by a past examination problem which asked:

■ A piece of plumber's solder weighing 3.0 g was dissolved in dilute nitric acid, then treated with dilute H_2SO_4. This precipitated the lead as

$PbSO_4$, which, after washing and drying, weighed 2.93 g. The solution was then neutralized to precipitate stannic acid, which was decomposed by heating, yielding 1.27 g of SnO_2. What is the analysis of the solder in terms of percentage of lead and tin? Atomic weights: Pb = 207.2; Sn = 118.7; O = 16.00; S = 32.07; H = 1.00.

The formula weight of $PbSO_4$ is 303.27 (207.2 + 32.07 + 64.0), which means that (207.2/303.27) × 2.93 = 2.00, or that there is a total of 2 g of lead in the 2.93 g of $PbSo_4$.

The formula weight of SnO_2 is 150.7, which means that the SnO_2 contains (118.7/150.7) × 1.27 = 1.00 g of tin.

The original 3 g of solder then contained 2.00 g of lead and 1.00 g of tin and therefore contained 66.7 percent lead and 33.3 percent tin.

A different type of problem is illustrated by the following example:

■ An Orsat analysis of flue gases yields the following volumetric analysis: CO_2 = 12.5 percent; CO = 0.5 percent; O_2 = 6.4 percent; N_2 = 80.6 percent. Convert this analysis to an analysis by weight. Atomic weight: C = 12; O = 16; N = 14.

All that is needed here is to apply the universal gas law. Molecular weight of CO_2 = 44. R = 1544/mol. wt = 35.1.

$$V = \frac{RT}{P} = \frac{35.1 \times (460 + 32)}{14.7 \times 144} = 8.16 \text{ ft}^3/\text{lb}$$

at standard conditions, or w = 0.1227 lb/ft³.

For each cubic foot of flue gas there will be 0.125 ft³ of CO_2, which will weigh 0.125 ft³ × 0.1227 lb/ft³ = 0.01535 lb. Similarly, at standard conditions there will be

0.005 ft³ × 0.0781 lb/ft³ = 0.00039 lb CO/ft³ of flue gas
0.064 ft³ × 0.0893 lb/ft³ = 0.00572 lb O_2/ft³ of flue gas
0.806 ft³ × 0.0781 lb/ft³ = 0.0630 lb N_2/ft³ of flue gas

The weight of 1 ft³ of flue gas at 1 atm pressure and 32°F is then 0.08446 lb, and the weight percentages of the different constituent gases are CO_2 = 18.1 percent; CO = 0.5 percent; O_2 = 6.8 percent; N_2 = 74.6 percent.

The weight percentage may be calculated more easily by determining the weight of each gas in a mole volume of 359 ft³. Since the volume percent would also be the mole percent, we would have, for 1 mole of flue gas:

Gas	Volume, %	Molecular weight	lb/mole of flue gas	Weight, %
CO_2	12.5	44	5.50	18.2
CO	0.5	28	0.14	0.44
O_2	6.4	32	2.05	6.76
N_2	80.6	28	22.57	74.6
Total			30.26	

7-7. COMBUSTION

Combustion need not necessarily be an oxidation; there are many things that burn with intense heat without combining with oxygen (e.g., the reaction in an atomic hydrogen welding torch which produces a very high temperature, $H + H \rightarrow H_2 +$ heat). Oxidation is the commonest type of burning, however, and is the one most frequently encountered. An example follows:

- One molecular weight of methyl alcohol, CH_4O, is burned in 10 percent excess air. Write the combustion equation and determine the following:

  a. Pounds of air required

  b. The respective partial pressure of each of the products for a total pressure of 15 psia, assuming that the water vapor has not condensed

  c. Volume occupied by the products at 240°F and 15 psia

  d. Volume of air at 15 psia and 60°F needed to burn 1 ton of methyl alcohol per hour under the above conditions

First, write the equation:

$$\_\_CH_4O + \_\_O_2 \rightarrow \_\_CO_2 + \_\_H_2O$$

Since we are burning with 10 percent excess air, we assume that the end products will be carbon dioxide and water vapor. Only the oxygen of the air will enter into the reaction; the nitrogen will not. Balancing the equation gives:

$$2CH_4O + 3O_2 \rightarrow 2CO_2 + 4H_2O$$

One pound mole of CH_4O weighs 32 lb and 32 lb of methyl alcohol combines with 48 lb of oxygen for complete combustion. Ten percent excess air gives a total of 52.8 lb of oxygen. Since air consists of 23.2 percent oxygen by weight, a total of

$$\frac{52.8}{0.232} = 228 \text{ lb}$$

of air would be required.

The products (of the combustion) would include CO_2, H_2O, the excess O_2, and the N_2, which took no part in the process. The complete equation, assuming the air to consist of 23.2 percent O_2 and 76.8 percent N_2 by weight (21 percent O_2 and 79 percent N_2 by volume), would be

$$CH_4O + O_2 + N_2 \rightarrow CO_2 + H_2O + N_2 + O_2$$

We are disregarding the approximately 1 percent of CO_2 and inert gases present in the air, but this will not give any appreciable error, and their inclusion would multiply the complexity of the solution needlessly.

The weights of the different components would be:

$$CH_4O + O_2 \quad + N_2 \quad \rightarrow CO_2 \quad + H_2O \quad + N_2 \quad + O_2$$
$$32 \text{ lb} + 52.8 \text{ lb} + 175.2 \text{ lb} \rightarrow 44 \text{ lb} + 36 \text{ lb} + 175.2 \text{ lb} + 4.8 \text{ lb}$$

The partial pressures will be in the same ratios as the volumes, so let us calculate part (c) before part (b).

From $PV = RT$, the specific volume of the CO_2 is

$$V = \frac{RT}{P} = \frac{1544}{44} \times (460 + 240) \times \frac{1}{15 \times 144} = 11.37 \text{ ft}^3/\text{lb} \qquad \text{for } CO_2$$

Similarly for H_2O, $V = 27.8$ ft^3/lb; for N_2, $V = 17.85$ ft^3/lb; for O_2, $V = 15.6$ ft^3/lb.

The volume occupied by all the products at 240°F and 15 psia would then equal

$$44 \times 11.37 + 36 \times 27.8 + 175.2 \times 17.85 + 4.8 \times 15.60$$
$$= 500 + 1000 + 3140 + 75 = 4715 \text{ } ft^3$$

The answer to part (b) would then be, for CO_2,

$$P = \frac{500}{4715} \times 15 = 1.59 \text{ psia}$$

for H_2O, $P = 3.18$ psia; for N_2, $P = 10$ psia; and for O_2, $P = 0.24$ psia.

Since it requires 228 lb of air to burn 32 lb of methyl alcohol, it would require $(2000/32) \times 228 = 14{,}250$ lb of air to burn 1 ton. At 15 psia and 60°F, the specific volume of air would be

$$V = \frac{53.3 \times 520}{15 \times 144} = 12.83 \text{ ft}^3/\text{lb}$$

The required volume of air would then be

$$14,250 \times 12.83 = 183,000 \text{ ft}^3/\text{hr}$$

We could also have calculated these answers by using mole volumes.

Gas	Weight, lb	Molecular weight	Moles	Volume, %	Partial pressure, psia
CO_2	44	44	1	10.63	1.59
H_2O	36	18	2	21.26	3.19
N_2	175.2	28	6.26	66.52	9.98
O_2	4.8	32	0.15	1.59	0.24
Total...	.....	...	9.41	100.00	15.00

Total volume = 9.41 moles.

$$9.41 \times 359 \times \frac{14.7}{15} \times \frac{460 + 240}{460 + 32} = 4710 \text{ ft}^3$$

One mole of CH_4O requires 1.5 moles of O_2 for complete combustion. Air is 21 percent O_2 by volume, so this would mean $1.5 \times 0.79/0.21 = 5.64$ moles of N_2. Adding 10 percent gives 1.65 moles of O_2 plus 6.21 moles of N_2, or a total of 7.86 moles of air, which would occupy

$$7.86 \times 359 \times 14.7/15 \times {}^{520}/_{492} = 2920 \text{ ft}^3 \text{ air/mole } CH_4O$$

One ton of CH_4O contains $2000/32 = 62.5$ moles of CH_4O.

$$62.5 \times 2920 = 182,500 \text{ ft}^3/\text{hr}$$

7-8. WATER SOFTENING

The degree of hardness is usually expressed in terms of the equivalent amount of $CaCO_3$ (calcium carbonate) in parts per million (ppm) by weight.

Hardness is caused by the presence in the water of bivalent ions of calcium, magnesium, or iron. These ions form insoluble precipitates when the water is boiled or when soap is added. Softening of the hard water may be accomplished by chemical precipitation of the undesirable ions before use or by replacing them with sodium ions which form soluble compounds. Softness may also be obtained by the formation of complex soluble ions.

An illustration of the substitution method of softening is afforded by the problem:

■ A water-treating plant processes 2000 gal of raw water per day. Analysis shows that the raw water contains 60 ppm (parts per million by weight) of calcium bicarbonate, $Ca(HCO_3)_2$. In passing through the zeolite softening process, the calcium ions are exchanged for the sodium ions, so that the softened water contains sodium bicarbonate, $NaHCO_3$, instead of $Ca(HCO_3)_2$. How many parts per million of $NaHCO_3$ does the softened water contain? Substituting Z for the zeolite radical, write the balanced chemical reaction. The formula for sodium zeolite is Na_2Z. Formula weights: Na, 23.0; H, 1.0; C, 12.0; Ca, 40.1; O, 16.0.

The equation of the process is $Na_2Z + Ca(HCO_3)_2 \rightarrow CaZ + 2NaHCO_3$. The formula weight of $Ca(HCO_3)_2$ is 162.1, and that of $NaHCO_3$ is 84, so 162.1 lb of calcium bicarbonate will form 168 lb of sodium bicarbonate. Then, 60 ppm of calcium bicarbonate would give $60 \times (168/162.1) = 62.2$ ppm of sodium bicarbonate. Actually, the 1,000,000 lb of water containing 60 lb of calcium bicarbonate theoretically becomes 1,000,002.2 lb of water containing 62.2 lb of sodium bicarbonate, but the slight change in total weight would have no effect on the calculations.

7-9. ELECTROCHEMISTRY

There is a direct relationship between a chemical balance and an electrical balance. That is, chemical ions are created by the addition, or subtraction, of electrons from atoms or molecules in exact numbers. A hydrogen ion, H^+, for example, is created by the removal of one electron from a hydrogen atom. A sulfate ion, SO_4^{2-}, is created by the addition of two electrons to an SO_4 molecule.

One gram-mole of a substance contains one Avogadro number of molecules. One gram-atom of an element contains one Avogadro number of atoms. The Avogadro number is 6.024×10^{23}.

A faraday equals 96,500 coulombs and is the amount of electricity that will deposit one gram-equivalent weight of a substance at an electrode. One gram-equivalent weight of an oxidizing agent is that number of grams of the substance that picks up one Avogadro number of electrons. One gram-equivalent weight of a reducing agent is that amount of the substance, in grams, that gives up one Avogadro number of electrons. Thus, one gram-equivalent of any oxidizing agent will react exactly with one gram-equivalent of any reducing agent.

It would, then, require one Avogadro number of electrons (one faraday), or 96,500 coulombs, to deposit one gram-equivalent weight of a positively ionized substance at the cathode of an electrolytic cell, or one gram-equivalent weight of a negatively ionized substance at the anode of an electrolytic cell.

One coulomb equals one ampere-second.

■ As an example, how many ampere-seconds (coulombs) would be required to produce 100 standard cubic feet of chlorine from an NaCl solution?

Chlorine has an atomic weight of 35.457 and a molecular weight of 70.914. The weight equivalent of 100 standard cubic feet would equal

$$W = \frac{P \times \text{vol}}{RT} = \frac{(14.7 \times 144) \times 100}{(1544/70.914)(460 + 32)} = 19.76 \text{ lb}$$

or, since 1 lb-mole occupies 359 ft^3 at standard conditions, the weight of 100 standard cubic feet would equal

$$(100/359) \times 70.914 = 19.75 \text{ lb}$$

19.75 lb corresponds to $19.75 \times 454 = 8966$ g, which equals $8966/70.914 = 126$ g-moles or 252 gram-equivalent weights of chlorine, since a chlorine atom has a valence of -1 and there are two atoms per molecule.

The release of this amount of chlorine gas, Cl_2, at the anode of the electrolytic cell would require 252 faradays, or $252 \times 96,500 = 24.32 \times 10^6$ coulombs, or 24.32×10^6 ampere-seconds. This would require a current of 200 amperes for 33.8 hr.

The chlorine would be produced from a solution of sodium chloride, NaCl. The NaCl would ionize into Na^+ and Cl^-. The anode reaction would be

$$2Cl^- \rightarrow Cl_2 \uparrow + 2e$$

The sodium ions would be simultaneously attracted to the cathode. But there would be H^+ ions in the solution as well. And less electric energy is required to deposit hydrogen than is required to deposit sodium. So hydrogen would be given off at the cathode and the sodium ions would remain in solution.

The reaction at the cathode would be

$$2HOH + 2e^- \rightarrow H_2 \uparrow + 2OH^-$$

A total of 252 gram-equivalent weights would be released, or $252 \times 1.008 = 254$ g of hydrogen gas.

SAMPLE PROBLEMS

7- 1   A 5.82-g silver coin is dissolved in nitric acid. When sodium chloride is added to the solution all the silver is precipitated as AgCl. The AgCl precipitate weighs 7.20 g. Determine the percentage of silver in the coin. Atomic weights: H = 1.008; N = 14.008; O = 16.000; Ag = 107.88; Cl = 35.457.

7- 2   In the electric-furnace method for producing phosphorus, the raw materials are phosphate rock, silica, and coke. The products are calcium silicate, phosphorus, and carbon monoxide.

a. Finish the following chemical equation:

$$?Ca_3(PO_4)_2 + ?SiO_2 + ?C = ?CaSiO_3 + 2P + ?CO$$

b. How many tons of $Ca_3(PO_4)_2$ are needed to make a ton of phosphorus? Atomic weight of P = 30.98; formula weight of $Ca_3(PO_4)_2$ = 310.2.

7-3 How much $PbSO_4$ is formed when a lead storage battery supplies 200 ampere-hours of its capacity? Atomic weights: Pb = 207.2; S = 32; O = 16. One faraday of charge = 96,494 coulombs.

7-4 Caustic soda (NaOH) is an important commercial chemical. It is often prepared by the reaction of soda ash, Na_2CO_3, with slaked lime, $Ca(OH)_2$.

a. How many pounds of caustic soda can be obtained by treating 11.023 lb of soda ash with slaked lime?

b. What weight in pounds of lime (CaO) would be required? Assume complete reaction. Use approximate atomic weights: Ca = 40.1; C = 12; Na = 23; O = 16; H = 1.

7-5 In a certain chemical process two liquids enter a mixing chamber (Fig. 7-P-5) and are thoroughly mixed before being discharged at 80°F and 50 gpm. Liquid (1) enters at 140°F and has a specific heat of 10 Btu/(gal)(°F). Liquid (2) enters at 65°F and has a specific heat of 8.33 Btu/(gal)(°F). Assume that there is no chemical reaction between liquid (1) and liquid (2), and that there is no heat lost or gained to the system.

a. What is the flow for liquid (1) and liquid (2)?

b. What is the specific heat of the mixed liquid?

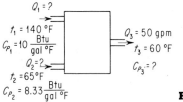

Figure 7-P-5

7-6 A producer gas has the following percentage composition by volume at standard conditions of T and P: carbon dioxide = 5.8; carbon monoxide = 19.8; hydrogen = 15.1; methane = 1.3; oxygen = 1.3; and nitrogen = 56.7. Calculate the volume of dry air at the same conditions of T and P required for the complete combustion of 100 ft³ of this gas.

7-7 a. How many pounds of 93 percent by weight H_2SO_4 can be obtained per ton of zinc sulfide (ZnS) ore of 50 percent by weight purity?

b. For part (a), how many cubic feet of air at standard conditions, 1 atm and 32°F, will be required per ton of ore?

$$2Zns + 3O_2 = 2ZnO + 2SO_2$$
$$2SO_2 + O_2 = 2SO_3$$
$$SO_3 + H_2O = H_2SO_4$$

7- 8   What weight of calcium carbonate ($CaCO_3$) will react with excess hydrochloric acid (HCl) to produce 10 liters of carbon dioxide, CO_2, at 25°C and 770 mm pressure? Atomic weights: Ca = 40; C = 120; O = 16; Cl = 35.

7- 9   Iron pyrite (FeS_2) is roasted in the presence of air so that sulfur dioxide (SO_2) is driven off in the form of a gas. The SO_2 is collected and combined with water to form sulfuric acid (H_2SO_4). How many pounds of 60 percent sulfuric acid (60 percent H_2SO_4 and 40 percent H_2O) can be produced per ton of iron pyrite, assuming no losses in the process?

7-10   The following is the equation for the disintegration of uranium 238. Explain the significance of the equation with respect to both superscripts and subscripts.

$$^{238}_{92}U = ^{234}_{90}Th + ^{4}_{2}He$$

7-11   Pure silver (Ag) is combined with nitric acid (HNO_3) to yield silver nitrate ($AgNO_3$), nitric oxide (NO), and water (H_2O).

a. Write the balanced chemical equation for this reaction.

b. How many pounds of silver would be required to liberate 100 ft^3 of nitric oxide at 60°F and 1 atm pressure? The gas constant for nitric oxide is 49.5 ft-lb/(lb)(°F). Formula weights: Ag = 107.9; N = 14.0; H = 1.01; O = 16.0.

7-12   Soda ash (Na_2CO_3) is produced commercially by the Solvay process, which may be represented by the following chemical equations:

$$CO_2 + NH_3 + NaCl + H_2O \rightarrow NaHCO_3 + NH_4Cl$$
$$2NaHCO_3 + heat \rightarrow Na_2CO_3 + H_2O + CO_2$$

In order to produce 2.2 kg of soda ash, how many grams of salt, NaCl, are required? Use atomic weights: Na = 23; C = 12; O = 16; Cl = 35.5

7-13   Limestone containing 60 percent by weight of $CaCo_3$, and 40 percent of $MgCO_3$, when heated in a kiln, undergoes the following reactions:

$$CaCO_3 \rightarrow CaO + CO_2$$
$$MgCO_3 \rightarrow MgO + CO_2$$

Given that 1 mol. wt of a gas occupies 380 ft^3 at the conditions of production from 1 ton of limestone,

a. How many pounds of MgO are produced?

b. How many cubic feet of CO_2 are produced?

The atomic weights are Ca = 40.1; Mg = 24.3; C = 12; O = 16.

7-14   If a current of 1 amp passes through a solution of NaCl for 1 hr, what are the products and what are the weights of the products formed at the cathode?

7-15   The complete combustion of propane gas is represented by the following skeleton equation:

$$C_3H_8 + O_2 = CO_2 + H_2O$$

a. Balance the equation.

b. How many cubic feet of air (air is 21 percent O_2 by volume and 79 percent N_2 by volume) measured at 25°C and a pressure of 760 mm Hg would be required to burn 10,500 ft^3 of propane gas measured at the same conditions of temperature and pressure?

7-16 A high-grade phosphate rock containing phosphate equivalent to 85 percent "bone-dry phosphate," $Ca_3(PO_4)_2$, is treated with an excess of sulfuric acid which converts 95 percent of the phosphate to phosphoric acid. How much phosphoric acid would be produced from 100 tons of rock? Assume the reaction involved is given by the following equation:

$$Ca_3(PO_4)_2 + H_2SO_4 \rightarrow H_3PO_4 + CaSO_4$$

Atomic weights: Ca = 40; P = 31; O = 16; H = 1; S = 32.

7-17 One of the principal scale-forming constituents of water is calcium carbonate, $Ca(HCO_3)_2$. Explain how this substance may be removed from water:

a. By boiling

b. By adding calcium hydroxide, $Ca(OH)_2$

Write the chemical equations.

7-18 A gas analysis by volume gives the following: CO_2, 12 percent; H, 4 percent; CH_4, 5 percent; CO, 23 percent; N_2, 56 percent.

a. Find the percentages by weight.

b. Write the combustion equation.

c. How much air is required for complete combustion?

Atomic weights: C = 12; O = 16; H = 1; N = 14.

7-19 For preparation of ferrous sulfide (FeS) three parts by weight of iron fillings are heated with two parts by weight of sulfur. The product from such a process is found to contain 60 percent by weight of FeS.

Calculate the complete analysis of the product in percent by weight, assuming that no sulfur is lost and no other sulfides are formed.

7-20 What is the composition in volume percent of the gases obtained from burning FeS_2 with 50 percent excess air, assuming that all the iron goes to Fe_2O_3 and all the sulfur to SO_2?

7-21 In a gaseous mixture collected over water at 14°C, the partial pressures of the components are hydrogen, 300 mm; ethane, 100 mm; oxygen, 50 mm; ethylene, 189 mm. The aqueous tension at 14°C is 12 mm.

a. What is the total pressure of the mixture?

b. What is the volume percent of hydrogen?

7-22 In the electric-furnace method of producing phosphorus, the raw materials are phosphate rock, silica, and coke. The products are calcium silicate, phosphorus, and carbon monoxide. Balance the following chemical equation:

$$Ca_3(PO_4)_2 + SiO_2 + C \rightarrow CaSiO_3 + P + CO$$

7-23 One of the principal scale-forming constituents of water is calcium bicarbonate, $Ca(HCO_3)_2$. This substance may be removed by treating the water with lime, $Ca(OH)_2$, in accordance with the following reactions:

$$?Ca(HCO_3)_2 + Ca(OH)_2 \rightarrow ?CaCO_3 + ?H_2O$$

a. Determine the number of pounds of lime required to remove 1 lb of calcium bicarbonate.

b. Determine the pounds of each of the products of the reaction per pound of calcium bicarbonate.

7-24 Gasoline may be represented approximately by the formula C_8H_{18}.
 a. Write the reaction for the complete combustion of C_8H_{18} with oxygen.
 b. Determine the weight of oxygen required per pound of fuel, and the weight of each of the products of combustion.

7-25 a. Chlorine may be prepared by the electrolysis of an aqueous solution of sodium chloride. If the decomposition efficiency of NaCl in the electrolytic cell is 50 percent, what weight of chlorine would be obtained per ton of salt electrolyzed?
 b. What would be the volume in cubic feet of the chlorine gas at standard conditions of temperature and pressure?
 c. Explain how application of chlorine to a municipal water supply makes the water bacteriologically safe for drinking.

7-26 A pound of cane sugar ($C_{12}H_{22}O_{11}$) is burned with the theoretical amount of air to give complete combustion. What is the composition of the products of combustion expressed as percent by volume?

7-27 An ultimate analysis of a Southern coal is as follows:

Carbon	0.7161
Hydrogen	0.0526
Oxygen	0.0979
Nitrogen	0.0123
Sulfur	0.0074
Ash	0.1137
	1.0000

Atomic weights: C = 12; H = 1; O = 16; N = 14; S = 32.
 a. Compute pounds of air per pound of fuel theoretically required for combustion.
 b. Compute the flue gas weight per pound of fuel at theoretically perfect combustion.
 c. What is the percentage of CO_2 in the wet gas?

7-28 Write the chemical formula for and calculate the weight of water and copper formed by the reaction of hydrogen with 10 g of cupric oxide.

MULTIPLE-CHOICE PROBLEMS

For each question select the correct answer from the five given possibilities.

7M- 1 One mole of carbon will combine with 1 mole of oxygen to form:
 (a) 1 mole of carbon monoxide
 (b) 1 mole of carbon dioxide
 (c) 2 moles of carbon dioxide
 (d) 1 mole of carbon dioxide and 1 mole of carbon monoxide
 (e) none of these

7M- 2 Hardness in water supplies is primarily due to the solution in it of:
(a) carbonates and sulfates of calcium and magnesium
(b) alum
(c) soda ash
(d) sodium sulfate
(e) sodium chloride

7M- 3 The group of elements containing bromine, iodine, and chlorine is known as the:
(a) reactants   (d) disaccharides
(b) ceramics   (e) halogens
(c) colloids

7M- 4 Which of the following statements is true?
(a) Equal volumes of gases at the same temperature and pressure have equal numbers of molecules.
(b) The viscosity of a gas is decreased by an increase in temperature if the pressure remains constant.
(c) All gases and vapors may be treated as ideal gases.
(d) All gases have the same specific heat at constant pressure.
(e) Gases are not soluble in water.

7M- 5 Which of the following chemical formulas is incorrect? (The valence of the various elements is as follows: $Al = 3+$; $Ba = 2+$; $SO_4 = 2-$; $NO_3 = 1-$; $O = 2-$)
(a) Al_2O_3 (d) $Al_2(SO_4)_3$
(b) $Al(NO_3)_3$ (e) $BaNO_3$
(c) $BaSO_4$

7M- 6 The process in which a solid changes directly to the gaseous state is called:
(a) sublimation   (d) vaporization
(b) homogenization   (e) distillation
(c) crystallization

7M- 7 The number of molecules in 22.4 liters (under standard conditions) of a substance in its gaseous state is called:
(a) Dulong's number   (d) Gay-Lussac's number
(b) Petit's number   (e) Graham's number
(c) Avogadro's number

7M- 8 One equivalent weight of H_2SO_4 is equal to:
(a) 98.06 g
(b) 2 g
(c) 49.03 g
(d) 96.06 g
(e) 32.06 g
Atomic weights: H = 1.00   S = 32.06   O = 16.00

7M- 9 H_2, Cl_2, and HCl are in equilibrium in a sealed box. What would be the effect on the concentration of the Cl_2 if more H_2 were injected into the sealed box?

(a) Increase

(b) Decrease

(c) Remain the same

(d) Equal the concentration of the HCl divided by the concentration of the H_2

(e) Equal the concentration of the H_2 minus the concentration of the HCl

7M-10 Two moles of sodium react with 2 moles of water to produce:

(a) 1 mole of sodium hydroxide and 1 mole of hydrogen

(b) 2 moles of sodium hydroxide and 2 moles of hydrogen

(c) 2 moles of sodium hydroxide and 1 mole of hydrogen

(d) 1 mole of sodium hydroxide and 2 moles of hydrogen

(e) none of these

8

ENGINEERING ECONOMICS

The principal emphasis has been on comparing the costs of alternative types of equipment purchase, payout periods, and costs of replacement of existing equipment. Included with these general subjects have been the complementary subjects of depreciation, interest, sinking funds, etc.

8-1. INTEREST

The simplest form of interest is termed, appropriately enough, "simple interest." This is the amount charged in dollars per year for the use of $100 of capital; it is similar to yearly rental. Thus, if someone charged $3 interest for a loan of $60 for a period of 8 months, the simple interest rate would be $(^3/_{60}) \times (^{12}/_8) = 7.5$ percent. Simple interest is seldom used for periods of longer than 1 year except when the interest is paid to the owner of the capital instead of being added to the capital. If the money is deposited in an account of some type, the interest is ordinarily added to the account when it is due, and for the next period interest is calculated on the total amount, principal plus accrued interest. This is known as compound interest.

By this method the amount in the account at the end of the first interest period would be $S_1 = \text{principal} + i \times \text{principal} = P(1 + i)$; at the end of the second interest period the sum would equal $S_2 = S_1 + i \times S_1 = S_1(1 + i) = P(1 + i)^2$. Similarly

$$S_3 = S_2 + i \times S_2 = S_2(1 + i) = P(1 + i)^3$$

For n periods, $S_n = P(1 + i)^n$, which is known as the compound interest law. Restating this we have, "The total amount of money at the end of n periods resulting from the investment of a principal P at an interest rate of i per period can be determined from the equation $S_n = P(1 + i)^n$." Note that n is the number of investment periods and i is the interest rate per period. If interest should be compounded quarterly, the number of periods would be four times the number of years and the interest rate i would be one-quarter the yearly or nominal rate. Similarly, n would equal twice the number of years the money was invested if the interest were compounded semiannually, and i would be one-half the nominal or yearly rate.

8-2. NOMINAL AND EFFECTIVE INTEREST

This leads to the distinction between nominal and effective rates of interest. Interest rates are usually quoted on an annual, or yearly, basis, and such an annually based rate is termed the "nominal" interest rate. Interest payments (or charges) are, however, frequently made semiannually, quarterannually, or monthly. In such a case part of the yearly interest can also earn interest during part of the year; the equivalent total yearly interest is termed the "effective" interest rate. Thus an interest rate of 5 percent compounded semiannually would give a nominal interest rate of 5 percent but an effective interest rate equal to $(1 + 0.025)^2 - 1 = 0.0506$, or 5.06 percent. Similarly, 5 percent compounded quarterly would given an effective rate of interest of 5.09 percent, though the nominal rate would still be 5 percent.

A past examination problem illustrating this point was as follows:

- You may purchase government bonds at $750 each which mature in 10 years and have a face value of $1000 at the end of 10 years. Determine:
  - *a.* The average nominal interest rate earned by the purchase price of $750, assuming annual compounding.
  - *b.* Compare the "nominal" rate of interest with the "effective" rate. Under what conditions will the "nominal" and "effective" rates of interest differ?

Annual compounding would mean ten interest periods, so $1000 = 750(1 + i)^{10}$, giving $i = 2.92$ percent nominal interest rate. The effective rate would be the same in this case, since the nominal and the effective rates will be different only when the interest periods are shorter than a year.

Effective annual interest rate $= [1 + (i/m)]^m - 1$, where i is the nominal interest rate and m is the number of interest periods per year.

The actual rate of interest paid (or earned) is an important factor in comparing alternate investments or methods of borrowing money. A slightly more complex interest problem is illustrated by the following:

- A household finance company lends a young engineer $120 cash for which he agrees to pay the company $8.72 per month for 18 months. He receives the $120 cash, begins payments at the end of the first month, and continues the monthly payments until 18 payments have been made to complete the contract. For this particular capital-recovery-with-return schedule, determine:

a. The nominal annual rate of interest earned by the company's capital.

b. The effective annual rate of interest earned by the company's capital.

The nominal yearly rate of interest is that equivalent rate which would have been charged on a yearly basis had not the payments been required monthly.

The total amount paid by the young engineer was

$$18 \times \$8.72 = \$157$$

The simple interest rate was then $(157 - 120)/120 = 30.8$ percent, but the young engineer had use of only half the money for the whole period, so the true simple interest was twice this, or 61.6 percent for a period of 18 months, or $1\frac{1}{2}$ years. The nominal or equivalent yearly rate i can be determined from the compound interest law: $0.616 = (1 + i)^{3/2} - 1$, which gives $i = 37.7$ percent.

When the young engineer pays $8.72 at the end of each month for 18 months, he is actually setting up a sinking fund, the present value of which is $120, since that is the amount he received in return for his agreement to pay $8.72 per month for 18 months. Theoretically the company can reinvest the $8.72 as it is received from the young engineer and receive interest on each payment from then on. Assuming 18 interest periods, each of 1 month's time, with a payment at the end of each period, the sum of money which would be available at the end of the 18 months would equal

$$S = 8.72(1 + i)^{17} + 8.72(1 + i)^{16} + \cdots + 8.72$$

since the first payment would earn interest for the 17 remaining periods, the second payment for the 16 remaining periods, etc. The last payment would be made at the end of the 18 periods and would earn no interest.

If we multiply both sides of the equation by $(1 + i)$ and then subtract the original equation from the new one, we obtain

$$
\begin{aligned}
S(1 + i) &= 8.72(1 + i)^{18} + 8.72(1 + i)^{17} + \cdots + 8.72(1 + i) \\
(-)S &= \phantom{8.72(1 + i)^{18} + {}} 8.72(1 + i)^{17} + \cdots + 8.72(1 + i) + 8.72 \\
S \times i &= 8.72(1 + i)^{18} - 8.72
\end{aligned}
$$

which can be generalized to give

$$S \times i = R[(1 + i)^n - 1]$$

where n = number of periods
i = interest rate per period
R = amount invested at the end of each period
S = total amount in the sinking fund at the end of n periods

If the young engineer had invested the \$120 for the 18 periods of 1 month each at an interest rate of i per period, he would have had a total of $S = 120(1 + i)^{18}$ at the end of the 18 months. The more general relationship is $S = P(1 + i)^n$, where S is the amount at the end of n periods and P the amount initially invested. Substituting this equivalence for S in the previous equation and solving for R gives

$$R = P\frac{i(1 + i)^n}{(1 + i)^n - 1}$$

where R = amount to be invested in the sinking fund at the end of each period
P = present worth of the sinking fund, or the amount which could be invested in a lump sum now to give an amount equal to that in the sinking fund at the end of n periods

The amount at the end of the n periods would be S, where $S = P(1 + i)^n$, or $S = R[(1 + i)^n - 1]/i$.

In the problem being considered, $R = \$8.72$, $P = \$120$, and $n = 18$. This problem cannot be solved directly for i; it would have to be solved by trial and error, assuming an i, solving the relationship $[i(1 + i)^n]/[(1 + i)^n - 1]$, and plotting the results against the value of i until the desired value of R/P was obtained. This expression, however, has already been calculated for many values of i and n and is listed in interest tables as the capital recovery factor. $R/P = 8.72/120 = 0.0727$, for the example being considered. From a capital-recovery-factor table with $n = 18$, the CRF = 0.0727 for $i = 3$ percent and 0.0697 for $i = 2\frac{1}{2}$ percent. The interest rate per period would then be 3 percent, and the effective annual rate of interest earned by the company's capital would equal $(1 + 0.03)^{12} - 1 = 1.426 - 1$, or 42.6 percent. Using the equation given previously, effective annual interest $= [1 + (i/m)]^m - 1$, we can again calculate the nominal rate i rom the effective rate.

$$0.426 = \left(1 + \frac{i}{12}\right)^{12} - 1$$

$$1.426^{1/12} = 1.031 = 1 + \frac{i}{12}$$

$$i = 12 \times 0.030 = 0.36, \text{ or } 36.0 \text{ percent nominal interest}$$

which is approximately the same as the value calculated previously by a different method.

8-3. PRESENT WORTH

The present worth of an amount of money required at some future date can be illustrated by means of an example:

■ A young man is about to purchase a new car. A roadster just right for his needs now costs $1000. The addition of a rumble seat for his prospective in-laws will cost $150 at some future date. A roadster with a rumble seat costs $1100 now. With interest at 8 percent, determine how soon he must get married to justify the purchase of the $1100 car.

Since the $1000 would be invested anyway, the question is, "When would $100 invested now be worth $150 if interest is at 8 percent?" The only question is whether to spend the $100 now or, theoretically, to invest it at 8 percent compounded annually. We have, then, that $150 = 100(1 + 0.08)^n$, which gives $n = 5.27$, or $100 invested today at 8 percent compounded annually would amount to $150 in 5.27 years. This means that the young man would have to get married before a lapse of 5.27 years to gain by the investment now in the rumble seat.

8-4. ANNUITY

Another important concept in economics is the annuity which might be termed a "sinking fund in reverse." That is, instead of depositing a certain amount at the end of each period so as to have available a large lump sum at the end of that time, the annuity does just the opposite. A lump sum is deposited in an account at a particular rate of interest and then a certain amount may be withdrawn from that account at the end of each period for a prescribed number of periods. This is the way pension plans are established. The equation which is used to determine how much must be deposited to provide a particular amount at the end of each period for a specified number of periods with a given interest rate is the equation for the present value of a sinking fund.

$$R = P\frac{i(1 + i)^n}{(1 + i)^n - 1} \quad \text{or} \quad P = R\frac{(1 + i)^n - 1}{i(1 + i)^n}$$

where R = amount to be paid out at end of each period
P = amount to be deposited at start of annuity
n = number of periods, or life of annuity
i = interest rate per period

An illustration of the application of this equation is afforded by the following problem:

- An engineer learns that for each additional year she works her pension will increase by $30 per month. The pension will be available to her when she reaches age 65. How much is the added pension worth to her per year at age 40? Age 45? Age 50?

At age 65 she would have a life expectancy of approximately 14 years. So the question becomes: How much would the engineer have to invest today to buy a 14-year annuity of $30 per month at age 65?

The value of the annuity at age 65, assuming a yearly interest rate of 7% throughout the life of the annuity, would equal

$$P = R\frac{(1 + i)^n - 1}{i(1 + i)^n}$$

$$30 \times \frac{1.005833^{168} - 1}{0.005833 \times 1.005833^{168}} = \$3207$$

P = value of annuity at age 65
R = $30 per month
n = 12 × 18 = 168 periods of one month each
i = $7/12$ percent = 0.5833 percent per month

If it is assumed that the engineer could invest her money at a net 6 percent per year while she was still actively employed (6 percent after paying income tax on the interest she obtains), the present value at age 40 of $3207 at age 65 (25 years hence) would equal

$$\text{Value at age 40} = \frac{3207}{1.06^{25}} = \$747$$

$$\text{Value at age 45} = \frac{3207}{1.06^{20}} = \$1000$$

$$\text{Value at age 50} = \frac{3207}{1.06^{15}} = \$1338$$

8-5. COST FACTORS

Interest is an expense. It is an expense charged if we borrow, and it is an expense if we purchase a piece of equipment, in that we do not receive the interest we would otherwise get if we invested the money in an interest-producing enterprise. This loss of interest due to investment in a machine is a cost attributable to the machine and must be taken into account in any consideration of the advisability of purchasing a given machine. There are, of course, other cost factors involved, and any final decision as to whether to purchase must be based on a consideration of all the contributing factors. These can best be reviewed with the aid of a few examples.

■ A company may furnish a car for use of its salesperson, or the company may pay him or her for the use of a car at the rate of 11 cents per mile. The following estimated data apply to company-furnished cars: A car costs $1800; it has a life of 4 years and a trade-in value of $700 at the end of that time. Monthly storage cost for the car is $3 and the cost of fuel, tires, and maintenance is $0.028 per mile. What annual mileage must a salesperson travel by car for the costs of the two methods to be equal if the interest rate is 8 percent?

The best way to handle such a problem is to itemize all the individual expenses of each of the alternative methods and add them to determine the total costs of the two possibilities. One method is a flat 11 cents per mile. The other method includes costs of operation, storage, depreciation, and loss of interest due to the invested capital.

The type of depreciation has not been specified, so we shall use straight-line depreciation and average interest. This would give an annual depreciation cost of $(1800 - 700)/4 = \$275$. The cost of depreciation is assessed at the end of each year; the amount invested throughout the first year is $1,800. Similarly, the amount invested throughout the second year is $1525, the third year $1250, and the fourth year $975. The average interest lost would then be $(144 + 122 + 100 + 78)/4 = \$111$ per year. This could also have been calculated by adding the first year's interest loss to the last year's and dividing by two, or

$$i_{\text{avg}} = \frac{144 + 78}{2} = \$111$$

per year. This can be generalized to give:

$$i_{\text{avg}} = \frac{\text{first cost} + \text{salvage value} + \text{yearly depreciation}}{2} \times i$$

Tabulate all the costs and add them to obtain total cost.

Depreciation $275
Loss of interest.............. 111
Storage (12 × $3)............. 36
Operating costs............... 0.028 × miles
　　Total yearly costs..........$422 + 0.028 × mileage

The problem asks when this would be equal to a unit cost of 11 cents per mile, so we have $0.11 \times m = 422 + 0.028 \times m$, giving a yearly mileage of 5146 miles.

8-6. COSTS OF NONOPERATION

Using this problem we can illustrate another concept, the cost of not operating something that is already owned. In this case, using the above figures and assuming that the salesperson owns his or her own car, our salesperson is confronted with the need of deciding, on a purely economic basis, whether to drive to a town 400 miles distant or whether to take the train. The ticket agent says that the ticket for the trip will cost only $14, which is considerably less than the $44 it would cost to drive the car at a cost of 11 cents/mile. We must recognize, however, that our salesperson already owns a car. The question is not whether he or she should buy a car, but rather, whether he should use a car he now owns. The costs of depreciation, loss of interest, and storage go right on whether the salesperson drives the car or not. We might, then, say that it costs him $(275 + 111 + 36)/5146 =$ $0.082 per mile to not drive the car, so the total cost of the train trip would be $14 + 400 \times 0.082 = 46.80, which is more expensive than driving. Another, and perhaps more realistic, way of looking at this question is to compare the additional out-of-pocket cost of driving vs. the cost of the train ride. Depreciation, loss of interest, and storage are what might be termed "fixed costs." A fixed cost is one which does not vary over the operational life span of the equipment being considered. The yearly operating cost of the car varies, depending upon how many miles the car is driven. The fixed costs might be considered costs of owning a car; the operating cost is only the cost of driving a car. The actual cost of driving the car on the proposed trip would be $400 \times 0.028 = 11.20, which again is $2.80 less than the cost of the train ride.

This is not a discussion of the pros and cons of car vs. train. These items are only incidental to the concept of fixed costs, which continue whether the equipment is operated or not; they help to illustrate the costs of nonoperation. It should be easy to see, with the aid of the illustration, why it is often more profitable—at least for a while—to operate a plant at a loss than to shut it down altogether.

As another example, let us consider the problem:

- Machine A cost $4400 five years ago. The book value of the machine today is $2400. The highest offer for it today is $1000. In five more years the salvage value will be $40.00. The annual operational costs have been $1300 and are expected to continue in that amount for the next 5 years. Machine B costs $7200 and has a life of 10 years, at which time it will be worth $900. Estimated annual cost is $300. With money worth 8 percent, what is the annual cost of each?

8-7. BOOK VALUE

Note that the book value of the machine is $2,400, while the best price it can be sold for is $1000. The concept of book value is confusing and causes many errors in investment judgment. Book value is a fictitious value and depends upon the accuracy of the estimated depreciation. It has no real significance and should be ignored when considering the cost of replacement. Regardless of what the book value may be, the machine is worth only what it can be sold for, and that is $1000. It may help to consider the case in which the book value is less than the salvage value. In this case you would certainly not dispose of the machine for the book value, you would sell it for the best price you could get, and that is the figure you would use in estimating replacement cost, not the lower book value. The same is true when the book value is higher than the salvage value; you would still use the sale price and not the book value in the determination of replacement cost.

The yearly costs of the machines will be depreciation, loss of interest on the capital invested, and the operating costs. Tabulate these and compare them:

Machine A now represents a capital investment of $1000 and at the end of 5 years will be worth only $40. The average yearly depreciation would then be $(1000 - 40)/5 = \$192$. The yearly operating cost would be $1300, and the average interest loss would equal $(1000 + 40 + 192)/2 \times 0.08 = \$49.28$. The comparative costs would be:

Cost item	Average yearly cost over next 5 years, machine A	Average yearly cost over next 10 years, machine B
Depreciation (straight-line)	$ 192.00	$ 630.00
Operating cost	1,300.00	300.00
Average interest	49.28	349.20
Total annual cost	$1,531.28	$1,279.20

It would pay to replace machine A with the new machine B. The average costs for machine B have been calculated for a 10-year period. If the average interest had been calculated for a 5-year period, the estimated costs would have been higher, as would the depreciation costs. The best information available, however, indicates that the new machine will last 10 years, so the average annual cost of operation has been estimated on a 10-year basis.

Another problem in an examination was:

■ An asset has a first cost of $13,000, an estimated life of 15 years, and a salvage of $1000. For depreciation, use the sinking-fund method, with interest at 5 percent compounded annually, and find:

a. The annual sinking-fund annuity or depreciation charge.

b. The balance in the sinking fund, i.e., the amount accumulated toward depreciation of the asset at the end of 9 years.

c. If the asset were to be sold for $4000 at the end of 9 years, what would be the net *book-value gain or loss?*

Since the salvage value at the end of 15 years is $1,000 and $13,000 will be required for replacement, there must be a total of $12,000 in the sinking fund at the end of the 15-year period. We need to determine, then, what amount we must deposit at the end of each year to amount to $12,000 at the end of 15 years with interest at 5 percent.

The sinking-fund formula previously derived is

$$R = \frac{Si}{(1 + i)^n - 1} = \frac{12,000 \times 0.05}{(1.05)^{15} - 1} = \$556$$

to be deposited at the end of each year for 15 years.

This could also have been determined with the aid of compound interest tables. Under the heading of "Sinking-Fund Factor" we find a value of 0.04634 for 5 percent compound interest and 15 periods. This is equal to $i/[(1 + i)^n - 1]$ and need only be multiplied by the value of S to give R. $R = 12,000 \times 0.04634 = \$556$.

We could also obtain the same value from the column headed "Compound Amount Factor" or "Amount of Annuity." This column gives the amount in a sinking fund at the end of n periods for a sinking-fund deposit amount of $1, or $R = 1$. The value in this column corresponding to $i = 5$ percent and $n = 15$ is 21.5785, or $1 deposited at the end of each year for 15 years would provide an amount equal to $21.5785 at the end of that time. Since we need $12,000, $R = 12,000/21.5785 = \$556$.

The annual sinking fund or annuity charge (a) is $556. This is also called the "sinking fund deposit factor."

At the end of 9 years there would be an amount equal to

$$S = R\frac{(1 + i)^9 - 1}{i} = 556\frac{1.05^9 - 1}{0.05} = \$6,130$$

The answer to part (b) is $6130.

If the asset were sold at this time for $4000, this would give a total of 6130 + 4000 = $10,130 available to balance the replacement cost of $13,000.

The book value at this time equals

$$13,000 - 6,130 = \$6,870$$

or the first cost less the amount in the sinking fund. The actual value is only the market value of \$4000, so there is a "book-value loss" of \$2870.

This brings up another interesting point. If we had used straight-line depreciation, the book value at the end of 9 years would have been

$$13,000 - 9 \times \frac{13,000 - 1000}{15} = \$5800$$

which would have given a book-value loss of \$1800.

8-8. DEPRECIATION

If we had used the "fixed-percentage-on-diminishing-balance" method of depreciation, the book value at the end of 9 years would be \$2,790, giving a "book-value gain" of \$1210. These three types of depreciation curves plus a curve for the sum-of-the-years-digits method of depreciation are shown in Figure 8-1.

The fixed-percentage-on-diminishing-balance method depreciates an asset by a fixed percentage of its value at the beginning of the year. The book value at the end of the first year is equal to Cost $\times$ $(1 - D)$; at the end of the second year,

$$C(1 - D)(1 - D) = C(1 - D)^2$$

at the end of the third year, $C(1 - D)^2(1 - D) = C(1 - D)^3$; at the end of nth year, $C(1 - D)^n$.

For the case considered, $C = \$13,000$, and the book value at the end of the fifteenth year is \$1000, so $13,000(1 - D)^{15} = \$1000$; $D = 0.1572$, and the depreciation allowance each year is 15.72 percent of the book value of the asset at the beginning of that year. The book value at the end of the first year would then be

$$13,000 \times 0.8428 = \$10,950$$

at the end of the second year, $13,000 \times 0.8428^2 = \$9,280$, or $10,950 \times 0.8428$; at the end of the fifth year,

$$13,000 \times 0.8428^5 = \$5,530$$

at the end of the ninth year, $13,000 \times 0.8428^9 = \$2,800$; at the end of the fifteenth year, $13,000 \times 0.8428^{15} = \$1,000$.

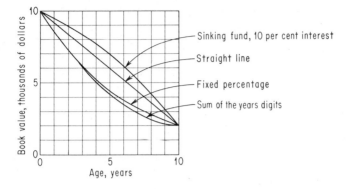

First cost... $10,000
Salvage value at end of ten years $2,000
Straight-line depreciation ...... (10,000 − 2,000)/10 = $800 per year
Book value .... $10,000 − 800$n$
Fixed percentage........................ 10,000$(1 − D)^{10}$ = 2,000
$$1 − D = 0.852$$
Book value .... 10,000 × 0.852^n
Sinking fund, interest at 10 percent

Amount in sinking fund .... $S = R \dfrac{(1 + i)^n - 1}{i}$

Book value .... $10,000 - 502 \times \dfrac{1.10^n - 1}{0.10}$

There is no simple relationship that will give the book value at the end of each year for the sum-of-the-years-digits method of depreciation, so the data for this method are tabulated below:

Year	Sum-of-the-year digits book value at end of year
0	$10,000
1	8,545
2	7,236
3	6,072
4	5,054
5	4,181
6	3,454
7	2,872
8	2,436
9	2,145
10	2,000

Figure 8-1

Another method of depreciation authorized by U.S. tax law is the sum-of-the-years-digits. With this method the sum of the digits corresponding to the estimated years of life are added together. For example, if the estimated life is 10 years, the sum-of-the-years digits $(10 + 9 + 8 + \cdots + 1)$ would equal 55. The depreciation the first year would equal $^{10}/_{55}$ times the depreciable cost. The second-year depreciation would equal $^{9}/_{55}$ times the depreciable cost, etc. (The tenth-year depreciation would equal $^{1}/_{55}$ times the depreciable cost.)

For the example used previously—first cost of $13,000, estimated life of 15 years, and salvage value of $1000—the first-year depreciation would equal $(^{15}/_{120}) \times 12,000$, or $1,500, and the book value would equal $11,500. The ninth-year depreciation would equal $(^{7}/_{120}) \times 1200$, or $700, and the book value at the end of the year would equal $3100. The fifteenth-year depreciation would be $100 and the book value at the end of the year would equal $1000.

The only difference in these cases is the method of keeping the books. Since sinking funds are seldom if ever actually set up in practice and exist only as accounts in the books of the organization, these cases would be just the same in so far as the actual investment procedure and recommendations were concerned.

Depreciation schedules are set up primarily for purposes of taxation and rate setting. They also give an estimate of the capital which may be required for replacement at any given time, but since replacement capital is ordinarily borrowed, long-term forecasts of this factor are seldom of any value. The tax picture is important however, and for this particular expense it can be seen that the fixed-percentage method will give the greatest benefit if the asset has to be sold before the end of its expected useful life. The book-value loss reflects a tax loss but should not enter into any investment decision.

Another example dealing with present value is afforded by the problem:

■ An engineer is currently paying $125 per month house rent. He or she can build a house for $12,000 which should have a life of 30 years. Taxes, insurance, and repairs will amount to $550 per year. This engineer has capital which is currently earning 6 percent on a mortgage which will be coming due very soon. Assuming that the capital can be reinvested at 6 percent and using straight-line method for depreciation, allowing $3000 salvage, calculate:

   a. The equivalent present worth of 30 years' rent at $125 per month using 6 percent interest.

b. The equivalent present worth of 30 years' owning the house.

c. Comparing present-worth values, which plan is more economical?

Part (*a*) asks, What is the present value of an annuity in which $125 is deposited at the end of each period with interest at $\frac{1}{2}$ percent per period, for 360 periods.

The annuity, or sinking-fund, formula used previously is

$$R = P \frac{i(1 + i)^n}{(1 + i)^n - 1}$$

which gives as the answer to (*a*)

$$P = 125 \frac{(1.005)^{360} - 1}{0.005(1.005)^{360}} = 125 \times 166.8 = \$20,850$$

This can also be calculated with the aid of a compound interest table. Split the total number of periods into four sets of 90 periods each. From the "Present Worth Factor" column the present worth of a 90-term annuity for $1 deposits with interest at $\frac{1}{2}$ percent per period is $72.33. In the "Compound Interest" column the present value of $1 invested at $\frac{1}{2}$ percent for 90 periods is $0.6383. If we invested $1 per period for 90 periods with interest at $\frac{1}{2}$ percent per period, the present value of the future sum is $72.33. However, the present value of the second annuity, $1 invested at the end of each period from period 91 through period 180, is $72.33 \times 0.6383 = \$46.17$, etc. We can build up a table to show this.

Period	Present value of annuity		
1–90	72.33×1	=	$ 72.33
91–180	72.33×0.6383	=	46.17
181–270	72.33×0.6383^2	=	29.47
271–360	72.33×0.6383^3	=	18.81
Present value of all four annuities . . .$166.78			

This is the same as the value obtained with the formula.

The equivalent present worth of 30 years' rent at $125 per month is, then, $20,850 with interest at $\frac{1}{2}$ percent per month.

This is the same as saying that if we invested $20,850 at an interest rate of $\frac{1}{2}$ percent per month, at the end of 30 years (360 months) it would equal the same amount that we should have if we invested $125 per month for 360 months (30 years) in an annuity with interest at $\frac{1}{2}$ percent per month.

$$\$20,850 \times 1.005^{360} = \$125,500$$

$$\$125 \times \frac{1.005^{360} - 1}{0.005} = \$125,500$$

An alternative method would be to figure the annuity based on 30 end-of-the-year payments of $1500 (12 × 125). The present worth of such an annuity would be

$$\$1500 \times \frac{1.06^{30} - 1}{0.06 \times 1.06^{30}} = 1500 \times 13.76 = \$20,640$$

If the house were owned, $550 would have to be spent every year for taxes, insurance, and repairs. This would amount to an annuity with a present value of

$$\$550 \times 13.76 = \$7570$$

This means that of the $20,640 present worth of the rent annuity, $7570 would have to be used to set up a fund to return $550 each year for 30 years for expenses.

Another way of looking at this is to say that of the $1,500 per year rental income $550 would have to be used for expenses, leaving a net of $950 to be deposited in the annuity at the end of each year. Either method will give the same value.

$$\$20,640 - \$7570 = \$13,070$$

$$\$950 \times 13.76 = \$13,072$$

In addition to the present worth of the net rental income annuity, there is also the estimated $3000 value of the house 30 years hence. The present worth of $3000 is

$$\frac{\$3000}{1.06^{30}} = \$522$$

The equivalent present worth of 30 years' owning the house—the answer to (b)—is then

$$\$13,076 + \$522 = \$13,598$$

The present worth of $12,000 is, of course, $12,000. Comparing these two figures we can see that the answer to part (c) is that it would be beneficial to build the house.

We can also determine this by calculating the values of the two alternatives at the end of 30 years. For the $12,000 invested at 6 percent we should have

$$\$12,000 \times 1.06^{30} = \$68,920$$

The rent money less expenses which could be invested in an annuity at the end of each year would equal $950. At the end of 30 years this would amount to

$$\$950 \times \frac{1.06^{30} - 1}{0.06} = 950 \times 79.058 = \$75,100$$

To this should be added the $3000 which the house will be worth in 30 years, giving a total of $78,100. This is $9180 more than the value of $12,000 thirty years hence. This should also equal the value of the difference of the present worths 30 years hence, or $1598 $\times$ 1.06^{30} = $9,180, which checks.

Another facet of engineering economics is illustrated by the past problem:

- A contract is let for $10,000. It is to be completed in 4 weeks' time, subject to a penalty of $240 per day if that time is exceeded. At the end of 3 weeks the work is 60 percent complete. Labor costs were initially estimated to come to $4000 on the basis of 40 hours per week. If overtime labor must be compensated at double time:

  a. Would it be cheaper to take the penalty or to complete the job by working overtime? (Give cost figures.)

  b. By what percentage would the estimated labor cost be increased by the cheaper plan?

Since the work is 60 percent complete, there is 40 percent to go. Working at the present rate will then require

$$(^{40}/_{60}) \times (3 \times 40) = 80 \text{ hours}$$

or 10 more days. With 1 week left this would mean 16 hour-days, but the problem implies that this would be all right. The cost of labor is 4000/4 = $1000 per week, or $200 per day. Figure costs of each method.

Penalty Method:

$$
\begin{array}{ll}
\text{10 days' labor @ \$200} \dots\dots\dots\dots\dots & \$2000 \\
\text{5 days' penalty @ \$240} \dots\dots\dots\dots\dots & \underline{1200} \\
\quad \text{Total cost labor plus penalty} \dots\dots & \$3200
\end{array}
$$

Overtime Method:

$$
\begin{array}{ll}
\text{5 days' straight time @ \$200} \dots\dots\dots\dots & \$1000 \\
\text{5 equivalent days' double time @ \$400} \dots\dots & \underline{2000} \\
\quad \text{Total cost} \dots\dots\dots\dots\dots\dots\dots\dots\dots & \$3000
\end{array}
$$

It would be cheaper to complete the job by working overtime.

With the cheaper plan the total labor cost would be the $3000 spent during the first 3 weeks plus the $3000 spent the last week, giving a total labor cost of $6000 as compared with the $4000 estimate, or a 50 percent increase.

An example of the type of problem encountered in recent examinations is as follows:

■ It is desired to construct a bridge over a marshy area in a commercial recreation park. A wooden bridge would cost $8000 and would last an estimated 8 years. Maintenance costs would equal $700 per year. There would be no salvage value. The interest rate is 8 percent.

1. What would be the yearly cost using the sinking-fund method?
   (a) $1875
   (b) $2250
   (c) $1680
   (d) $2090
   (e) $1920

2. What would be the yearly cost using straight-line depreciation and average interest?
   (a) $1800
   (b) $1660
   (c) $2060
   (d) $2200
   (e) $1980

3. What would be the first-year depreciation if the sum-of-the-years-digits method were used?
   (a) $1008
   (b) $1576
   (c) $1778
   (d) $1876
   (e) $1248

4. How much would have to be deposited in an account to pay for the $700 per year maintenance costs over the 8-year life?
   (a) $4022
   (b) $3785
   (c) $5600
   (d) $4080
   (e) $3628

5. If the bridge is reinforced at the end of 5 years, the usable life could

be extended an estimated 3 years (to 11 years total life). How large an expense could be justified to reinforce the bridge? (The same yearly maintenance costs would apply. Use sinking fund cost.)

(a) $3026

(b) $2848

(c) $2972

(d) $2756

(e) $2932

6. An alternative method of construction would be to make the bridge of prestressed concrete construction. The estimated life of such a structure would be 20 years. The first cost would be $15,000 and it would have a salvage value of $2000 at the end of its useful life. The yearly maintenance cost would be $200. What would be the total yearly cost using the sinking-fund method?

(a) $1684

(b) $2098

(c) $1875

(d) $1586

(e) $1714

7. What would be the yearly cost of the prestressed concrete bridge using straight-line depreciation and average interest?

(a) $1722

(b) $1488

(c) $1684

(d) $1624

(e) $1556

8. What would be the depreciation for the third year using the sum-of-the-years-digits method of depreciation?

(a) $1216

(b) $1114

(c) $1162

(d) $1098

(e) $1052

9. What would be the depreciation for the third year if the fixed-percentage-on-diminishing-balance method were used?

(a) $1175

(b) $1150

(c) $1200

(d) $1135

(e) $1298

10. For what rate of interest would the yearly costs of the different structures in parts 1 and 6 be the same?

(*a*) 6 percent
(*b*) 12 percent
(*c*) 10 percent
(*d*) 14 percent
(*e*) 16 percent

(1) The salvage value of the wooden bridge at the end of its useful life is zero, so there would have to be a total of $8,000 plus the lost interest, or $8000(1.08)^8$, at the end of 8 years. The present value of the sinking fund is $8,000 so the end-of-the-year deposits would equal

$$R = \frac{Si \times (1 + i)^n}{(1 + i)^n - 1} = \frac{8000 \times 0.08 \times 1.08^8}{1.08^8 - 1} = \$1392.12$$

To this must be added the maintenance cost of $700 per year, giving a total yearly cost of $2092.12. Answer (*d*) is the correct one.

(2) Using straight-line depreciation, the yearly depreciation would equal $8000/8 = \$1000$ per year. The average yearly interest cost would equal

$$\frac{8000 + 1000}{2} \times 0.08 = \$360 \text{ per year}$$

Total yearly costs would equal $1,000 + $360 + $700 = $2,060. The correct answer would be (*c*).

(3) The sum of the digits of an even number of years equals $(n + 1) \times n/2$, so the sum of the digits for 8 years would equal 9×4, or 36. The sum of the digits for an odd number of years is obtained with the same equation, but is more easily understood if rearranged as $(n + 1)/2 \times n$.

The sum-of-the-years digits for a life of 40 years, for example, equals 41×20, or 820. For 45 years the sum equals $(46/2) \times 45$, or 1035.

The sum of the years digits for 8 years is 36. The first-year depreciation would then equal

$$(8/36) \times 8,000 = \$1,778$$

Answer (*c*) would be the correct answer.

(4) The amount required would be the amount necessary to set up a $700 per year annuity with a life of 8 years (eight payments). This is the present value of a sinking fund.

$$P = R\frac{(1 + i)^n - 1}{i(1 + i)^n} = 700\frac{(1.08)^8 - 1}{0.08(1.08)^8} = \$4,022$$

Answer (a) is the correct one.

(5) The capital-recovery-plus-a-return yearly payment for the original $8,000 was found to equal $1,392.12 in part 1 of the problem. This is the amount which would be paid each year for three additional years to pay for the original bridge plus the reinforcement.

At the end of 5 years the amount in the sinking fund would equal $1,392.12 \times (1.08^5 - 1)/0.08$, or $8,167.01. The original investment of $8,000 would have appreciated to $8,000 \times 1.08^5$, or $11,754.62, if it had been invested at 8 percent; so there would still remain $3,587.61 to be paid off. That is, the book value of the bridge at the end of 5 years would equal $3,587.61. The cost of reinforcement would be added to this to give a value of $(3,587.61 + R)$ for the bridge at the end of 5 years, after it had been reinforced. This would appreciate to $(3,587.61 + R) \times 1.08^6$ after six more years. This must equal the additional amount in the sinking fund resulting from deposits made at the end of the sixth through eleventh years. Thus,

$$(3,587.61 + R) \times 1.08^6 = 1,392.12 \times (1.08^6 - 1)/0.08 = 10,212.49$$
or
$$R = (10,212.49/1.08^6) - 3,587.61 = \$2,847.99$$

which is the amount which could be justified for the reinforcement of the bridge. This assumes that the interest rate would remain constant at 8 percent for 11 years. The correct answer is (b).

To check this number, $8,000 would be invested for 11 years and $2,847.99 would be invested for 6 years. The total value to which these investments would have appreciated at the end of the life of the reinforced bridge would equal

$$9,000 \times 1.08^{11} + 2,847.99 \times 1.08^6 = \$23,172.51$$

This should equal the value of the sinking fund at the end of 11 years, or

$$1,392.12 \times (1.08^{11} - 1)/0.08 = \$23,172.51$$

which checks.

(6) The depreciable cost equals $13,000. This is the capital which must be recovered. The $2,000 would remain tied up throughout the 20-year life and would be recoverable at the end of that time. The present value of the sinking fund would then equal $13,000. The total cost would equal the capital-recovery-plus-a-return yearly payment for the $13,000 depreciable cost plus the interest lost on the $2,000 which would be tied up for 20 years plus the yearly maintenance cost of $200.

$$R = P\frac{i(1 + i)^n}{(1 + i)^n - 1} = 13,000\frac{0.08 \times 1.08^{20}}{1.08^{20} - 1} = \$1,324.08$$

The total yearly cost would equal

$$1324.08 + 0.08 \times 2,000 + 200 = \$1,684.08$$

The correct answer is (a).

(7) The depreciable cost is \$13,000, so the straight-line depreciation would equal (13,000/20), or \$650 per year.

The average interest would equal

$$\frac{15,000 + 2000 + 650}{2} \times 0.08 = \$706.00 \text{ per year}$$

The total yearly cost would then equal

$$650.00 + 706.00 + 200.00 = \$1556.00$$

Answer (e) is correct.

(8) The sum of the years digits would equal $(20 + 1)(20/2)$, or 210. The first-year depreciation would equal $(20/210) \times 13,000$ or \$1,238. Second year depreciation would equal $(19/210) \times 13,000$, or \$1,176. And depreciation for the third year would equal $(18/210) \times 13,000$, or \$1114. Answer (b) is correct.

(9) The first cost equals \$15,000 and the salvage value at the end of 20 years would equal \$2,000.

$$\text{(First cost)} \times (1 - D)^n = \text{salvage value}$$
$$15,000 \times (1 - D)^{20} = 2,000$$
$$1 - D = 0.1333^{1/20} = 0.9042, \text{ giving } D = 0.0958, \text{ or } 9.58 \text{ percent}$$

The first-year depreciation would equal

$$0.0958 \times 15,000 = \$1,437.00$$

The amounts for depreciation for the next few years are listed below.

Year	Book Value	Depreciation
1	\$15,000.00	\$1,437.00
2	13,563.00	1,299.34
3	12,263.66	1,174.86
4	11,088.80	1,062.31
5	10,026.49	960.54

The correct answer is (a).

(10) The total cost for part 1 was found to equal

$$8000 \times \frac{i(1 + i)^8}{(1 + i)^8 - 1} + 700$$

The total cost for part 6 was found to equal

$$13,000 \times \frac{i(1 + i)^{20}}{(1 + i)^{20} - 1} + 2000 \times i + 200$$

A table can be set up to compare the total costs of the wooden and concrete bridges for different interest rates.

%	Cost of Wooden Bridge per year	Cost of Concrete Bridge per year
8	$2,092	$1,684
9	2,145	1,804
12	2,310	2,180
14	2,424	2,442
13.75	2,410	2,410

Answer (d): 14 percent is the closest to the interest rate at which the costs of the two different types of bridge would be equal, so (d) is the correct answer.

8-9. CASH FLOW

Cash flow is the flow of money (cash) into and out of a business. It includes the money received from the sale of goods and services, and money paid out for business expenses. Accounts receivable represent assets, but they cannot be considered in the cash flow until they are collected. This is the reason for the existence of "factors," those who buy accounts receivable (at a discount) and then collect the money due. Sale of the accounts receivable provides cash to the company originally holding the paper so that it can pay its bills, salaries, and taxes. Depreciation is a legitimate business expense, but it is not a cash item, so it is not included in the cash flow.

As an example, assume that the Koe Co. can save $5,000 per year in materials-handling expenses by rearranging the storage area and materials-handling equipment in its warehouse. Rearrangement of the facilities will cost $37,000 and it is estimated that the savings will be realized for a total of 12 years. What would be the return on the investment?

Since no new equipment would be purchased, and no existing equipment would be repaired or modernized, there would be no depreciation charges. The $37,000 would constitute an operating expense and the expense would all be incurred during the first year. The Koe Co. pays a 50 percent tax on its income.

The cash flow the first year, assuming that a $5000 saving would be realized during the first year, would include the following:

Cash outlay for improvements	−$37,000
Savings in handling	+ 5,000
Net cash outlay before income tax	−$32,000
Savings on income tax	+$16,000
Cash flow for the first year after income tax	−$16,000

The second year there would be no additional expense and a savings of $5000 would be realized. Of this savings, 50 percent, or $2500, would have to be paid in taxes. The cash flow after income taxes for the second through the twelfth years would then equal +$2500 per year.

Thus the question becomes: If $16,000 is invested now for a return of $2500 per year for 11 years, what would be the rate of return? Or, if an annuity is purchased for $16,000 which will pay $2500 at the end of each year for 11 years, what is the rate of interest?

$$16,000 = 2500 \frac{(1 + i)^{11} - 1}{(1 + i)^{11} \times i}$$

which gives an interest rate of 10.3 percent per year.

SAMPLE PROBLEMS

8- 1 To obtain a certain service two plans are being considered. (1) Plan A will require the purchasing of certain special equipment costing $12,000. The salvage for this special equipment will be $500 regardless of the length of time it may have been used. Annual labor costs will average $3500. (2) Plan B simply involves $5000 labor cost per year. All other incomes and expenses under the two plans will be the same. Allow 5 percent interest on capital, 5 percent compounded annually on amortization or sinking fund for depreciation, and determine how many years it will take for the special equipment of Plan A to pay for itself. (Hint: Annual cost tabulation and a final interpolation give an exact answer.)

8- 2 A young engineer has estimated that her annual earnings should average $6000, $10,000, and $15,000 per year in succeeding decades from the time she takes her first job after graduation. Allow 3 percent interest compounded annually each for cost of money and return, and determine:

a. Present worth (at graduation) in cash of the 30 years' earnings
b. Equivalent uniform annual value of the 30 years' estimated income

8- 3 A piece of earth-moving equipment was purchased at a cash price of $25,000. The life of this equipment was estimated at 6 years with no salvage. However, at the end of 4 years the machine had become so inefficient, because of wearing of parts, that it was replaced. Depreciation was allowed on the company books by the sinking-fund method with 4 percent interest. Determine:

a. Estimated annual depreciation
b. Balance in the depreciation sinking fund at the end of the 4 years
c. Sunk cost at the time of replacement

8- 4 A certain type of automatic milling machine can be purchased for $6,500. Expert estimates indicate a life of 10 years and $500 salvage for this particular machine. Use the sinking-fund method for depreciation charges and calculate total *annual capital recovery* charges, which include return on capital on this machine, when interest is (a) 4 percent, (b) 6 percent.

8- 5 An old light-capacity highway bridge may be strengthened at a cost of $9,000, or it may be replaced by a new bridge of sufficient capacity at a cost of $40,000. The present net salvage value of the old bridge is $13,000. It is estimated that the old bridge, when reinforced, will last for 20 years, with a maintenance cost of $500 per year and a salvage value of $10,000 at the end of 20 years. The estimated salvage value of the new bridge after 20 years of service is $15,000. The maintenance on the new bridge will be $100 per year. If the interest is 6 percent, determine whether it is more economical to reinforce the old bridge or replace it. Use straight-line depreciation, plus average interest.

8- 6 A lathe costs $10,000 now. It has a life expectancy of 20 years and an estimated salvage value of $2000 at the end of 20 years.

At 7 percent interest, what is the annual "capital-recovery-with-a-return" cost of the lathe on the basis of the above estimates? Use straight-line depreciation plus average interest in your calculations.

8- 7 A company buys a machine for $12,000, which it agrees to pay in five equal annual payments, beginning 1 year after the date of purchase, at an interest rate of 4 percent per annum. Immediately after the second payment, the terms of the agreement are changed to allow the balance due to be paid off in a single payment the next year.

a. What is the yearly payment for the first 2 years?
b. What is the final payment?

8- 8 A manufacturing plant has been purchasing the energy required for plant operation. It is considering building a power plant to supply a load estimated at 1,000 kW (24 hours per day, 365 days per year). Purchased power will cost

$0.023 per kWh. The cost of the required plant is $1,000,000, and total operating expense including interest on investment is estimated at $95,000 per annum. Assuming 8 percent sinking-fund depreciation, 15-year life, and zero scrap value, determine whether energy should be purchased or supplied by the new power plant.

8- 9 A manufacturer is planning to produce a new line of products which will require the buying or renting of new machinery. A new machine will cost $17,000 and have an estimated value of $14,000 at the end of 5 years. Special tools for the new machine will cost $5000 and have an estimated value of $2500 at the end of 5 years. Maintenance costs for the machine and tools are estimated to be $200 per year. What will be the average annual cost of ownership during the next 5 years if interest is 6 percent? Assume that the annual cost is equal to the straight-line depreciation, plus average interest.

8-10 How much would the owner of a building be justified in paying for a sprinkler system that will save $500 per year in insurance premiums? The system will have to be renewed every 20 years and has a salvage value of 10 percent of its initial cost. Interest is at 5 percent.

8-11 The owners of a concrete batching plant have in use a power steam shovel, which, if repaired at a cost of $2000, would last another 10 years. Maintenance costs, operating costs, taxes, and insurance would be $2500 per year. The present salvage value of the power steam shovel is $6000; if it is repaired and used for 10 years, the estimated salvage value at the end of that time is estimated to be $500. A new power shovel could be purchased at a cost of $30,000, and its estimated salvage value at the end of 10 years would be $20,000. The maintenance costs, operating costs, taxes, and insurance would be $1500 per year. If the interest rate is 6 percent, determine whether it is more economical to repair and use the old steam shovel or to buy the new one. Use straight-line depreciation, plus average interest.

8-12 A debt of $10,000 with interest compounded annually at the rate of 4 percent is to be paid in a lump sum at the end of 5 years. To create a fund with which to pay the debt, the debtor decides to deposit equal sums at the end of each 6 months with a savings association where the interest is compounded semiannually at an annual rate of 3 percent. What should be the amount of the semiannual deposit?

8-13 The state will furnish an automobile to certain employees or will permit them to use their personal car and will pay 10 cents per mile. A car costs $2350 and has a life of 5 years, with a trade-in value of $600; it costs $10 per month for incidentals, the cost of tires and repairs is $4 per month, and fuel and oil cost 2 cents per mile.

   a. It is necessary for an employee to travel how many miles per year in order that the cost of the two methods be equal? Interest is computed at 6 percent; depreciation is computed by the straight-line formula.

   b. If the individual travels 10,000 miles per year, what will be the financial advantage or disadvantage?

MULTIPLE-CHOICE PROBLEMS

For each question select the correct answer from the five given possibilities.

8M- 1 An individual wishes to deposit a certain quantity of money so that at the end of 5 years, at 4 percent interest, compounded semiannually, he will have $500. He must deposit:
(a) $609.50   (d) $410.15
(b) $451.35   (e) none of these
(c) $337.80

8M- 2 Money is invested at a nominal rate of interest of 5 percent per annum compounded semiannually. What is the effective rate per annum?
(a) 10 percent   (d) 5.06 percent
(b) $5^{1}/_{4}$ percent   (e) none of these
(c) 5 percent

8M- 3 When a cost analysis of an engineering project is figured on a straight-line depreciation and an average interest basis, the average interest is:
(a) the interest for the first period plus the interest for the last period, divided by two.
(b) the interest figured on half the principal.
(c) half the annual interest.
(d) half the interest calculated at the end of the first period.
(e) the sum of the interests for each period divided by two.

8M- 4 The uniform annual-end-of-year payment to repay a debt (the lender's investment) in n years, with an interest rate of i is determined by multiplying the capital recovery factor by the:
(a) average investment
(b) initial investment, plus total interest
(c) average investment, plus interest
(d) initial investment, plus first year's interest
(e) initial investment

8M- 5 The straight-line depreciation, plus average interest, is used to calculate the yearly cost of the investment in a machine because:
(a) It is a measure of the likelihood of bankruptcy.
(b) It is required by the Bureau of Internal Revenue in figuring income taxes.
(c) It is more accurate than the sinking-fund formula.
(d) It is easily understood and fairly accurate.
(e) It takes into account the "sunk cost" of the machine.

8M- 6 The formula for determining average annual interest is

(a) $$(P - L)\frac{i}{2}\left(\frac{n + 1}{n}\right) + Li$$

(b) $$(P - L)\left(\frac{1}{2}\frac{n + 1}{n}\right)$$

(c) $$\left(\frac{P + L}{2}\right)i$$

(d) $$(P - L)\left[\frac{1}{(1 + i)^n - 1}\right] + Li$$

(e) $$(P - L)\left[\frac{1}{(1 + i)^n - 1}\right]$$

8M- 7 The sum which must be paid yearly to pay a total sum at the end of a given number of years and at a given rate of interest is the:

(a) sinking fund   (d) compound interest rate
(b) present worth   (e) discount rate
(c) capital recovery

8M- 8 In the interest formula $A = P(1 + i/12)^{12n}$ the interest is compounded:

(a) daily   (d) semiannually
(b) monthly   (e) annually
(c) quarterly

8M- 9 A sum of money invested at 4 percent compounded semiannually will double in amount in approximately:

(a) $15\frac{1}{2}$ years   (d) $18\frac{1}{2}$ years
(b) $21\frac{1}{2}$ years   (e) $19\frac{1}{2}$ years
(c) $17\frac{1}{2}$ years

8M-10 The present value of $5000.00 ten years hence with interest at 7.5 percent compounded annually most nearly equals

(a) $2525.00   (d) $2500.00
(b) $2385.00   (e) $2400.00
(c) $2425.00

ANSWERS TO SAMPLE PROBLEMS

1- **1.** 0.5 radians/min, or 28.7°/min

1- **2.** $x = 0.850D$, $y = 0.526D$

1- **3.** 433 ft

1- **4.** $y = 2x^2 + x - 1$

1- **5.** 13 in.²/sec

1- **6.** $3y = 14x + 18$

1- **7.** $dr/dt = -0.1193$ in./sec

1- **8.** 40,100 ft²

1- **9.** 44.4°

1-**10.** 8.58 gal

1-**11.** -1

1-**12.** *a.* 5.34 mph closing the gap
   b. 10.7 mph separating
   c. 1.30 hr, 20.0 miles apart

1-**13.** $wl/4 - w/6$

1-**14.** $\sin^{-1}(0.25l - 0.75) - 3.989$ radians

1-**15.** *a.* $x = -8.87$, $y = 9.913$ and $x = 7.02$, $y = -0.0683$; angle between line and tangents is 78.5°
   b. $2 \tan x$

1-**16.** 1,842 ft

1-**17.** $3y = 4x + 8$

1-**18.** *a.* slope $= 0, -1$
   b. $x = 4a$, $y = -2.67a + 1.25$, and $x = 0$, $y = \frac{5}{4}$
   c. $x = 4.83a$,
   $y = -2.27a + 1.25$, and
   $x = -0.83a$,
   $y = -0.393a + 1.25$

1-**19.** 453 cu units

1-**20.** 500 ft

1-**21.** a. 100 ft × 100 ft, and 50 ft × 50 ft
   b. 10,000 sq ft, and 2,500 sq ft.

1-**23.** $\pi/4$

1-**24.** $a/2(e^{x_{1/a}} - e^{-x_{1/a}})$

1-**25.** *a.* 45°
   b. 68.3 yd

1-**26.** 1.194 fpm

1-**27.** *a.* $x = 0$
   b. $x = -\pi/2$

1-**28.** 21.4 miles

1-**31.** *a.* $34x^2/15$
   b. $1/(16y^8z^{12})$
   c. $1/(a + x)$
   d. $2.5^{1/6}$
   e. $x = 7$

1-**32.** 19.3°; 125.4°; 118.4 ft

1-**33.** 111.5°

1-**34.** 5.043

1-**35.** 1.605 hr

1-**36.** y increases as x increases, no maximum. y approaches infinity as x approaches infinity. $y_{min} = -10^{1/3}$ at $x = \frac{1}{3}$

1-**37.** *b.* $y = -0.50x - 2.5$
   c. 25.0 sq units

1-**38.** $D = 23.4$ ft; $L = 23.3$ ft

1-39. 374 ft

1-40. $\frac{1}{2}$

1-41. $\frac{1}{2}$

1M- 1. d

1M- 2. a

1M- 3. e

1M- 4. c

1M- 5. d

1M- 6. a

1M- 7. a

1M- 8. c

1M- 9. d

1M-10. c

2S- 1. 500 lb, 300 lb (coefficient of friction = 0.50)

2S- 2. 13.33 ft; 4,170 lb

2S- 3. 4.33 in.

2S- 4. a. 2 lb b. 2.67 lb

2S- 5. U_3U_4, 15,000 lb compression; U_4L_4, 13,330 lb tension

2S- 6. AB, 618 lb compression; AC, 425 lb tension; AD, 816 lb tension

2S- 7. 10.15 lb

2S- 8. a. 732 lb, b. 250 lb, c. 688 lb

2S- 9. $\bar{x}$ = 3.45 in., $\bar{y}$ = 3.46 in. from lower left corner

2S-10. 6.4 in.

2S-11. CH = 1020 lb tension; CD = 4160 lb compression

2S-12. AC = 840 lb tension; BC = 840 lb tension; CD = 2500 lb compression

2S-13. a. $F = W \tan \alpha$;
b. vertical row, force = F
horizontal row, force = W
slanting row, force = $W/\cos \alpha$
$= F/\sin \alpha$
c. $\alpha = 5.7°$

2S-14. a. L_1U_2 = 2.23 k compression
b. U_0U_1 = 5.98 k compression

2S-15. a. 11.48 ft, b. $\theta = 50.4°$,

c. F_x = 1,657 lb, d. BN = 6.93 ft

2S-16. a. T = 4520 lb
b. M = 75,000 lb-ft.

2S-17. a. $\bar{x}$ = 5.84 in., $\bar{y}$ = 2.00 in.
b. Required I = 225.3 in.⁴

2S-18. a. I_{x-x} = 1551 in.⁴
b. P = 8620 lb

2S-19. a. 16,500 lb tension
b. 10,000 lb tension

2S-20. A = 20,000 lb tension
B = 30,000 lb compression

2S-21. a. X = 2 × weight of tractor
b. X = 0.966 × weight of tractor

2S-22. a. A_x = −1.06P B_x = −1.06P
A_y = 0.353P B_y = 0.353P
b. AE = 0; AD = 0.707P tension
BF = 0.750P compression;
DD = 0.350P compression

2S-23. $P = (W/r)\sqrt{2rh - h^2}$
$\alpha = \sin^{-1} (1/r)\sqrt{2rh - h^2}$

2S-24. 5.7 in. up from base;
I = 380 in.⁴

2S-25. a. 160 lb-in. 9 ft length, 120 lb in 12-ft length
b. 143 lb in. cable

2S-26. 19.1 lb to right, 47.1° down from horizontal; from lower left corner, x = 30.9 in., y = 2.77 in.

2S-27. L_0U_2 = 11,440 lb compression;
L_0L_2 = 22,860 lb tension
U_2U_4 = 20,000 lb compression;
U_2L_4 = 4050 lb compression

2S-28. 2.67 ft to left and 4.0 ft below

2S-29. 17.67 lb

2S-30. I = 327 in.⁴

2S-31. 140 lb

2S-32. b. 20 in.

2S-33. a. 2.73 lb/in. b. 3.33 in.

2S-34. 10.18 ft.

2S-35. a. 40.4 lb b. 63.2°
c. 44.8 lb

2S-36. a. 16,000 lb tension
b. 16,000 lb compression
c. 8000 lb tension

2SM- 1. *b*

2SM- 2. *a*

2SM- 3. *c*

2SM- 4. *e*

2SM- 5. *d*

2SM- 6. *c*

2SM- 7. *d*

2SM- 8. *c*

2SM- 9. *d*

2SM-10. *a*

2D- 1. 1.868 sec

2D- 2. 4.76 in.

2D- 3. zero

2D- 4. 2.72 lb/ft

2D- 5. 35 ft/sec

2D- 6. 20.3 ft/sec

2D- 7. 0.795 radians/sec

2D- 8. $v = t^3 - 6t^2 + 8t$
$a = 3t^2 - 12t + 8$
b. momentary at $t = 0.844$ sec
$v \to \infty$ as $t \to \infty$
c. $t = 2$ to $t = 4$ sec

2D- 9. 0.534 ft   108.7 mph

2D-10. 15.7 hp

2D-11. 56,500 lb

2D-12. 3.52 ft/sec

2D-13. 56,600 ft high
227,000 ft horizontally

2D-14. *b*. 213.7 mph
c. N 7.6°W

2D-15. *a*. 208,800 rpm
b. 2,294,000 ft/sec²

2D-16. $= -(V_1 - V_2)^2/2a$

2D-17. $a = (19.2e^{4t} + 16)/(1.20e^{4t} - 1)$

2D-18. 19.7 lb

2D-19. *a*. $\omega = 0.00116$ rad/sec
b. 5.94×10^8 ft-lb

2D-20. $A_y = 96.4$ lb, $B_y = 103.6$ lb,
$P = 82.2$ lb, $A_x = 32.1$ lb, $B_x = 0$

2D-21. *a*. 14.2 ft
b. Maximum moment $= 296,000$
lb-ft at rear wheels

2D-22. *a*. $v = 6.15$ ft/sec
b. $h = 369$ ft

2D-23. $\mu = 0.433$

2D-24. 1.74 in.

2D-25. *a*. 3.22 ft/sec
b. 1.93 ft from edge

2D-26. 81.7 rpm

2D-27. 4.25 in.

2D-28. 47.4 mph, 13.5°

2D-29. 17 ft/sec, 25 ft-lb

2D-30. 18.8 in.

2D-31. 5.5 revolutions

2D-32. 379 ft

2D-33. 25.0 psi

2D-34. 324 lb

2D-35. *a*. 115.08 ft; $a = 7.5$ ft/sec² at
time $t = 2$ sec; at $t = 5$ sec, $a =$
9.2 ft/sec², $v = 46$ ft/sec

2D-36. *a*. 322 lb
b. 0.0394 ft/ft

2D-37. 3500 ft/sec

2D-38. 635 ft/sec

2D-39. 13.0 ft/sec to right, 15° down
from horizontal

2D-40. 9100 ft-lb

2D-41. *a*. 256 ft-lb
b. 240 ft-lb
c. 16 ft-lb

2D-42. *a*. 16.1 ft/sec²
b. 193.2 ft

2D-43. 31.6 ft/sec; 31,000 ft-lb

2D-44. 1.90 sec; 5220 ft-lb

2D-45. *a*. 48.9 sec, 3230 ft
b. 6610 ft
c. 2 hrs. 6 min.

2D-46. 2.08 ft/sec and 5.68 ft/sec

2D-47. $11.4\sqrt{R - r}$ (includes rotational
KE of ball)

2D-48. *a*. 15.1 ft/sec
b. 13.55 in.

2D-49. $v = 25.5$ ft/sec to right at angle of
45° up from horizontal; $a = 134$
ft/sec² to right, 10.3° above horizontal

2D-50. 9.85 in.

2D-51. $F_1 = 4$ lb; $F_2 = 117.6$ lb

2D-52. $b/c = 2.13$

2D-53. *a*. 6.44 ft/sec²   *b*. 9.84 ft/sec

      c. 153 lb-sec   *d*. 600 lb

2D-54. 0.875 fpm

    b. 11.8 min. from start of movement

2DM- 1. *d*

2DM- 2. *b*

2DM- 3. *e*

2DM- 4. *a*

2DM- 5. *c*

2DM- 6. *e*

2DM- 7. *d*

2DM- 8. *a*

2DM- 9. *a*

2DM-10. *a*

3- 1. 561 lb

3- 2. 40.7 ft/sec

3- 3. 40,400 gal/hr

3- 4. 0.66 sp gr

3- 5. 7.13 ft

3- 6. *a*. 250 psig

    b. 3.47 cfs

3- 7. With the figures given, the suction pressure is calculated to be -316 psig, which is impossible. If the flow rate is taken as 60 gpm, the suction pressure is -7.92 psig.

3- 8. 41.6 ft

3- 9. 9.54 ft from the toe

3-10. 361.9 hp

3-11. 231 gpm

3-12. 18,750 lb

3-13. *a*. 99.1 ft

    b. 67.5 ft

    c. 39 gpm

    d. 78.0 percent

3-14. *a*. 13.5 ft from toe

    b. 5250 psf at toe, 2850 psf at heel

    c. 19.8 ft from toe

    d. Yes

3-15. 87.0 lb

3-16. Problem not solvable with data given. If 10.5 psig changed to 8.0 psig, the discharge would be 335 gpm.

3-17. *a*. 4.11 kW

    b. 5.20 hp

    c. 2.72 hp

    d. 52.4 percent

3-18. 51,400 lb to right, 60° down from horizontal, if flow is to right

3-19. *a*. $x = 2.11$ ft

    b. $F = 12,000$ lb

3-20. 263 cfs

3-21. *a*. 1180 lb force

    b. 56.4 hp

3-22. *a*. 196 ft

    b. 590 cfs

3-23. 608 lb lift

3-24. 452 lb

3-25. *a*. 6.9 cm

    b. Above

3-26. 1.37 cfs

3-27. *a*. 147.5 ft

    b. 137.5 ft

    c. 10 ft

3-28. a. 7660 lb force

    b. 0.187 in. below horizontal center line

3-29. 3.25 ft

3-30. 6.50 min

3-31. 936 psig (assuming $f = 0.017$)

3-32. 194 ft/1000 ft

3-33. *a*. 3.82 ft/sec

    b. 42.2 percent efficiency

    c. 0.0299

3-34. 28.2 lb/ft

3-35. 201 centipoises

3-36. Assuming free discharge at 3 and taking $p_3 = 0$,

    $p_1 = 43.7$ psig, $v_1 = 46.6$ ft/sec

    $p_2 = 10.7$ psig, $v_2 = 62.1$ ft/sec

    $p_3 = 0$, $v_3 = 93.2$ ft/sec

3-37. Schedule 40³/₄-in. pipe

3-38. *a*. 5 cm

    b. 4900 dynes/cm²

3-39. 42 min, assuming a discharge coefficient of 0.60

3-40. 115.7 psig; 102.5 hp

3-41. *a*. 16.0 ft/sec

    b. 2.40 ft/sec

3-42. 40 psi

3-43. 32.9 g

3-44. 1.8 ft

3-45. 6.54 ft

3-46. 0.166b

3-47. 16.9 cu cm

3-48. a. 77.2 lb to right, 85.9° down from horizontal

    b. 870 lb to right, 32.1° down from horizontal

    c. 916 lb to right, 36.0° down from horizontal

3-49. −290 psf at upstream edge of base (in tension)

3-50. 3,540 lb

3-51. 17.8 psi

3-52. 53.4 cfs through 24-in. diameter line

    6.6 cfs through 12-in. diameter line

3-53. a. 1.385 cfs

    b. 4.46 psi

3M- 1. e

3M- 2. c

3M- 3. c

3M- 4. e

3M- 5. b

3M- 6. c

3M- 7. c

3M- 8. c

3M- 9. a

3M-10. b

4- 1. a. 28.5 percent

    b. 125.4°F

    c. 494 hp

    d. 548,000

4- 2. a. 53.4 cu ft

    b. 60°F

    c. −266,000 ft-lb

    d. zero

4- 3. 120 lb

4- 4. 2.18 cu ft

4- 5. 1106 Btu

4- 6. a. $U = 0.356$ Btu/(hr)(sq ft)(°F)

    b. Brick face $h = 2.85$ Btu/(hr)(sq ft)(°F); concrete face $h = 1.815$ Btu/(hr)(sq ft)(°F)

4- 7. 54 cents

4- 8. 423,000 ft-lb

4- 9. Partial pressures:

    $N_2 = 14.0$ psia, $CO_2 = 4.0$ psia, $CO = 2.0$ psia

    Weight analysis: $N_2 = 62.8$ percent, $CO_2 = 28.2$ percent, $CO = 9.0$ percent

4-10. 1.13×10^7 kWh

4-11. 300 hp

4-12. a. 59 grains/lb dry air

    b. 31.9 percent

    c. 51.7°F

4-13. a. (Assuming 20 psig) 1270°F, 1965 psig

    b. 7800 ft-lb

4-14. 151.3 hp

4-15. 7.05 in.

4-16. 0.179 in./hr

4-17. 54 g of ice; 8 g of steam

4-18. 334 hp

4-19. a. $n = 1.25$

    b. 345,600 ft-lb

    c. 172 Btu added to gas

4-20. b. State 2: $p = 1.069p_1$, $V = V_1$, $T = 1.069T_1$

    State 3: $p = 0.848p_1$, $V = 1.18V_1$, $T = T_1$

    State 4: $p = p_1$, $V = V_1$, $T = T_1$

    c. 1265 joules

    d. 72 liters

4-21. 4.95 qt

4-22. 88.6 percent quality

4-23. 7.83 Btu/(mole)(°R)

4-24. 2215 lb

4-25. 6.46 cu ft/cu ft

4-26. 119.7 hp

4-27. 0.296 Btu/(lb)(°F)

4-28. a. 636,000 Btu/hr supplied; 382,000 Btu/hr rejected

    b. 40 percent

4-29. 282.4 knots

4-30. 75.62 cm

4-31. *b*. 780 cal added. No change in internal energy over a complete cycle.

4-32. *a*. 134.63 psia; 29.6 percent

b. 47.21 psia; 100 percent

4-33. 76.3 Btu (nonflow process)

4-34. $t = 363.13°F$; $v = 5.696$ cu ft; $h = 1195.0$ Btu/lb; $s = 1.5646$ Btu/(lb)(°F); $u = 1111.1$ Btu/lb

4-35. $DB = 81°F$; $WB = 60°F$; 52 percent relative humidity

4-36. 3000 lb/hr

4-37. bmep = 236 psi; brake torque = 123 lb-ft; brake thermal efficiency = 23 percent; efficiency based on cold-air standard = 40.7 percent

4-38. 92.9 psi

4-39. 388 psia; 1110°F

4-40. *a*. 9480 cfm

b. isothermal, lower; adiabatic, higher

4-41. *a*. 67.2 Btu/sec

b. 0.0163 cfs

4-42. *a*. 234 Btu/lb

b. Efficiency = (usable work out of turbine)/(heat added in combuster)

4-43. 176 Btu

4-44. *a*. 58°F

b. 10.025 Btu

4-45. *a*. −46°F

b. −127.7 Btu

c. 59,400 ft-lb

d. 0.1036 Btu/°F

e. 51.2 Btu given off

4-46. *a*. 24.5 percent

b. 0.479

c. 0.545 lb/hp-hr

4-47. *a*. 174.8 psig

b. 48.5 percent methane 51.5 percent ethane

4-48. $x = 98.83$ percent

4-49. 1106°F; 420 psia

4-50. $U = 0.836$ Btu/(sq ft)(hr)(°F)

4-51. 12 lb

4-52. *a*. 57 grains/lb

b. 51.5°F

c. 30 percent relative humidity

4-53. 8.05 percent

4-54. 187°F

4-55. *a*. 2680 lb thrust

b. 3300 hp

c. 13 percent

4-56. 2.11 percent

4-57. 21.4 cu ft

4-58. *a*. 2510 ft/sec

b. 1955 ft/sec

4-59. *a*. 6.50 percent hydrogen, 93.50 percent carbon (assuming complete combustion)

b. 48.2 percent

c. 255 cu ft

4-60. 12 cu ft (assuming 45 psig)

4-61. *a*. 11.15 Btu

b. 9.54 Btu

c. 14.7 percent

d. 62 percent

4-62. *a*. 82.2 lb/min

b. 0.028 Btu/(lb)(°F)

4-63. *a*. 3.71

b. 131 Btu/lb

c. 487 Btu/lb

d. 12.3 lb/min

e. 1.27 hp/ton

f. 498 lb/min

4-64. 0.192 cu ft

4M- 1. *c*

4M- 2. a

4M- 3. *b*

4M- 4. *c*

4M- 5. *d*

4M- 6. *d*

4M- 7. *e*

4M- 8. *d*

4M- 9. *a*

4M-10. *b*

5- 1. 261,000 lb-in.

5- 2. $\dfrac{dy}{dx} = \dfrac{w}{EI}\left(\dfrac{Lx^2}{4} - \dfrac{x^3}{6} - \dfrac{L^3}{24}\right)$

5- 3. a. Each copper wire holds 266 lb, steel wire 468 lb.

    b. 119.6°F

5- 4. 50,000 psi

5- 5. (Maximum shear stress)/(average shear stress) = $^3/_2$. (See Fig. 5-P-105.)

5- 6. b. M_{max} = 363,000 lb-in., 14 ft (See Fig. 5-P-106.)

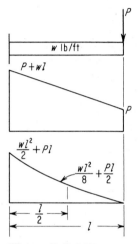

Figure 5-P-105

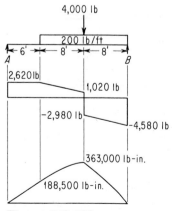

Figure 5-P-106

5- 7. 47,200 psi (disregarding weight of cable

5- 8. 7.28 in.

5- 9. a. 0.00216P

    b. 2.25P

5-10. 394,000 lb load permissible

5-11. See Fig. 5-P-111.

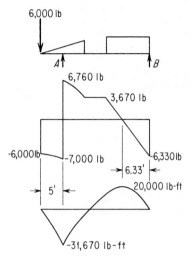

Figure 5-P-111

5-12. a. 225 in.-lb

    b. 40 lb

5-13. a. See Fig. 5-P-113

    b. 0.96 cu in.

5-14. Maximum shear = −12,000 lb; maximum moment = −90,000 lb-ft; both located 10.0 ft from left end of beam (see also Fig. 5-P-114)

5-15. M_{max} = 10,230 lb-ft at point 3.25 ft from left end; S_{max} = −5,400 lb

5-16. M_{max} = 288,000 lb-ft, M_{min} = 192,000 lb-ft; S_{max} = 10,000 lb S_{min} = zero

5-17. 2,000.778 ft

5-18. Use $2^3/_4$-in.-diameter shaft

5-19. 549 lb cement, 1296 lb sand, 1894 lb aggregate, 292 lb water

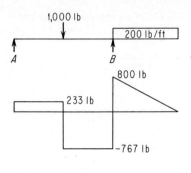

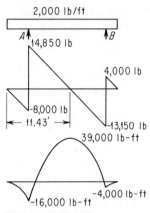

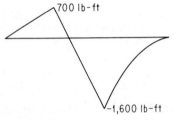

Figure 5-P-113

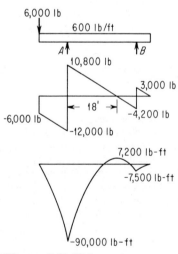

Figure 5-P-114

5-20. $M_{max} = 39,000 \; ft \times lb +$ 11.43 ft from left end. (See also Fig. 5-P-120.)

Figure 5-P-120

5-21. 5440 lb/ft
5-22. 8400 psi shear stress and 12,600 psi crushing stress
5-23. 9870 lb shear; 11,200 lb-ft bending moment
5-24. See Fig. 5-P-124.
    c. 13,500 psi

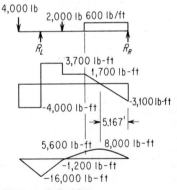

Figure 5-P-124

5-25. 11 ft from left end of beam; 188,000 lb-ft
5-26. $-58°F$
5-27. See Fig. 5-P-127.
5-28. The deflections are approximately equal at $2.66 \times 10^{-4} \times F$.

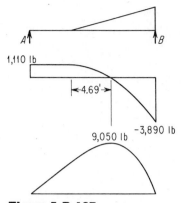

Figure 5-P-127

5-29. Steel, 200 lb; copper, 100 lb
$x = 7.0$ ft.

5-30. 7500 psi

5-31. *a*. 750 psi
b. 500 psi

5-32. 8020 psi; 0.895°

5-33. 18,700 psi

5-34. *a*. No, 93.8 psi shear horizontally
b. No, it is 1,685 psi bending

5-35. *a*. 2.34 in.
b. Small shaft twist = 1.71 × large shaft twist

5M- 1. *b*

5M- 2. *c*

5M- 3. *d*

5M- 4. *a*

5M- 5. *a*

5M- 6. *b*

5M- 7. *a*

5M- 8. *b*

5M- 9. *b*

5M-10. *d*

6- 1. *a*. 5500 volts
b. 201 watts

6- 2. 10.4 ohms

6- 3. 0.35

6- 4. 15 ohm, 1.35 amp; 25 ohm, 0.811 amp; 60 ohm, 1.66 amp; 200 ohm, 0.50 amp

6- 5. *a*. 4.0 amp
b. 89.4 percent
c. 14.1 hp

6- 6. Voltages equal at 53.3 volts
$Q_1 = 320 \ \mu$coulombs,
$Q_2 = 640 \ \mu$coulombs

6- 7. *a*. 10.67 ohms
b. 14.67 ohms
c. 0.728
d. 43.3°
e. current lags
f. 10.05 ohms
g. Inductive

6- 8. 104.4 volts

6- 9. 656 μF

6-10. 21.7 kvar

6-11. *a*. 245 amp
b. 459 kvar

6-12. 45.3 watts

6-13. Open, $V_L = 89.7$ volts,
$V_{R_1} = 25.6$ volts,
$V_{R_3} = 67.2$ volts;
closed, $V_L = 83.8$ volts,
$V_{R_1} = V_{R_3} = 49.5$ volts

6-14. *a*. 39.6 ohms
b. 2.78 amp
c. 72.3°
d. 0.303
e. 92.7 watts

6-15. *a*. 14,040 volts
b. 14,800 volts
c. 12,900 volts

6-16. 0.455 amp

6-17. 0.002 coulomb, 1.0 joule, 20,000 volts/meter; 250 volts, 0.250 joule

6-18. $I_1 = 15.65$ amp;
$I_2 = 7.5$ amp; $I_3 = 12.25$ amp

6-19. *a*. 52.5 volts
b. 63.475 volts
c. 21.9 percent

6-20. 1.0 volt; 2.5 ohms

6-21. *a*. 886 watts
b. 0.943 leading

6-22. *a*. 0.192 volts per conductor

b. $E_{avg} = 0.637 E_{max}$
$E_{rms} = 0.707 E_{max}$
for sine×wave voltages

6-23. *a.* Inductive reactance is greater.
b. 24.0 amp

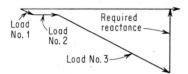

Figure 6-P-130

6-24. *a.* 4.48 ohms
b. 85.8 percent
c. 2510 watts
6-25. 125 volts; 100 volts
6-26. *a.* 30.3 ohms
b. 0.031 amp
c. 0.089 amp
6-27. 257-kVA transformers; 290 hp at 93 percent efficiency and 90 percent power factor
6-28. *a.* 604 coulombs
b. 120.8 amp
c. 129.6 amp
6-29. 1.5 amp through 9-volt battery; 0.5 amp through 6.5-volt battery
6-30. *a.* See Fig. 6-P-130
b. 118.15 kvar capacitance
6-31. *a.* 2000 kW
b. Same
6-32. 3.33 ohms
6-33. $I_1 = 3.23$ amp, $I_2 = 3.66$ amp, $I_3 = 0.43$ amp
6-34. 20 amp
6-35. *a.* 25.4 ohms
b. 61.7 volts

c. 309 watts
d. 0.786
6-36. 1.705 amp
6-37. *a.* 31 percent full-load torque
b. 196.6 volts
6-38. 14.2 ohms
6-39. 9.9 hp
6-40. *a.* 18 amp
b. 13.7 amp at 39.8° lagging
6-41. 0.106 henry
6-42. *a.* 50 amp
b. 5 volts
c. 250 watts
d. 25 amp
6-43. *a.* 1,746 rpm
b. 602 lb-ft
c. 253 amp
6-44. *a.* 2.13 volts
b. 19.36 amp
6-45. 10 volts
6-46. *a.* See Fig. 6-P-146.
b. 1440.5 kW
6-47. 1227 rpm (assuming full-load armature current of 58 amp)
6-48. 55 kVA
6-49. *a.* 217 volts at 8.7° lagging
b. See Fig. 6-P-149.
6-50. *a.* 53.4 amp
b. 15.35 amp
c. 3.68 kVA
d. 1.06 kW
6-51. *a.* 200 watts
b. 141.4 volts
c. 127.2 volts
d. 1.11
6-52. 125.25 volts
6-53. 10 ohms, 2.21 amp; 30 ohms, 0.73 amp; 50 ohms, 1.96 amp; 100 ohms, 0.98 amp

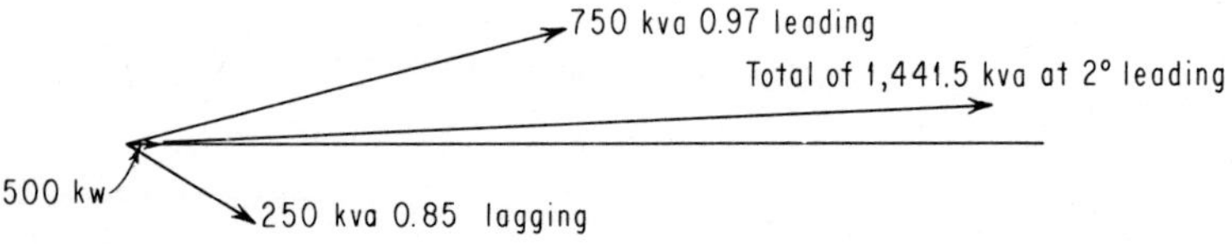

Figure 6-P-146

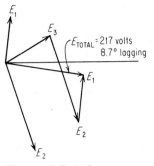

Figure 6-P-149

6-54. 0.944
6-55. *a.* 55 volts
    b. 0.11 amp
    c. 0.055 amp

6M- 1. *c*
6M- 2. *a*
6M- 3. *a*
6M- 4. *b*
6M- 5. *d*
6M- 6. *e*
6M- 7. *e*
6M- 8. *e*
6M- 9. *d*
6M-10. *b*

7- 1. 93.2 percent silver
7- 2. *a.* $Ca_3(PO_4)_2 + 3SiO_2 + 5C \rightarrow$
                    $3CaSiO_3 + 2P + 5CO$
    b. 5.0
7- 3. 2.48 lb
7- 4. *a.* 8.33 lb
    b. 5.83 lb
7- 5. *a.* $Q_1 = 8.62$ gpm, $Q_2 = 41.38$ gpm
    b. 8.62 Btu/(gal)(°F)
7- 6. 89.4 cu ft dry air
7- 7. *a.* 1082 lb
    b. 35,000 cu ft
7- 8. 41.5 g
7- 9. 5460 lb
7-11. *a.* $3Ag + 4HNO_3 \rightarrow 3AgNo_3$
                $+ NO + 2H_2O$

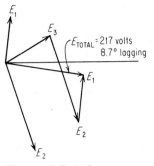

    b. 88.7 lb
7-12. 1990 g
7-13. 382 lb MgO; 8160 cu ft
7-14. 0.0373 g hof hydrogen
7-15. *a.* $C_3H_8 + 5O_2 \rightarrow 3CO_2 + 4H_2O$
    b. 250,000 cu ft
7-16. 51 tons
7-17. *a.* $Ca(HCO_3)_2 \rightarrow CaCO_3\downarrow$
                  $+ H_2CO_3$
                     $\rightarrow H_2O + CO_2\uparrow$
    b. $Ca(HCO_3)_2 + Ca(OH)_2 \rightarrow$
                    $2CaCO_3\downarrow + 2H_2O$
7-18. *a.* 18.70 percent CO_2, 0.28 percent
    H_2, 2.82 percent CH_4, 22.80 per-
    cent CO, 55.40 percent N_2
    b. $(CO_2) + H_2 + CH_4 + CO + (N_2)$
      $+ 3O_2 \rightarrow 3H_2O + 2CO_2 + (N_2) +$
                    (CO_2)
    c. 1.10 cu ft/cu ft
7-19. 60 percent FeS, 18.1 percent S,
    21.9 percent Fe
7-20. 10.6 percent SO_2, 82.1 percent N_2,
    7.3 percent O_2
7-21. *a.* 651 mm Hg
    b. 46.1 percent
7-22. $Ca_3(PO_4)_2 + 3SiO_2 + 5C \rightarrow$
                $3CaSiO_3 + 2P + 5CO$
7-23. *a.* 0.457 lb
    b. 1.235 lb $CaCO_3$, 0.222 lb H_2O
7-24. *a.* $2C_8H_{18} + 25O_2 \rightarrow 16CO_2$
                     $+ 18H_2O$
    b. 3.51 lb O_2, 3.09 lb CO_2, 1.42 lb
    H_2O
7-25. *a.* 606 lb
    b. 3060 cu ft
7-26. 17.6 percent CO_2, 16.1 percent
    H_2O, 66.3 percent N_2
7-27. *a.* 9.60 lb air
    b. 10.490 lb flue gas (excluding ash)
    per lb fuel
    c. 17.1 percent by volume
7-28. $CuO + H_2 \rightarrow Cu + H_2O$; 7.98 g Cu,
    2.27 g H_2O

7M- 1. *b*

7M- 2. *a*

7M- 3. *e*

7M- 4. *a*

7M- 5. *e*

7M- 6. *a*

7M- 7. *c*

7M- 8. *a*

7M- 9. *a*

7M-10. *c*

8- 1. 10 years, 2 months

8- 2. *a.* $185,600

    b. $9460

8- 3. *a.* $3770

    b. $16,000

    c. $9000

8- 4. *a.* $760

    b. $846

8- 5. Reinforce old bridge. $2078 per year vs. $3038 per year

8- 6. $834

8- 7. *a.* $2700

    b. $7780

8- 8. Purchased, $201,000 per year; power plant $131,900 per year

8- 9. $2490

8-10. $6480

8-11. Repair old shovel; $3,528 vs. $4030

8-12. $1135

8-13. *a.* 7720 miles

    b. $183 advantage to driving own car

8M- 1. *d*

8M- 2. *d*

8M- 3. *a*

8M- 4. *e*

8M- 5. *d*

8M- 6. *c*

8M- 7. *c*

8M- 8. *b*

8M- 9. *c*

8M-10. *c*

INDEX